CONRAD BURRI

PETROCHEMICAL CALCULATIONS

based on equivalents

(Methods of Paul Niggli)

Translated from German

Israel Program for Scientific Translations
Jerusalem 1964

Published in the U. S. A. by:
DANIEL DAVEY & CO., INC.
257 Park Avenue South, New York, N. Y

© 1964 Israel Program for Scientific Translations Ltd.

This book is a translation of

PETROCHEMISCHE BERECHNUNGSMETHODEN
AUF ÄQUIVALENTER GRUNDLAGE
(Methoden von Paul Niggli)

By Conrad Burri
Published in 1959 by
© Birkhauser Verlag, Basle, Switzerland

Translated by A. Katz, M. Sc.

IPST Cat. No. 2401

Printed by Sivan Press, Jerusalem
Binding: K. Wiener

Dedicated in gratitude to the memory of
PAUL NIGGLI
the creator of the Equivalent Norm

ABOUT THE AUTHOR

Born 22 May 1900 in Zurich. Graduated from the Municipal High School, Bern. Studied at the Eidgenössischen Technischen Hochschule (ETH) in Zurich and at the Universities of Zurich and Bern. Member of the faculty at ETH 1929. Associate Professor ETH 1932; full professorship ETH and Zurich University 1954.

Publications, among others, on petrochemistry and petrographic provinces of young igneous rocks, as well as crystal optics. Since 1953 President of the Foundation "Vulkaninstitut Immanuel Friedlaender".

TABLE OF CONTENTS

	English page	German page
PREFACE	1	7
A. INTRODUCTION	3	13
I. The Significance and Representation of Rock Chemistry in General	3	13
1. Chemical rock analysis	3	13
2. Calculation and projection methods	7	18
II. Methods of Graphical Representation	8	19
1. Illustration of the behavior of a single variable ..	9	19
2. Illustration of the interdependence between two variables	17	28
3. Illustration of the mutual relations between two, three, and four variables having a constant sum ...	20	32
a) Two variables: The concentration line	20	33
b) Three variables: The concentration triangle .	21	33
c) Four variables: The concentration tetrahedron	26	38
III. Aids for the Calculation	31	44
B. PETROCHEMICAL CALCULATION METHODS BASED ON EQUIVALENCE ACCORDING TO P. NIGGLI	34	47
I. The Niggli-Values si, al, fm, c, alk, k, mg, ti, p, etc, and Their Applications	34	47
1. The calculation of the Niggli-values	34	47
a) Niggli-values and weight percentages	34	47
α) Calculation of the Niggli-values from the weight percentages	34	47
Examples	37	50
β) Calculation of weight percentages from given Niggli-values	38	52
Examples	39	53
b) Niggli-values and cation percentages	41	55
α) Definition and calculation of the cation percentages	41	55
Examples	42	55
β) Relationships between cation percentages and Niggli-values	43	56
γ) Calculation of the cation percentages from given Niggli-values	43	57
Examples	44	57

	English page	German page
δ) Calculation of the Niggli-values from given cation percentages	44	58
Examples	45	58
c) Niggli-values and molecular equivalent percentages (molecular percentages)	46	59
α) Calculation of the Niggli-values from the molecular equivalent percentages	46	59
Example	46	60
β) Calculation of the molecular equivalent percentages from given Niggli-values	47	60
2. The representation of fundamental petrochemical relationships by means of the Niggli-values	47	60
a) The quartz number qz as an expression of saturation with respect to SiO_2	47	60
b) The alumina excess	49	63
c) Derivation of a simplified normative mineral composition from the Niggli-values	50	64
3. Application limits of the Niggli-values	52	66
4. Further applications of the Niggli-values	55	70
a) Compositional and quantitative relations of the normative feldspars	55	70
b) The state of silication	61	75
c) The al-fm-c-alk-tetrahedron and the representation of the igneous field	62	77
d) Further representations of the igneous field	64	80
e) The variation diagram	67	83
f) Schematic review of the mutual dependence between si and alk for certain idealized normative mineral compositions	73	88
g) Distribution of SiO_2 among normative leucocratic and melanocratic components	75	90
5. Magma-types	76	92
a) The concept of a magma-type	76	92
b) Review of magma-types (C. Burri and P. Niggli, 1945)	78	94
II. The Equivalent Norm	84	100
1. General consideration of experiments for the calculation of the normative mineral composition	84	100
2. The principle of Niggli's equivalent norm	87	103
3. The basis and the calculation of the basis components	90	106

	English page	German page
Examples	93	109
4. Basis and cation percentages	95	112
a) Relationships between basis bonds and cation percentages	95	112
b) Calculation of the basis from given cation percentages	95	112
Examples	95	112
c) Calculation of the cation percentages from the basis	97	114
Examples	97	114
5. Derivation of normative mineral compositions from the basis	99	116
a) The standard katanorm and its derivation from the basis	104	121
Examples	110	128
b) The formation of variants of the standard katanorm	114	133
6. The standard katanorm and cation percentages	116	135
a) Relationships between cation percentages and the standard katanorm	116	135
b) Calculation of the standard katanorm from given cation percentages	117	136
Examples	117	136
c) Calculation of the cation percentages from the standard katanorm	118	138
Examples	119	139
7. Standard katanorm and Niggli-values	120	140
a) Calculation of the standard katanorm from given Niggli-values	121	140
Examples	121	141
b) Calculation of the Niggli-values from the standard katanorm	124	144
Examples	125	145
8. The application of the equivalent norm to the study of heteromorphic relations	127	147
a) The vaugnerite from Vaugneray (Dép. du Rhône, France)	128	148
b) Selected examples of the heteromorphic possibilities of alkali-gabbroidal magmas	136	157
α) Maifraite, luscladite, and berondrite	136	157
β) The Fasinite	142	164

	English page	German page
9. Equivalent norm, weight and volume percentages	147	170
a) Equivalent percentages and weight percentages	147	170
b) Equivalent percentages and volume percentages	153	176
c) Volume percentages and weight percentages	155	179
10. Graphic representation of the chemism of the rock based on the basis components	156	181
a) The basic-group values Q, L, M and the QLM-triangle	156	181
b) The KNaCa- and MgFeCa-triangles and the determination of normative feldspar relations	161	186
c) The method of the remainder triangle	169	194
11. The direct calculation of the basis group values Q, L, M from Niggli-values	172	198
12. The application of the equivalent norm to the study of metamorphic rocks	175	200
a) General	175	200
b) The standard katanorm of metamorphic rocks	179	206
Examples	181	207
c) The epinorm of metamorphic rocks	187	214
α) General	187	214
β) The calculation of a standardized epinorm	192	219
Examples	199	226
γ) Variants of the calculated modus of the standard epinorm	203	230
Examples	205	233
δ) The calculation of epinorm variants adapted to the modus	208	236
d) Additional selected examples for the calculations of metamorphic rocks	210	238
C. THE SUGGESTIONS OF T.F.W. BARTH AND P. ESKOLA FOR THE CONSIDERATION OF ANIONS IN PETROCHEMICAL CALCULATIONS, AND THEIR RELATIONSHIP TO NIGGLI'S EQUIVALENT NORM	230	260
I. General	230	260
II. The Calculation of Rock Analyses with Consideration of the Anions	230	260
1. Calculation starting from the weight percentages	231	261
2. Calculation starting from the Niggli-values	232	262
3. Calculation involving the presence of additional anions	233	263
III. Barth's Standard Cell	233	263

	English page	German page
1. Quartz-diorite, Spanish Peak, California (calculation from the cation percentages)	234	264
2. Quartz-diorite, Spanish Peak, California (calculation from the weight percentages)	234	265
3. Quartz-diorite, Spanish Peak, California (calculation from the Niggli-values)	235	265
4. The norm of the standard cell	236	266
5. The application of Barth's standard cell for the representation of the material balance of iso-volumetric metasomatic processes	237	267
D. APPENDIX	241	272
I. Alphabetical Index of Used Basis- and Equivalent-Normative Components	241	272
II. Compilation of Important Reaction Relations Between Basis- and Equivalent-Normative Components	243	274
III. Tables of the Molecular and Atomic Equivalent Numbers, Multiplied by 1000, for the Important Rock-Forming Oxides	249	280
E. BIBLIOGRAPHY	286	318
F. AUTHOR INDEX	291	323
G. SUBJECT INDEX	293	325

All cross references in this book allude to the original German page numbers indicated in the left-hand margin

PREFACE

The works dealing with the chemical aspects of the world of rocks occupy a special place in the astonishingly many-sided life-work of the late Paul Niggli, Professor of Mineralogy and Petrography at the Züricher Hochschule, who died in 1953. This subject was already of interest to him early in his career, and even in later years, when he was mainly occupied with the theory of crystal structure, he returned to it on numerous occasions. While initially concerning himself with problems dealing with the chemical characterization of rock types and the comparison of rock series and rock provinces — his unpublished doctorate thesis, the subject of which was suggested by his teacher, Ulrich Grubenmann, in 1911, dealt with this subject — he later became mainly interested in the problem of the relation between the chemistry of rocks and their mineralogical composition. This problem was successfully and ingeniously solved by his introduction of the equivalent norm (initially termed "molecular norm"). In comparison with similar earlier attempts, his new calculation method, which is applicable to both igneous and metamorphic rocks, constitutes a great step forward. "Congruent formula units" ("übereinstimmende Formelgrossen"), composed of equal numbers of cations, were introduced and used for the chemical formulation of rock-forming minerals. This allows for the transformation of a once-calculated percentile distribution or mineral facies to other facies types by means of simply-formulated reactions, without resorting to the recalculation of the sum to 100. An additional advantage of the method is that chemical compositions of rock-forming minerals, as obtained from chemical analysis, may be introduced into the calculation, both together with, and instead of, the ideally formulated mineral compositions. Thus, this method provides a much-needed tool for handling problems of rock metamorphism and heteromorphism of igneous rocks. This method, in similarity to the earlier methods of Niggli, was supplemented by clear graphical representations, being the products of his eminent geometrical talent which generally manifests itself throughout his entire work.

However, it was not easy for those interested in the subject to adapt themselves to Niggli's train of thought. Basic papers were scattered over a number of not always easily accessible publications which, after the war, were mostly out of print; applications of this method, in the form of worked-out and discussed examples, were only available in dissertations of Niggli's students, thus not reaching any further stage of development. A comprehensive representation, as repeatedly suggested by many of his colleagues, was put off time and again. Not until the latter years of his life did Niggli agree to, and start planning, a joint publication with the undersigned. However, with the exception of the first disposition, nothing was ready by the 13 January, 1953, the date of his untimely death.

The undersigned was privileged to follow the development of Niggli's petrochemical system in all its phases from close approach. Therefore, the author regarded the publication of the planned work as an obligation in honor of the memory of his teacher. Thus, he decided to carry out this work, fully realizing that other works, very close to his heart, would have to be delayed in consequence. The inclusion of completely worked-out examples applying to various fields was very much stressed. The author hopes that the book will thereby gain in usefulness.

The author wishes to express his thanks to many individuals for their generous help and support: Prof.Dr.H. Pallmann, President of the Swiss Education Council, kindly covered the drawing costs of 41 text-figures from the 1955 anniversary fund of the ETH. An additional number of figures was made available by the endowment of the "Vulkaninstitut Immanuel Friedlaender" in Zürich, from an earlier publication. The molecular and equivalent tables of the important rock-forming oxides, given in the appendix, were recalculated by W. Oberholzer, Dr.Sc.Nat., M.Weibel, Dr.Sc.Nat., along with R. Steiger, dipl.ing.geol. ETH, and R. Jakob, Dr.Phil., checked the calculations cited as examples. My wife, Mariette Burri, Dr.Phil., prepared the index. Mr. R. Steiger and K. Soldatos, Dr.Sc.Nat., as well as Miss J. Marquardt assisted me greatly in reading the proofs. I recall with special pleasure many discussions with colleagues, of whom I here will mention only Prof. T.W.F. Barth, Oslo, who critically perused part of the proof sheets. I also wish to thank the publisher for his constant cooperation and willing compliance with my wishes and for the care taken in publishing the book in accordance with the traditions of his publishing house.

A few brief comments are possibly merited here on the use of the terms "petrochemistry" and "petrochemical", as recently (from about 1942 onward) these terms were not used in connection with rock chemistry, but with the chemistry of petroleum and products of natural gas. However, it must be stated that priority should be given to the rock-chemistry sense of petrochemistry. This fact was recently also recognized by the nomenclature commission of the department of organic chemistry of the American Chemical Society. This commission recommended that "petrochemistry" and "petrochemical" (as adjective) should be used in scientific publications in the "rock-chemistry" sense only. On the other hand, it is also recognized that "petrochemical(s)" as a noun has a wide range of use as a popular term indicating chemicals derived from petroleum and natural gas in the daily and commercial press. The same also applies to its use in the adjectival form, by producers. The usage of the term in this sense in non-scientific publications of the American Chemical Society should therefore be tolerated (Chem. A. Eng. News 32, 1954, 3111 and 33, 1955, 70).

Zürich, the Institute of Crystallography and
Petrography of the Eidgenössischen Technischen Hochschule

15 March 1959   Conrad Burri

A. INTRODUCTION

I. The Significance and Representation of Rock Chemistry in General

1. Chemical Rock Analysis

In the study of rocks their chemistry is exceedingly significant. Without this knowledge many questions concerning the exact identification and comparison of the rock, its genesis and affinities, as well as its technical characteristics and possible use, can only be solved partially, if at all. Moreover, the mutual relationship between the chemistry and mineral composition is a point of great significance as far as crystalline rocks are concerned. However, though a given mineral composition uniquely determines the chemistry of the rock, the opposite is not necessarily true. Different mineral constituents (facies), which have identical chemical composition, may occur, depending on the conditions prevailing during the formation of the rock. This phenomenon which was named h e t e r o m o r p h i s m (Lacroix, 1920) is of great significance, especially as far as metamorphic rocks are concerned. Being comparable, in a sense, to the polymorphism of the individual minerals, its full meaning in rocks of igneous origin was recognized only relatively recently. Heteromorphism indeed constitutes the basic phenomenon of metamorphism, namely, the dependence of the mineralogical composition on the conditions of formation, as expressed in the zoning theory of Becke-Grubenmann or in the facies concept of Eskola.

The above-mentioned statement, i.e., that a chemical composition is fixed for a given mineralogical combination, should, however, not be accepted in general without certain qualifications. Only if the mineralogy is known both qualitatively and quantitatively can the chemical composition of a rock be given in more than general terms. The reason for this is that most, and strictly speaking even all, rock-forming minerals, with the exception of quartz, are mixed crystals. Their chemical composition can be arrived at only approximately by means of the idealized assumptions made during petrographic work resorting, frequently as the only means, to optical methods. Unfortunately, this is the case encountered by all petrographers even when considering plagioclases of relatively simple composition, which rank among the most widely known minerals. We are still obliged to be content with the approximate determination of the anorthite content, and nothing definite can be said about the potassium feldspar constituent which is always present, and, still less, about the amount of carnegieite which may be present. These difficulties apply to an even greater extent to the complex mineral species, e.g. sesquioxide-bearing hornblendes and augites. Thus, it may be concluded that even when the mineralogical composition is known, as determined by optical methods, and the quantitative relations between the individual minerals are also known, the chemical analysis of the rock under consideration is by no means dispensable. The chemical analysis may often indirectly allow for a more precise chemical characterization of a complex component by using suitable calculations.

The "oxide form" is nowadays most preferred for the representation of the results obtained from a chemical analysis of rocks, i.e. the individual metal-atom species are given in the form of weight percentages of their oxides. It should be noted, however, that this mode of representation is by no means the only one possible. The weight percentages of the individual atom species, including oxygen, might be given, for example, and this form is of use for certain purposes. Some suggestions more recently proposed by T.F.W. Barth and P. Eskola will be discussed in greater detail further on. The reason for the preference of the "oxide form", as well as its presentation in general, stems from the basic crystallochemical and geochemical regularities. Oxygen is by far the most abundant element in the outer lithosphere, where the rocks under examination originate. This is shown on Table I.

TABLE I

Abundance of the important elements of the outer lithosphere

	Weight %	Atom %	Volume %
O	46.60	60.50	91.83
Si	27.70	20.45	0.83
Al	8.13	6.24	0.79
Fe	5.00	1.87	0.58
Ca	3.63	1.87	1.50
Na	2.83	2.54	1.64
Mg	2.09	1.79	0.58
K	2.59	1.40	2.19
Ti	0.44	0.19	0.05
	99.01	96.85	99.99

Oxygen percentages are as follows: 46.60 by weight, 60.50 as atom percentages and 91.83 by volume, assuming heteropolar bonds. It is followed by Si and Al which are, however, not nearly as abundant. Therefore it is understandable that the silicates and aluminosilicates, respectively, are the most important rock-forming minerals. In these minerals the elements Si, Al, Mg, Fe, Ca, Na, and K are in most cases surrounded by and bound to oxygen. The same also applies to less abundant elements, such as C, P, N, Ti, Mn, Li, B, Cr, etc, as well as to many non-silicates. Consequently, the great majority of rock-forming silicates may hypothetically be broken up into oxides, a procedure which often provides a clear formulation, as for example

Orthoclase	$6\,SiO_2 \cdot Al_2O_3 \cdot K_2O$
Albite	$6\,SiO_2 \cdot Al_2O_3 \cdot Na_2O$
Anorthite	$2\,SiO_2 \cdot Al_2O_3 \cdot CaO$
Nepheline	$2\,SiO_2 \cdot Al_2O_3 \cdot Na_2O$
Diopside	$2\,SiO_2 \cdot (Mg, Fe)O \cdot CaO$
Aegirine	$4\,SiO_2 \cdot Fe_2O_3 \cdot Na_2O$
Olivine	$SiO_2 \cdot 2\,(Mg, Fe)O$

It is therefore reasonable to utilize these same oxide components for the representation of a chemical analysis of rocks. The following components are generally given in common rock analyses carried out for petrographical purposes: SiO_2, Al_2O_3, Fe_2O_3, FeO, MnO, MgO, CaO, Na_2O, K_2O, P_2O_5, TiO_2, H_2O. The determination of Fe_2O_3, in addition to FeO, is of importance here, as it represents the degree of oxidation of the iron. Unfortunately, this determination is still often neglected. The so-called constitutional "H_2O+" driven out from the rock at temperature above 110° (red-heat), is distinguished from mere moisture "H_2O-", which is determined by drying the sample at 110°. More detailed analyses also give the rarer elements, such as BaO, SrO, Li_2O, as well as the anions Cl, F, S, SO_3, and B_2O_3. The determination of additional trace elements by special methods is mainly carried out when geochemical problems are dealt with. F and S often occur in place of oxygen. When they are present in appreciable amounts, an equivalent amount of O must be deducted from the sum of the analysis. This is necessary, as in the formulation of the analyzed components it is assumed, a priori, that all electropositive elements are bound to oxygen, e.g., the total iron, including that portion of it which is combined in pyrite or other sulfides. Carbon dioxide, CO_2, should always be determined when carbonates (primary or secondary) are present.

It is not intended to provide a more detailed description of analytical technique, since this subject is beyond the scope of this book. Brief mention will only be made of the fact that a completely new development in the field of rock analysis is beginning to take shape. Rapid methods, mainly spectrographic and spectrometric, are being suggested (L. Shapiro and W. W. Brannock, 1956; W. W. Brannock and S. M. Berthold, 1953; L. H. Ahrens, 1954), in addition to the so-called classical methods — gravimetric and titrimetric, and in some instances also colorimetric, which were almost exclusively used until very recently. The determinations carried out by these rapid methods are, however, somewhat less accurate as compared to the classical methods when used by an experienced and conscientious analyst. However, this drawback is more than offset by the possibility of carrying out a much greater number of determinations in the same time, thereby facilitating the application of statistical methods during the interpretation. L. Shapiro and W. W. Brannock (1956) have proposed a working plan whereby eight rock analyses may be accomplished within a period of three days.

Regarding the criteria for assessing the reliability of a rock analysis and the precision of an individual determination, mention should be made, in addition, to the well known work of H. S. Washington (1917a, 9-26), of the experimental work carried out at an international scale, by H. W. Fairbairn in collaboration with the U.S. Geological Survey, in which two rocks, a granite from Westerley, Rhode Island (named "G-1") and a diabase from Centerville, Virginia (named "W-1"), were analyzed by 34 analysts from different countries (H.W. Fairbairn, et al., 1951, 1953; H.W. Fairbairn and I.F. Schairer, 1952; W.G. Schelcht, 1951; L.H. Ahrens, 1954). The detailed report on this work, including the statistical evaluation of the results (H.W. Fairbairn et al., 1951), is of the greatest importance to all those actively engaged in the field of rock chemistry. The study of this work is much recommended.

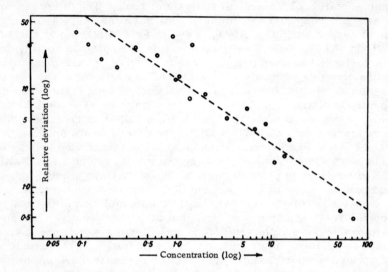

FIGURE 1. Accuracy of the determination of the individual components in granite G-1 and diabase W-1, as expressed by the relative deviation (percentage standard deviation) as a function of the concentration (both variables represented logarithmically). After L.H. Ahrens (1954)

As proved by L.H. Ahrens (1954a), the results of this test have also allowed for an estimation of the errors innate to the separate determinations of the rock analysis as carried out by the classical methods. In sensu strictu, these conclusions only apply to the granite G-1 and the diabase W-1. They should, however, also hold true for other rocks of similar composition or for rocks of compositions intermediate between G-1 and W-1.

A remarkable linear correlation is obtained between the accuracy of determination and the concentration of a given component. This is achieved, as suggested by H.W. Fairbairn, by plotting the logarithm of the relative deviation c, i.e. the standard deviation in per cent (cf. next chapter) as a measure of accuracy, against the logarithm of concentration, within a system of rectangular coordinates (Figure 1, after L.H. Ahrens, 1954a). A general regularity follows, namely, the precision in the determination of a given element decreases as the concentration of this element in the rock decreases. In Figure 2, also after L.H. Ahrens, the same points are given as in Figure 1, but the components and rock type are also indicated. The general statement outlined above is also applicable in this case for each individual component, with the exception of Al_2O_3. The decrease in accuracy caused by lowering of concentration is somewhat less marked in the case of the two colorimetrically determined components TiO_2 and MnO, than in the case of all the other components, which were determined gravimetrically and titrimetrically. The determinations of Fe_2O_3 show far greater discrepancies than the other determinations. This fact is of the greatest consequence in all petrochemical calculation methods. On the other hand, the determination of CaO is more accurate than the great majority of the other determinations, which fact is well known to every rock analyst. Ahrens' representation lucidly illustrates many interesting facts. By means of interpolation, it provides an estimate of the degree of error, expressed as a function of the concentration of an individual component, to be anticipated during rock analysis.

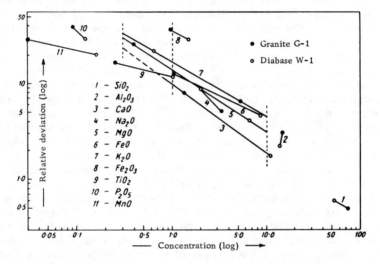

FIGURE 2. The same representation as in Figure 1, but with the individual determinations indicated. After L.H. Ahrens (1954)

2. Calculation and Projection Methods

During the chemical study of rocks, and especially in comparative research by means of rock analyses, the need is at once felt of simplification, and of improving clarity by reducing the number of components invovled. It is desirable to distinguish g r o u p s o f c o m p o n e n t s, as past experinece has shown that great difficulties arise when attempting to review a large number of analyses, each of which includes 14 or more highly variable components. The various petrochemical calculation methods subsequently suggested have provided a means of meeting this requirement. The restriction, common to all these methods, of the number of components by suitable grouping, also facilitates clear graphical representation. Owing to the restrictions imposed by the three-dimensional nature of space, the number of variables should be restricted.

However, the calculation methods have an additional aim. During the study of igneous rocks, and even more so in the case of metamorphic rocks, the relation between the chemical and mineralogical composition is one of the most fundamental problems. The ability to solve this problem may even be accepted as a direct measure of the usefulness of a particular petrochmical calculation method. It follows immediately, however, that such calculation methods must fulfill an important condition. The rock-forming minerals are chemical compounds and obey the stochiometric laws (with the necessary restrictions arising from the formation of mixed crystals). Thus, only those methods which utilize relations founded on a "molecular" i.e., an equivalent basis, instead of the weight percentage ratios, may be considered as suitable for revealing the relations between the chemical and mineralogical composition. It is advisable to avoid using the term "molecular" in this context, since this term, as originally defined, refers to a closed, finite particle configuration. As such, the term is unsuitable for the purely inorganic, mainly silicate and oxide crystal bonds considered here. Therefore, the following terms are introduced in the descriptions hereafter: "equivalent" instead of "molecular", "formula weight" instead of "molecular weight", and "formula unit" instead of "molecule". Thus, $6\,SiO_2 \cdot Al_2O_3 \cdot K_2O$ does not designate "1 molecule of orthoclase" but one "formula unit", and 556.49 does not represent the "molecular weight" of orthoclase, but its "formula weight". The necessity for using equivalent units exclusively in petrochemical calculations, repeatedly emphasized by P. Niggli, should nowadays be recognized universally. Only the works of T. F. W. Barth (1948, 1952, 1955), P. Eskola (1954) and P. Laffitte (1957) are mentioned here in this connection.

From the various calculation methods based on chemical equivalence, it was the system introduced by P. Niggli in 1917, and since continually elaborated, to which special attention has been paid, due not only to its great flexibility which in many instances permitted adequate solutions to be found also for special cases, but mainly due to its equal applicability to rocks of magmatic, sedimentary, and metamorphic origin. This point especially emphasizes the advantages of this method over the earlier systems, and undoubtedly constitutes a great step forward.

II. Methods of Graphical Representation

Since graphical representation will be much used in the following chapters, it is only appropriate to preface the subsequent explanations with some basic remarks. The following cases are of chief interest with regard to the problems of representing the chemistry of rocks:

1. Illustration of the behavior of a single variable.
2. Illustration of the interdependence of two variables.
3. Illustration of the variation of 2, 3, or 4 variables having a constant sum.

1. Illustration of the Behavior of a Single Variable

A single variable x is given, which may assume different values within certain limits. As an example we may consider the content of SiO_2 or some other oxide in the various rocks belonging to the rock series or petrographic province under research. A representation is being sought, which is able to provide a clear picture of the frequency of the different values of the variable, i.e. the so-called frequency distribution of the variable. This treatment is generally only appropriate when dealing with large-scale observational material. A comparative study of large provinces, the relations between which are established by 100 analyses, or more, may provide an example.

The individual values are tabulated in classes in a suitable manner. The number of values which fall within each class interval is noted and is either used as such or expressed as the relative values, e.g. the **percentage frequency values**. The suitably selected interval size is significant in this case. The entire variation range is, as a rule, divided into 15-25 intervals, or else, the intervals are chosen to be of the size of the standard deviation mentioned below.

The graphical representation of the tabulated values can be carried out in various ways. With regard to the detailed modes of representation, reference should be made to the usual methods used in statistics. They are partially known to petrographers from their application in sedimentary petrography, where they have been used for some time to illustrate the grain-size distribution of detrital sediments. The following modes of representation are useful:

a) **The histogram.** The class intervals are marked off on the abscissa and serve for bases of rectangles, the heights of which are proportional to the absolute, respectively relative frequency values (e.g. percentage values), whereby a clear picture is obtained of the prevailing frequency distribution. This constitutes the histogram.

b) **The frequency polygon and the frequency curve.** Ordinate values proportional to the frequencies are erected in the middle of the class intervals. A line is drawn through the end points of these ordinates, or else through the midpoints of the upper edges of the histogram rectangles. By this method the frequency distribution is visually depicted as a so-called frequency polygon. This polygon becomes a smooth curve when the number of the individual values become large and the class intervals are reduced. This is the frequency curve, which may replace the frequency polygon.

c) **The cumulative curve.** If the ordinates are not marked off to represent the absolute or relative frequency values of their respective abscissae, but rather signify the sum of all frequency values corresponding to the total of the smaller or larger abscissa values observed, that is, signify the so-called cumulative frequency values or **cumulative percentages**, then the so-called cumulative curves or lines are obtained.

All these methods are suitable for the illustration of frequency distributions of a single variable. The selection of a particular mode of representation is up to the choice of the author. Histograms or cumulative curves are generally applied in sedimentary petrography, whereas frequency polygons and curves are preferred in petrochemical work.

The saturation with respect to SiO_2 of the young volcanic rocks of the intra-Pacific suite (Hawaii Islands, Tahiti, Morea, Tubuai Manu, Cook Islands, Huahine, Raiatea, Tahaa, Bora-Bora, Maupiti, Austral Islands, Rapa, Pitcairn, Marquesas Islands, Carolines, Gambier Islands, Cocos, Galapagos, San Felix, San Ambrosio, Juan Fernandez, Easter Island) and those of the Atlantic (Canary Islands, Cape Verde, Madiera, Azores, Principe, Fernando Poo, S. Thome, Annobón, St. Helena, Ascension, Tristan da Cunha, Gough, Trinidad, Fernando Noronha, Bermuda) may serve as an example. 418 rock analyses are available from the first suite and 369 from the second. The "quartz number", qz, introduced by P. Niggli as a measure of the saturation and further exemplified in section BI 2a, will be used here. This number equals zero for rocks having an SiO_2-saturated norm, i.e., for rocks which have a norm exclusively consisting of feldspars and pyroxenes. When free quartz is present, the number becomes positive, and if olivine or foids* are present, the number will attain a negative value. Table II presents the absolute and percentage frequency numbers for the qz-values. Figure 3 illustrates the histograms and frequency polygons derived therefrom, and Figure 4 presents the cumulative curves.

TABLE II

The frequency distribution of the quartz number qz as a measure of the saturation with respect to SiO_2 for 418 young volcanic rocks from the intra-Pacific suite and 369 samples from the Atlantic suite

	qz	Pacific	%	Atlantic	%
	110–100.1	0	—	3	0.8
	100– 90.1	2	0.5	10	2.7
	90– 80.1	4	1.0	12	3.3
	80– 70.1	10	2.4	17	4.6
	70– 60.1	19	4.5	26	7.1
	60– 50.1	22	5.3	54	14.6
	50– 40.1	41	9.8	64	17.3
	40– 30.1	78	18.7	63	17.1
$-qz$	30– 20.1	101	24.2	54	14.6
	20– 10.1	58	13.9	27	7.3
	10– 0	46	11.0	9	2.4
	0 – 10	19	4.5	7	1.9
	10.1– 20	3	0.7	3	0.8
$+qz$	20.1– 30	2	0.5	5	1.4
	30.1– 40	1	0.2	4	1.1
	40.1– 50	1	0.2	2	0.5
	50.1– 60	1	0.2	0	—
	60.1– 70	0	—	2	0.5
	70.1– 80	2	0.5	2	0.5
	80.1– 90	0	—	0	—
	90.1–100	1	0.2	2	0.5
	100.1–110	1	0.2	1	0.3
	110.1–120	1	0.2	1	0.3
	120.1–130	2	0.5	1	0.3
	130	3	0.7	0	—
	Total	418	99.9	369	99.9

* [A term coined by A. Johannsen for the collective group of the felspathoids, of which word it is a condensation].

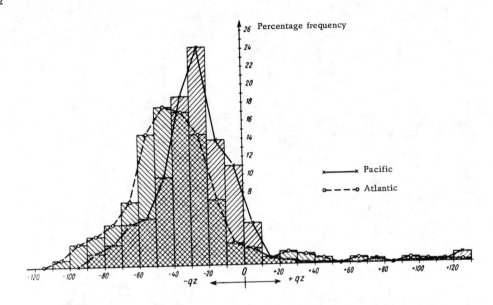

FIGURE 3. Frequency distribution histogram of the quartz numbers ± qz for 418 young volcanic rocks from the intra-Pacific suite and 369 young volcanic rocks from the Middle and South Atlantic suites

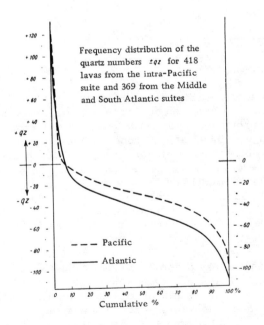

FIGURE 4. The same as Figure 3, but illustrated with cumulative curves (cumulative percentages)

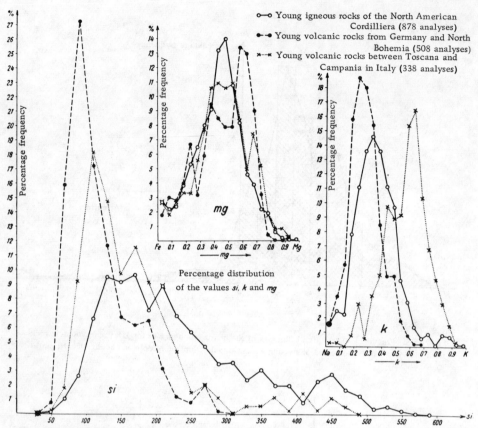

FIGURE 5. Percentage frequencies of the Niggli values si, mg and k for three selected main provinces. After C. Burri and P. Niggli (1945)

In contrast to the prevailing opinion that the volcanic rocks of the intra-Pacific and Atlantic suites are chemically similar, it is clearly shown that the Atlantic province, as an entity, has a lower silica content than the intra-Pacific province. It is beyond the scope of this book to explain the significance of this fact, which must be considered in every tectonic-petrographic general synthesis.

Figures 5-7 (after Burri and Niggli, 1945) should be considered as examples of distribution curves constructed from a larger number of analyses in order to characterize main provinces. The percentage-frequency distribution of the so-called Niggli-values (later explained in detail), si, al, fm, c, alk, k and mg, as calculated for three well-examined main provinces, are presented in these figures. These figures include the late Mesozoic and younger igneous rocks of the North American Cordillera (878 analyses, cf. also Burri, 1926), the young volcanic rocks of the Alpine foreland in Germany and North Bohemia (cf. Jung, 1928), and the young volcanic rocks between Toscana and Campania in Italy (338 analyses). The differences between these three main provinces, which at the same time represent Niggli's provincial types, the "Pacific", the "Atlantic" and the "Mediterranean", show up clearly, despite the arbitrary and random manner in which the analyzed examples were chosen.

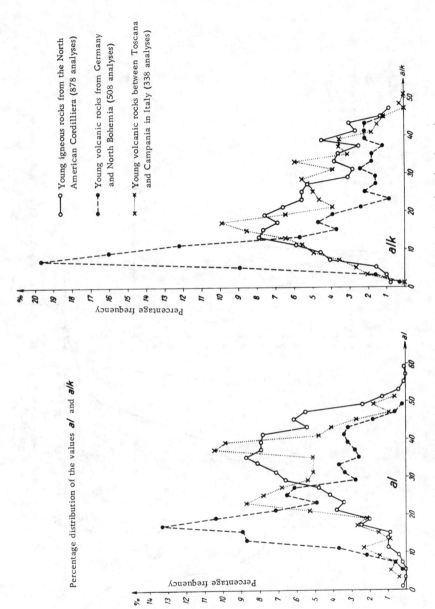

FIGURE 6. Percentage frequencies of the Niggli-values *al* and *alk* for three selected main provinces. After C. Burri and P. Niggli (1945)

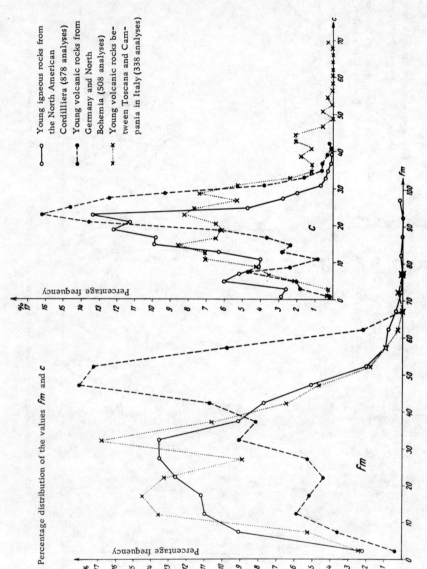

FIGURE 7. Percentage frequencies of the the Niggli-values fm and c for three selected main provinces. After C. Burri and P. Niggli (1945)

The characterization, by suitable numerical values, of the special qualities of frequency distributions or frequency polygons and curves has been attempted in various ways and is treated, in greater detail, in numerous papers on mathematical statistics, hereafter referred to (including A. Linder, 1950; G. U. Yule, and M. G. Kendall, 1950; L. H. C. Tippet, 1952; W. A. Wallis and H. V. Roberts, 1957). Of all these so-called statistical indices, intended for characterizing frequency distributions, the following should be mentioned first:

a) Mean values

a) The arithmetic mean or the average of several given values. If these values are designated as $x_1, x_2, x_3, \ldots, x_n$, their average value $\bar{x}$ will be

$$\bar{x} = \frac{1}{n} \sum_{i=1}^{n} x_i.$$

β) The median, central or middle value. In any distribtuion this will be the value for which there exist equal numbers of smaller and larger values of the variable. The plane, bounded by the abscissa and the frequency curve, is divided into two equal halves by the ordinate of the median value. This facilitates a simple graphical determination of the median value.

γ) The most frequent value or the mode. This corresponds to the maximum ordinate value on the frequency curve. The mean value, the median and the mode become equal in a symmetrical distribution; this does not hold true for asymmetrical distributions. In the case of moderately asymmetrical distributions, the following emprical rule can be applied with great approximation

Mode = Mean $-$ 3 (Mean value $-$ Median)

i. e. the median lies between the mean value and the mode, and is located at one third of the distance from the mean toward the mode.

b) Variability values

a) The variation range. Primary data concerning the variability is provided by the so-called variation range, i. e. the data on the minimum and maximum values taken on by the variable in the distribution under study.

β) The variance. The so-called variance states the distribution mode of the various values of the variable x about the mean $\bar{x}$.
If the deviation of the i-th x from the mean $\bar{x}$ is designated by (x_i), i. e. defining

$$(x_i) = (x_i - \bar{x}),$$

then the variance is given by

$$v = \frac{1}{n-1} \sum_{i=1}^{n} (x_i^2),$$

and the important widely used so-called mean or standard deviation σ is obtained

$$\sigma = \sqrt{v} = \sqrt{\frac{1}{n-1} \sum_{i=1}^{n} (x_i^2)}.$$

which is identical with Gauss's so-called mean square or mean error as applied to observation series.

For a series of time-saving and advantageous methods for the practical calculation of σ and v, as well as $\bar{x}$ from a large number of values, the reader is again referred to the above-mentioned works on mathematical statistics.

Sometimes the relative standard deviation or the so-called variation coefficient C

$$C = \frac{\sigma}{\bar{x}}$$

is also used. In contrast to the standard deviation, this is a pure number, generally expressed in percentages.

Further variability values deal with the characterization of the deviation of a distribution from a symmetrical form. This is expressed by the so-called skewness. A useful measure of this is given by

$$\varrho = \frac{\text{Mean} - \text{Mode}}{\text{Standard deviation}}$$

which is also a pure number, equalling zero for symmetrical distributions. For the purposes of our representation, the exact calculation of the statistical indices mentioned just now, frequently associated with a considerable amount of calculations, does not seem practical, if only because the sampling, that is, the choice of the rocks subjected to chemical analysis, very seldom answers the requirements of statistical research, but is always involved with a certain arbitrariness. It should be remembered, for example, that generally only a relatively small number of chemical analyses can be made available for petrographic research. Even though, generally, the main rock types do undergo chemical examination, the rarer rocks, occurring in subordinate amounts and only rarely described previously, are also frequently analyzed, thereby exaggerating their significance in the research. Important rock types, although recognized as such, are often discarded because they are not sufficiently fresh for analysis. On considering these facts it becomes apparent that a strict utilization of statistical methods at the present state of chemical-petrography research is out of question, their use being generally limited to a more illustrative comparison of the curves plotted.

However, it is advantageous for petrographers interested in petrochemistry to become acquainted with the methods of mathematical statistics and the possibilities of evaluation of a more extensive observation material offered thereby. As already mentioned, petrochemical research is about to enter a stage when the utilization of statistical methods will offer new possibilities. The rapid methods, mentioned previously, will provide for much more numerous data, so that, in the future, statistical interpretation will become the rule.

2. Illustration of the Interdependence Between Two Variables

In mathematics, the interdependence of two variables is known as a function. The simplest and best known way of graphically representing such a function is to compare the two variables in a system of rectangular coordinates. The functional dependence is then reflected by curves of various configurations, including the straight line. From such a curve one is able to read, for each value of one variable, one or more precisely defined values of the other variable. This dependence can also be expressed by an equation or by a table of values instead of by graphic representation. Such representations are also yielded by experimental research, provided that foreign influences are successfully excluded by suitable means. This is not the case with the material at hand in petrochemical research. This material has a different character, and the chemical qualities of the rocks under research are dependent on so many factors, the influence of which it is impossible to estimate, that no clear-cut functional relationship is to be expected, unless an exceptionally simple case is at hand. Therefore, the relations are much more of the type to be met with in **statistical investigation**. Here, as a rule, a larger number of observations is reflected not as a simple curve, but as a cluster of more or less scattered points. Instead of a functional association there arises the concept of a more or less complete correlation, and one deals therefore with **correlation diagrams**. The degree of interdependence of two variables can thus range from a complete functional dependence, through degrees of more or less complete dependence, to a total disorder (Figure 8).

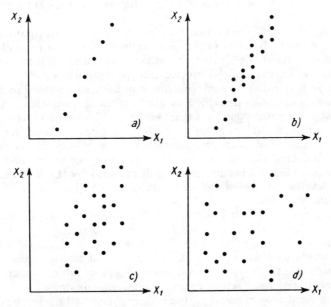

FIGURE 8. a) (linear) functional relationship between x_1 and x_2; b) fairly good (positive) correlation between x_1 and x_2; c) poor correlation between x_1 and x_2; d) total disorder.

If a very large number of individual values are available, and the points lie very close to one another in the correlation diagram, or even partially overlap, they can be dealt with as was explained for the case of the histogram, or the distribution curves, that is, suitably chosen intervals or classes are marked off on the abscissa and ordinate axes. Thereby the diagram may be broken down into individual squares. These squares may be inscribed with the respective numbers of points pertaining to them. Relevant examples will be cited later.

By erecting on top of the squares columns perpendicular to the $x_1 x_2$-plane, their heights being proportional to the respective number of points included in the square, a further illustrative representation is provided. In this case the correlation between the two variables is demonstrated in a three-dimensional structure. This is analogous to the frequency distribution of a single variable, as represented by a two-dimensional structure bounded by the frequency curve and the abscissa.

30 In this case, too, parameters were developed by the method of mathematical statistics, which enable one to define the degree of correlation numerically. The so-called c o r r e l a t i o n coefficient

$$r_{12} = \frac{\frac{1}{n-1} \sum_{i=1}^{n} (x_{1_i} x_{2_i})}{\sigma_1 \sigma_2} = \frac{\sum (x_{1_i} x_{2_i})}{\sqrt{\sum (x_{1_i}^2) \sum (x_{2_i}^2)}}$$

is most suitable for this purpose. It varies between -1 and $+1$. If $r=1$, a complete linear dependence results. x_1 increases with an increas in x_2 (so-called p o s i t i v e correlation) when $r = +1$, and decreases with the increase of x_2, when $r = -1$ (so-called n e g a t i v e correlation). Total disorder is expressed by $r = 0$, whereas intermediate degrees of correlation will have r values between 0 and $|1|$, being either p o s i t i v e or n e g a t i v e.

Reference should again be made to the introduction to mathematical statistics and to M. Ezekiel (1941) for further details on the relations between the correlation coefficient and other statistical indices, as well as for other formulations and geometric interpretations.

A more or less complete correlation can be also reduced to simple functions using methods provided by mathematical statistics. Thus, a field of scattered points can be replaced by an ideal mean curve. In many cases this is a straight line, the so-called r e g r e s s i o n l i n e. The assumption is thus made that the deviations are purely fortuitous, being caused by uncontrollable factors or by imperfect sampling, and that the ideal curve does indeed reflect the actual behavior pattern of the variables. For more details reference should again be made to the introduction to mathematical statistics.

For our present aims it is unnecessary, in most of the cases, as already explained, to apply these methods, frequently requiring lengthy and tedious calculations. If the correlation is of such a kind as to permit replacing a cluster of points by a simple curve, an intuitive drawing of the latter is generally sufficient. Nevertheless, the concept of the correlation coefficient is of decided importance with regard to later discussions.

Special difficulties arise during the investigation of the interdependence between two variables, if the correlation dealt with is not one of absolute magnitudes, but of ratios formed thereof. Special recognition is due to F. Chayes (1949), who stressed the significance of this point with respect to petrochemical problems, mentioning that misleading conclusions might well be arrived at if this distinction were ignored. This would be the case if, for instance, an attempt were made to infer from the correlation of ratios formed of absolute magnitudes, on the correlation of these absolute magnitudes themselves. Mathematical statistics has recognized this circumstance long ago (K. Pearson (1896-1897), a fact also recognized by F. Chayes. The following brief descriptions are mainly based on works of this author. For further details, reference should be made to the original treatise (F. Chayes, 1949).

If x_1, x_2, x_3 and x_4 are four absolute values (indices) and the ratios derived therefrom are

$$y = \frac{x_1}{x_2}, \qquad z = \frac{x_3}{x_4},$$

then, according to K. Pearson (1896-1897), the correlation coefficient with respect to y and z will be given by

$$r_{yz} = \frac{r_{13} C_1 C_3 - r_{14} C_1 C_4 - r_{23} C_2 C_3 + r_{24} C_2 C_4}{\sqrt{C_1^2 + C_2^2 - 2 r_{12} C_1 C_2} \sqrt{C_3^2 + C_4^2 - 2 r_{34} C_3 C_4}},$$

where C_i is the variation coefficient, defined as the ratio of the standard deviation and the arithmetic mean, as given previously. In deriving the formula, the assumption was made that C is so small that its third power can be neglected. It is clearly evident that the correlation coefficient r_{yz} is in itself a function of the correlation states existing between the four absolute values. As the denominator of the fraction will always be positive, the sign of r_{yz} will only depend on that of the numerator. The correlation coefficient r_{yz} will become zero (i.e., no correlation will exist between y and z) only when $r_{13} = r_{14} = r_{23} = r_{24} = 0$ (and obviously also in the case when the four terms of the numerator cancel out exactly. As the probability of such a situation arising is exceedingly small, this eventuality may safely be ignored).

Should one of the four mentioned correlation coefficients be different from zero, r_{yz} will also be different from zero, and a correlation will exist between y and z. In all events, correlation coefficients having values differing from zero will naturally designate different absolute amounts. Also, their signs may differ. Whereas only a single value of r_{yz} can result from a given combination of r and C values, widely varying combinations of r and C values may satisfy any r_{yz} value. Thus it follows that no unique conclusions can be drawn from confirmed correlations between ratios, with regard to the amount and sign of the correlation between the absolute values from which these ratios were derived.

The term r_{yz} becomes significantly simplified in special cases when the individual values of r become equal among themselves, or else become equal to either zero or unity, and individual C values become equal among themselves. Among the cases discussed by F. Chayes, the one in which y and z have a common denominator is of significance for the methods here concerned. This occurs, for example, when $x_2 = x_4$ and consequently

$$y = \frac{x_1}{x_2}, \qquad z = \frac{x_3}{x_2}.$$

Then $C_2 = C_4$, $r_{24} = 1$, $r_{12} = r_{14}$, $r_{23} = r_{34}$, and the general expression takes on the following simplified form:

$$r_{yz} = \frac{r_{13} C_1 C_3 - r_{12} C_1 C_2 - r_{23} C_2 C_3 + C_2^2}{\sqrt{C_1^2 + C_2^2 - 2 r_{12} C_1 C_2} \sqrt{C_2^2 + C_3^2 - 2 r_{23} C_2 C_3}}.$$

If no correlation exists between the absolute values x_1, x_2 and x_3, $r_{12} = r_{13} = r_{23} = 0$, and the same expression is further simplified as follows:

$$r_{yz} = \frac{C_2^2}{\sqrt{C_1^2 + C_2^2} \sqrt{C_2^2 + C_3^2}}.$$

Since the right hand side of the equation will always be >0, a positive correlation should always exist between y and z, although x_1, x_2 and x_3 are assumed as uncorrelated. This case, which is of great practical importance, corresponds to the so-called **spurious correlation** of K. Pearson. If, for example, x_1, x_2 and x_3 designate the absolute quantities in a rock series of three different minerals uncorrelated with respect to their quantities, a correlation, always positive, will nevertheless be formed from the ratios derived from x_1, x_2 and x_3, which have a common denominator, irrespective of which of the three is chosen as this denominator. Thus, the conclusion that a correlation exists between the absolute quantities of the individual minerals, drawn because of an existing ratio correlation, would be a gross mistake.

3. Illustration of the Mutual Relations Between Two, Three and Four Variables Having a Constant Sum

If, in general, there are n variables and their sum S is stipulated to have a constant value, $(n - 1)$ terms will suffice for their determination. If one assumes, as is customary, that $S = 100$, the variables will correspond to the conventional percentage values.

Therefore 2 variables with a constant sum S can be determined and depicted graphically from a single term or coordinate; 3 variables from two terms or coordinates; and 4 variables from three terms. The graphical representation leads to the well-known concepts of the **concentration line**, the **concentration triangle**, and the **concentration tetrahedron**. These were used in physical chemistry by J. W. Gibbs, in order to demonstrate the concentration ratios in binary, ternary and quaternary systems. This type of representation is no longer visually illustrative when dealing with more than 4 components (variables), because of the three-dimensional nature of our space, though this way of representation is valid also for multidimensional spaces. Such experiments were indeed tried, but will not be considered here.

a) **Two variables:** The concentration line. Two variables or a binary system, i.e. the percentage portion of two components A and B, where $A + B = S$ or $= 100$, may be represented by a straight line having

A and B as its end points (Figure 9). The end point A corresponds to 100% A and 0% B, and the end point B, conversely, to 0% A and 100% B. Thus, every point on the line AB is associated with a precisely defined composition $A_xB_{(S-x)}$ or $A_xB_{(100-x)}$ which can be easily read off from the diagram. The ordinate is available for indicating additional values as functions of the ratio $A:B$.

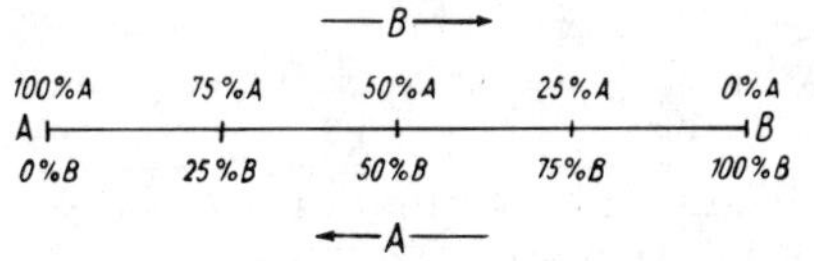

```
                        —B—→
    100%A      75%A      50%A      25%A       0%A
    A├─────────+─────────+─────────+──────────┤B
    0%B        25%B      50%B      75%B       100%B
                        ←—A—
```

FIGURE 9. Concentration line. Each point on the line corresponds to a definite ratio of $A:B$, and visa versa. The second dimension (ordinate) is available for indicating an additional variable (e.g. temperature, pressure, refractive index) as a function of the $A:B$ ratio

b) **Three variables:** The concentration triangle. The graphical representation of three variables or the quantitative relationships between three components A, B, and C, with a constant sum S, or 100, can be realized by means of the so-called concentration triangle, as introduced into physical chemistry by J. W. Gibbs. This triangle is usually equilateral, but need not necessarily be so.

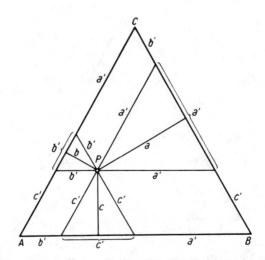

FIGURE 10. Concentration triangle for the illustration of the concentrations of three components A, B and C. Every point within the triangle corresponds to a definite ratio of $A:B:C$ and visa versa. Variables which are functions of concentration should be plotted on ordinates normal to the plane of the paper, thereby converting the representation to a three-dimensional framework.

In the triangle ABC (Figure 10) a, b, and c designate the perpendiculars from a point P on the three sides of the triangle. Lines parallel to the sides of the triangle are drawn through P. Three new triangles are thus formed, having the sides a', b', and c'. Since, according to Pythagoras, the ratio of the perpendicular height to any side in an equilateral triangle is always equal to $\sqrt{3} : 2$, it follows that

$$\frac{a}{a'} = \frac{b}{b'} = \frac{c}{c'} = \frac{\sqrt{3}}{2},$$

and

$$a' = \frac{2}{\sqrt{3}} a, \quad b' = \frac{2}{\sqrt{3}} b, \quad c' = \frac{2}{\sqrt{3}} c.$$

From the figure it also follows that $a' + b' + c' = s$, which is the total length of the sides of the triangle, so that:

$$a + b + c = \frac{s}{2} \sqrt{3} = h,$$

i.e. the sum of the three normals from a selected point P to the three sides of the triangle is equal to the height h of the triangle. Normals $a, b,$ and c are termed the triangular coordinates of the point P. Since only two coordinates are necessary in order to determine the location of a point in a plane, only two triangular coordinates are needed here, e.g. a and b. The third is obtained as follows: $c = h - (a + b)$.

Instead of a, b, and c the parallels to the sides of the triangle, a', b', and c' may be used as coordinates for determining the location of point P. In this case also, two parallels are sufficient, e.g. a' and b'. The third is obtained as follows: $c' = s - (a' + b')$.

The triangular coordinates, as introduced here, can also be used for defining points outside the triangle ABC (Figure 11), provided that the appropriate sign is indicated.

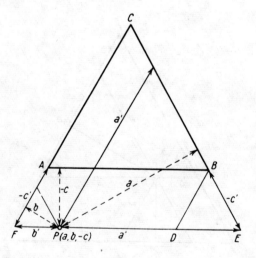

FIGURE 11. Concentration triangle, showing points outside the $\triangle ABC$. Since $\triangle BDE$ is equilateral, it follows for $P(a, b, -c)$, that $a' + b' - c' = AB = s$. In the case of $\triangle FEC$, $a + b$ is equal to the height of the triangle and, consequently, $a + b - c = h$, the height of $\triangle ABC$.

The coordinates a, b, and c or a', b', and c' should be designated by either positive or negative values, depending on the position of point P in relation to the corner of the triangle to which the coordinate is related. Figure 12 shows a scheme of the signs in the 7 areas bounded by the sides of the triangle or their extensions. The signs are shown in the order a, b, and c. From the figure it is quite evident that only points falling within the triangle will have all their coordinates positive, and that a situation never arises where all the coordinates are negative. This also follows from the equation $a + b + c = h$ or $a' + b' + c' = s$, since both h and s are always positive.

If either s or h is made to equal 100, the triangle ABC becomes suitable for the representation of the percentage distribution of three components. Each point within the triangle with the coordinates a, b, c or a', b', c' corresponds to a mixture of a or $a'\% A$, b or $b'\% B$, and c or $c'\% C$. The three sides of the triangle, AB, BC, and CA correspond to a distribution, for which C, or A or B respectively equal zero. They represent the concentration lines of the binary systems AB, BC or CA. The apexes A, B, and C correspond to the pure components A, B, and C.

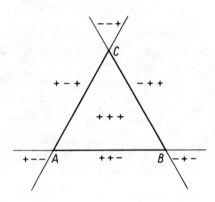

FIGURE 12. Sign scheme of the triangular coordinates for points within and outside $\triangle ABC$.

Lines parallel to the sides of the triangle, i.e. level lines with respect to one component, correspond to a distribution having a constant content with respect to the component situated at the opposite apex, e.g. the content A remains constant along a line parallel to BC but the ratio $B : C$ may vary from 0 to ∞. Lines drawn through the apexes correspond to distributions with a constant ratio between the other two components, e.g. the ratio of $A : B$ will remain constant, whereas C will vary along a line drawn through the apex C. In the special case when the line passing through the apex at the same time also corresponds to the height of the triangle, the constant ratio between the two components will be 1:1.

In practice, the representation of ternary systems or of three components with a constant sum of 100, is carried out using commercially available triangular coordinate paper with side lengths equal to 20 cm*. On this paper the sides of the triangle, as well as its heights, are divided into 100

* Produced by Aerni-Leuch (Bern, Switzerland), Schleicher and Schüll (Düren, Rheinland, West Germany) and others.

divisions (Figure 13). The use of this type of paper is extremely convenient. For any given percentage distribution of the three components, the corresponding point can be plotted immediately from its coordinates a, b, c or a', b', c'. In addition, the coordinates of any selected point may easily be read, thus providing the respective percentage distribution (Figure 14).

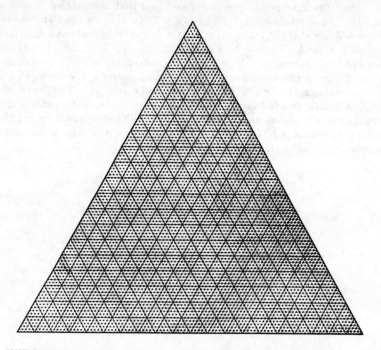

FIGURE 13. Commercially available triangular coordinate paper (side lengths = 20 cm)

If no triangular coordinate paper is available, a distorted right-angled isosceles triangle may be prepared on commonly-used coordinate (millimeter-) paper. Let us again make $S = 100$, and select a rectangular coordinate system XOY, (Figure 15). The quantity of A is plotted on the X-axis, for example, and that of B, on the Y-axis. Thus, it is immediately evident that points having definite ratios of $A:B:C$ (when $A + B + C = 100$) can only fall within the triangle OAB. Lines parallel to the X-axis correspond to a constant content of B, and those parallel to the Y-axis—to a constant content of A. The axes themselves represent $B = 0$ or $A = 0$. In a similar manner the side of the triangle AB corresponds to $C = 0$, and lines parallel to this side represent geometrical loci with a constant value of C, whereas the origin O corresponds to $C = 100$. Every point within the triangle thus corresponds to a certain ratio or a certain concentration A_x, B_y, $C_{[100-(x+y)]}$, which can be read directly therefrom. This mode of representation was used, among others, by F. Becke for his SiLU triangle, for representing chemism of metamorphic rocks (Becke, 1912).

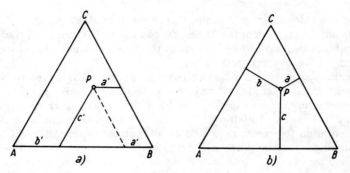

FIGURE 14. The plotting of a point P according to its coordinates (concentration, percentage contents) a) using the coordinates a', b', c'; b) using the coordinates a, b, c.

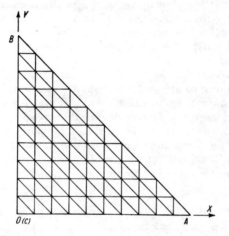

FIGURE 15. Rectangularly distorted concentration triangle ABC.

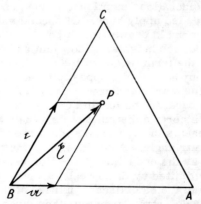

FIGURE 16. Plotting of a point in a concentration triangle on the basis of the vectorial addition of two coordinates considered as vectors.

These considerations also apply to oblique-angled triangles, for which, however, no detailed proof will be offered here. Oblique-angled triangles have a certain degree of usefulness in the later discussed method of the so-called remainder triangle.

If, in the absence of special coordinate paper, points are to be plotted for concentration triangles of any shape, whether equilateral, rectangular or oblique-angled, the following can be done. Starting from any apex, e.g. B (Figure 16), the quantities of the two other components, i. e. A and C, are indicated as two vectors $\mathfrak{a}$ and $\mathfrak{c}$, directed towards A and C respectively. The desired projection point will be given by the end point of the vector $\mathfrak{r}$, as obtained by vectorial addition of $\mathfrak{a}$ and $\mathfrak{c}$, i. e. by the diagonal of the parallelogram defined by $\mathfrak{a}$ and $\mathfrak{c}$. This construction is also valid for points located outside the triangle ABC, i. e. in cases where 1 or 2 components are considered to be negative. The reason for the correctness of the demonstrated construction lies in the fact already mentioned, that a system of three variables with a constant sum is already determined by two of them. Knowing the sum of two components, e. g. $A + B$, the third may be obtained from $C = [100 - (A + B)]$, the components of the vector $\mathfrak{r}$ corresponding to the coordinates a', b' or c'.

In order to read off the quantitative ratios of the three components A, B, and C corresponding to any selected points within or without the triangle, the inverse procedure may be used. Any apex of the triangle is connected to P by means of a vector. The vector is then resolved into two components parallel to the sides of the triangle, which intersect at the selected apex. The lengths of these components provide two of the desired concentrations. This construction, carried out by simple shifting of parallels, is especially useful in the case of oblique-angled concentration triangles.

c) Four variables: The concentration tetrahedron.

α) General. Four-component systems can be represented by means of a concentration tetrahedron in analogy to the representation of three-component systems or, in general, the distribution of three components, in triangular coordination. This method has also been long used in physical chemistry. The considerations mentioned in the previous section with regard to the triangle also apply here. Of special validity is the assumption that the sum of normals drawn from a point P within the tetrahedron to the four planes, i. e. the heights, is constant and equal to the height of the tetrahedron. By the term height of the tetrahedron h we refer to the normal drawn from any apex to the opposite face. Equally valid also is the assumption that the sum of the coordinates of P, parallel to the edges of the tetrahedron, is equal to the length of one edge. In the triangle the point characterizing a certain distribution is situated at the end point of a vector which represents the vectorial sum of two vectors directed parallel to the sides of the triangle. Analogously, in the tetrahedron the point P can be considered as the end point of a vector representing a diagonal of a parallelipiped delimited by three vectors, each parallel to one edge.

A further analogy to the established qualities of the triangle is the fact that at the corners of the tetrahedron the other three components have zero values. On the edges of the tetrahedron two components become equal to zero, whereas on the faces of the tetrahedron only one component becomes equal to zero. In sections which are parallel to a face of the tetrahedron, the concentration of one of the components, namely, the component corresponding to the opposite corner, remains constant. The plane of the section is an "equal level" plane with respect to this component. Section planes which include a tetrahedral edge will represent a constant quantitative ratio of the two components not associated with that edge.

In principle, the regular tetrahedron may also be replaced by a rectangular tetrahedron. In analogy to the right-angled concentration triangle, the percentage quantities of three components, for example, A, B and C, can be plotted on the three axes of a rectangular system of coordinates. If the consideration regarding the triangle are applied to the three dimensional relations, planes parallel to XOY will correspond to constant values of C, planes parallel to XOZ — to constant values of B, and planes parallel to ZOY — to constant values of A, the planes parallel to ABC correspond to constant values of the fourth component D, while at the origin O itself, $D = 0$. Occasional use will be made of the rectangular tetrahedron in the following sections.

β) **Representation of the concentration tetrahedron on a plane.** In contrast to the concentration triangle which represents a plane figure, the tetrahedron gives rise to the problem as to how it is to be represented on the plane of the drawing. Simple perspective images are only suitable for providing a general view and do not permit the plotting of points with reference to their coordinates or the derivation of the latter from indicated or constructed points. This problem can be solved by means of two methods: either by considering suitable section planes which are later swivelled into the plane of the drawing where they appear in their true proportions, or by projecting the tetrahedron onto one or more suitable planes. The first method was used, among others, by P. Niggli for illustrating the igneous field in the *al-fm-c-alk* tetrahedron. Therefore, this method will be clarified in the corresponding section. The projection of the concentration tetrahedron onto different planes was mainly used by H. E. Boeke, W. Eitel and F. Becke.

Only the two most important methods, which are especially suited for petrographic purposes, suggested by H. E. Boeke (H. E. Boeke, 1914, 1915, 1916, 1917; W. Eitel, 1923), will be given here, namely, the projection onto one of the tetrahedral faces, and the projection onto the faces of a cube circumscribing the tetrahedron, i. e. on the crystallographic axial planes of the tetrahedron. A modified version of the last method, given by F. Becke, 1925, will also be mentioned. For additional methods reference should be made to the previously quoted works of H. E. Boeke.

αα) **Projection onto a tetrahedron face.** A point P is plotted, corresponding to the quantities $a, b, c,$ and d of the respective component A, B, C and D, where $a + b + c + d = 100$. The tetrahedral face BCD is selected as the plane of projection. Preferably using triangular-coordinate paper (Figure 17), the length of the edge is set at $s = 100$. A section plane $B'C'D'$ is visualized, which includes the point P and is parallel to BCD. Coordinates $b, c,$ and d, parallel to the plane of projection BCD, appear in this plane in their true length. If the coordinates of P on the triangle BCD are marked $b'c'd'$, it follows from the figure that $b' + c' + d' = s$. Since, according to the original assumption, $a + b + c + d = s$, and $BB' = CC' = DD'$, the difference $s - (b' + c' + d') = a$ must be equally divided among the three coordinates, so that

$$b' = b + \frac{a}{3}, \qquad c' = c + \frac{a}{3}, \qquad d' = d + \frac{a}{3}.$$

41 Thus, the coordinates of P for quaternary mixtures or for four quantities with a constant sum can be indicated in the projection. Since all the points along a normal to the projection plane are represented by the same point P, it is necessary to project this point P onto a second face of the tetrahedron in order to obtain its exact location.

[40]

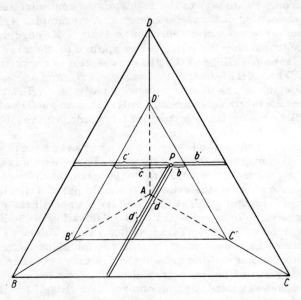

FIGURE 17. Projection of a concentration tetrahedron $ABCD$ onto the face BCD and the plotting of a point according to its coordinates.

$\beta\beta$) Projection on a face of a cube circumscribing the tetrahedron (crystallographic axial plane). In Figure 18 the cube face parallel to the tetrahedron edges AB and CD is selected as the plane of projection. A plane drawn through P, parallel to the plane of projection, will cut the tetrahedron, forming the rectangle $A'B'C'D'$. Both the coordinates a and b, parallel to the edges AB, as well as the coordinates c and d, parallel to CD, will appear in their true length. Introducing a new set of rectangular coordinates x, y, with their origin at O, and parallel to the edges AB and CD, we obtain

$$2(x+b) = a+b \quad \text{and likewise} \quad 2(y+c) = c+d,$$

and it thus follows that

$$x = \frac{a-b}{2}, \quad y = \frac{d-c}{2}.$$

42 The new rectangular coordinates in the plane of projection (x, y) are thereby expressed by the tetrahedral coordinates (a, b, c, d). At the same time attention must be paid to the signs. Given the orientation accepted for the system XOY, OA and OC, for instance, will correspond to its positive directions, and OB and OD—to its negative directions. Thus, a and c are to be considered as

positive and *b* and *d* as negative. Since in the considered use $|a|>|b|$ and $|d|>|c|$, *x* turns out to be positive and *y* negative.

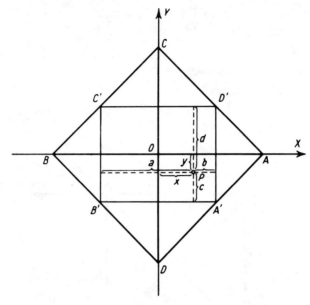

FIGURE 18. Projection of a concentration tetrahedron *A B C D* onto a face of the circumscribing cube and the plotting of a point according to its coordinates.

In order to avoid changing signs within the tetrahedron and to obviate the subtraction of arithmetic values, a modified form of representation was suggested by F. Becke (1925). Accordingly, the coordinate system is constructed parallel to the edges of the cube circumscribing the tetrahedron, instead of parallel to the tetrahedron edges; the origin of the coordinates is thus displaced to the bottom rear corner of the cube. In this way the entire tetrahedron comes to be located within the I octant. To this end, a transformation of coordinates is required, consisting of a displacement of the origin from O to $O'=B$, followed by a clockwise rotation of 45° (Figure 19). After the parallel translocation of the system XOY by 1/2 (edge length of the tetrahedron = 1) the new coordinates (x', y') of P will be given by

$$x' = x + \frac{1}{2}, \qquad y' = y.$$

After the 45° rotation of the system $X'O'Y'$ in a negative sense, the new coordinates (ξ, η) will be given by

$$\xi = x' \cos 45° - y' \sin 45°, \quad \eta = x' \sin 45° + y' \cos 45°$$

or, if these values are substituted for x', y' and noting that $\sin 45° = \cos 45° = 1/\sqrt{2}$ and $a/2 + b/2 + c/2 + d/2 = 1/2$, then

$$\xi = \frac{a+d}{\sqrt{2}}, \quad \eta = \frac{a+c}{\sqrt{2}}.$$

Since $1/\sqrt{2}$ = the edge w of the cube circumscribing the tetrahedron of an edge length of l, it follows that

$$\xi = w(a+d), \quad \eta = w(a+c).$$

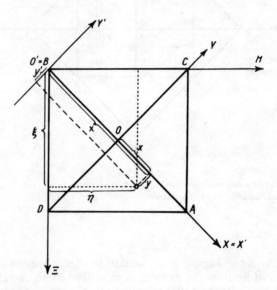

FIGURE 19. Displacement of the origin from the center of the tetrahedron to the bottom rear corner of the circumscribed cube, in order to avoid negative coordinates. After F. Becke (1925)

This method of representation, using a projection upon a single cube plane, is also not unequivocal, since all the points within the tetrahedron which fall on a given normal to the selected plane will coincide in the projection, i.e., they will all have the same coordinates (ξ, η). Therefore, in this case it is also necessary to introduce an additional projection on another of the cube's faces (Figure 20). If (001) is taken as the cube face used, the projection is obtained on (010), the side view, in which the tetrahedron is rotated by 90° (anticlockwise) about [100] = crystallographic $a = \Xi$. The coordinate ξ is retained, η is eliminated, and a new coordinate ζ is introduced, parallel to the cube edge (001) = crystallographic c. Its value is given by

$$\zeta = w(c+d).$$

The projection on the third cube face (100), i.e., the elevation, is obtained by rotation of the tetrahedron about [010] = crystallographic $b = H$. In this case the coordinate ξ disappears, and we obtain

$$\eta = w(a + c) \quad \text{and} \quad \zeta = w(c + d).$$

If the coordinates (ξ, η, ζ) are given, the quantities $a, b, c,$ and d can be calculated. If the length of cube edge $w = 1$ (thereby setting the tetrahedron's edge at $\sqrt{2}$), then, according to F. Becke,

$$a = \frac{1}{2}(\xi + \eta - \zeta), \quad c = \frac{1}{2}(-\xi + \eta + \zeta),$$

$$b = 1 - \frac{1}{2}(\xi + \eta + \zeta), \quad d = \frac{1}{2}(\xi - \eta + \zeta).$$

F. Becke (1925, 38-40) also considered projections on the diagonal planes of the cube (the tetrahedron symmetry planes, the faces of the rhombic dodecahedron). Utilization of Becke's projection is especially apparent in the works of the Viennese school, and among others, those of Ch. Bacon (1926, 164) and A. Marchet (1931, 490-537).

III. Aids for the Calculation

The petrochemical-calculation methods here described only call for exclusively elementary arithmetic operations. Although these operations are neither complicated nor time-consuming, it is nevertheless advantageous to mechanize them as far as possible, thus saving time and also avoiding errors.

The constantly recurring process of transforming weight-percentage analysis values to equivalent numbers by dividing by the corresponding formula weights, can best be carried out by resorting to tables. The tables prepared by H. S. Washington are commonly used. These tables, which were prepared by rounding off formula weights for the norm calculation of his well known collection of analyses (H. S. Washington, 1917), were later reprinted by various authors, including A. Holmes (1921) and A. Johannsen (1931).

The rounded-off formula weights used are given below (values considered exact at present are given in brackets for purpose of comparison):

SiO_2	60	(60.06)	CaO	56	(56.08)
Al_2O_3	102	(101.94)	Na_2O	62	(61.994)
Fe_2O_3	160	(159.68)	K_2O	94	(94.192)
FeO	72	(71.84)	TiO_2	80	(79.90)
MnO	71	(70.93)	P_2O_5	142	(141.96)
MgO	40	(40.32),	etc.		

Nowadays there is no longer any reason for using tables which are based on rounded-off values. Newer tables containing exact values are available, e.g., those of H. V. Philipsborn (1933). These include equivalent numbers to four decimal places for intervals of 0.01 wt %. This accuracy is mainly meant for calculations of mineral analyses. Intervals of 0.1 wt % and three decimal places are sufficient for petrochemical purposes, rendering the tables much clearer and handier. Such tables were recalculated by the author and published in 1945 (C. Burri and P. Niggli, 1945). They are presented in a somewhat enlarged form in the appendix. The tables also include, along with the so-called "molecular" equivalent numbers of the most important rock-forming oxides, such as SiO_2, Al_2O_3, Fe_2O_3, FeO, etc, the "atomic" equivalent numbers, corresponding to the metal atoms (cations). These are obtained by dividing the weight percentage of Al_2O_3 by 1/2 Al_2O_3, of Fe_2O_3 by 1/2 Fe_2O_3, etc. One of their uses is for the conversion of an analysis to the at present much used metal atoms or cation percentages. The tables are presented in a form which directly provides the (molecular and atomic) equivalent numbers for intervals of 0.1 wt %. Interpolation for intervals of 0.01 wt % is carried out by means of proportional tables, provided in the margin, as is the practice in logarithmic tables.

A small calculating machine, as used in commerce under different names, e.g. "Addiator", "Stima", etc, will be of advantage during the ensuing operations of addition and subtraction.

The logarithmic slide rule is the best-suited instrument for arithmetic operations involving multiplication and division. This applies to the recurrent calculations involving rule-of-three and proportions and to all the important calculations of a number of terms making up a sum of 100. The author assumes that the reader is familiar with the construction and use of the slide rule. Brief mention will only be made of how the last-mentioned operation, i.e., the calculation of any desired number of terms making up a sum of 100, is carried out to special advantage.

If $a + b + c + d + \cdots = S$ and $a' + b' + c' + d' + \cdots = 100$, the desired percentages will be obtained as follows:

$$a' = \frac{100\,a}{S}, \qquad b' = \frac{100\,b}{S}, \qquad c' = \frac{100\,c}{S} \quad \text{etc.}$$

The easiest way to arrive at these values is to set the slide rule to $1/s$ and then multiply this quotient by 100 *a*, 100 *b*, 100 *c*, etc. In order to minimize the moving of the slide rule, the two upper scales are used to advantage. The readings obtained are more accurate if one uses a large model, e.g. a rule with a scale length of 50 cm.

Moving the slide can be obviated altogether, i.e. all the percentage numbers required can be obtained from one setting, if a disk-shaped slide rule is used instead of the common slide rule. This corresponds in a way to a slide rule the start and end of which coincide, so that the point of reading will be of no significance. Similar advantages are provided by slide rules made by the firm Dennert and Pape (Hamburg). A second base scale, displaced by the value of π as in the model "Aristo Technica" for example, permits the immediate reading of any desired percentage value at one and the same setting of the slide.

Disk-shaped slide rules, such as have proved themselves for many years during the work carried out in the Zürcher Mineralogical-Petrographical Institute, are produced in various types with excellent finish by the firm Loga-Calculator in Uster (Zürich). The simpler models with two scales are fully sufficient for the present purposes. More elaborate models provide the same possibilities as the well-known slide-rule models "Rietz" or "Darmstadt". The disk-ruler pocket models with a diameter of 12 cm correspond in accuracy to rod-shaped slide rules of 30 cm scale length. The very handy, supported, so-called "bureau-model" of about 30 cm diameter corresponds to slide rules with 75 cm scale length and provides accordingly 3 to 4 value numbers*.

[43]

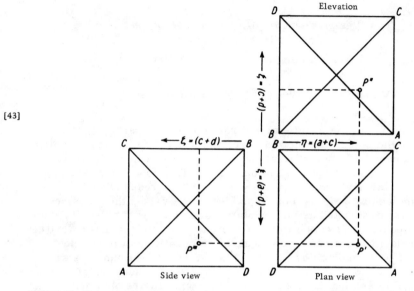

FIGURE 20. Representation of a point within the concentration tetrahedron $ABCD$ in plan view, elevation and side view. After F. Becke (1925).

---

* The pocket model which was previously described by the author (C. Burri, 1944, 310-344) is no longer in production. This model featured a staggered upper division for the transitive reading of 3 value numbers, as well as a possibility of introducing special values (for example formula weights) on the back side with the ability of transfer to the front side. Information concerning a large number of other models is provided in the catalogue of the manufacturers.

B. PETROCHEMICAL CALCULATION METHODS BASED ON EQUIVALENCE ACCORDING TO P. NIGGLI

I. The Niggli-Values *si, al, fm, c, alk, k, mg*, etc
and Their Applications

1. The Calculation of the Niggli-Values

a) Niggli-Values and Weight Percentages

α) Calculation of the Niggli-values from the weight percentages. According to the principle followed here, the weight-percentage data provided by the chemical rock analysis should be transformed into e q u i v a -
l e n t v a l u e s, i. e., to equivalent numbers. This is accomplished by dividing the numerical results of analysis by the corresponding formula weights ("molecular weights") of the oxides. The values thus obtained, which are also designated as "molecular proportions", "molecular numbers", "molecular quotients", or "m o l e c u l a r e q u i v a l e n t n u m b e r s", are inter-related in a manner identical to the quantities of the oxide equivalents ("oxide molecules"). The obtained equivalent numbers are generally multiplied by a factor of 1,000 to obviate the handling of fractions. As mentioned in section A III, the equivalent numbers are obtained from the tables provided in the appendix.

In order to reduce the number of the components for the sake of greater clarity, a g r o u p i n g of the components, with the exclusion of SiO_2, TiO_2 and P_2O_5, is made in the following manner:

I. Al_2O_3
II. $FeO + Fe_2O_3$ (calculated as FeO)* + MnO
III. CaO
IV. $Na_2O + K_2O$

Rarer elements or oxides, which are therefore often not dealt with in rock analyses carried out for petrographic purposes, should be added to their respective groups I-IV. In earlier days this addition was made in accordance with pure chemical considerations, i. e., with preference to the valency or to the position in the periodic table. According to the present state of knowledge of crystal chemistry, space requirement qualities constitute the prime factor governing the incorporation of elements into crystal structures. These are characterized by the so-called empirical ionic radii. Thus, changes must be introduced with respect to the earlier accepted rules. The following should be noted (C. Burri 1956):

Cr_2O_3 and V_2O_3 were previously added to Al_2O_3, because of their trivalency. They should, however, be added to Fe_2O_3, as shown by the

_____
* Since $1 Fe_2O_3$ is equivalent to $2 FeO$, the equivalent number of Fe_2O_3 should be multiplied by 2 before the summation.

empirical ionic radii: $Cr^{3+}-0.64$, V^{3+} 0.65, Fe^{3+} 0.67, as compared with Al^{3+} 0.57.

MnO can be grouped together with FeO, although Mn may replace Ca to some extent, e. g., in apatite. NiO and CaO are added to FeO, as is already conventional. The corresponding ionic radii are: Mn^{2+} 0.91, Fe^{2+} 0.83, Ni^{2+} 0.78, Co^{2+} 0.82 and Ca^{2+} 1.06. Until recently SrO and BaO were always grouped with CaO. Ba is, however, too large to replace Ca, and it more frequently replaces K, e. g., in the feldspars. Therefore it should be added to K_2O. It should be noted that 1 BaO is equivalent to only one half of K_2O. SrO may replace either Ca or K. Since SrO generally occurs in only limited amounts, for the sake of simplicity it can be treated like BaO. The corresponding ionic radii are: Ba^{2+} 1.43, Sr^{2+}-1.27, K^+ 1.33 and Ca^{2+}-1.06.

The rarely determined Rb_2O and Cs_2O, as well as the somewhat more frequently determined Li_2O, have till now been calculated as Na_2O and K_2O. This procedure need not be changed in the case of the two first-mentioned oxides, as Rb and Cs generally follow K (ionic radii: K^+ 1.33, Rb^+ 1.49, Cs^+ 1.65) in the late stages of crystallization despite their larger space requirements. In contrast, the far smaller Li^+ ion with radius size 0.78 replaces Mg^{2+} (0.78) in the most important Li-minerals — the Li-micas. It is the Mg, with equal space requirements, and not the far larger K, which is replaced, and thus Li^+ should be added to Mg^{2+}. It should be noted that 1 MgO is equivalent to one half Li_2O, i.e., 1 Li_2O is equivalent to 2 MgO.

All these elements occur only in such small amounts in crystalline rocks, that the suggested innovations with respect to the calculation procedure, as compared with the usual one, will have hardly any influence on the Niggli-values. The new findings should, however, be taken into consideration, in principle, even if their effect will be generally small.

The evaluation of the role of titanium encounters special difficulties. This element quantitatively constitutes 0.44 % (by weight) of the lithosphere and is the ninth most abundant element, being placed after O, Si, Al, Fe, Ca, Na, Mg and K. These difficulties are mainly due to the fact that, from the analytical point of view, it is very difficult to distinguish between the trivalent form and the quadravalent form. This is indeed impossible in the presence of divalent and trivalent iron.

On the one hand titanium is incorporated into independent Ti-minerals such as ilmenite $FeTiO_3$, rutile TiO_2 (rarely also anatase or brookite) and titanite $Ca[Ti(O, OH, F)SiO_4]$, which can be replaced by $CaTiO_3$ perovskite, when conditions of extremely low silication are prevalent. On the other hand, Ti also replaces Fe^{3+}, Al^{3+} or Mg^{2+}, e. g., in titanomagnetite or in Ti-bearing augites or hornblendes, to mention only a few important cases. The corresponding ionic radii are $Ti^{3+}-0.69$, $Ti^{4+}-0.64$, $Mg^{2+}-0.78$, $Fe^{3+}-0.57$ and $Fe^{2+}-0.67$. The possible substitution relation of Ti^{4+} for Si^{4+}, which was often considered in the past, is nowadays much less taken into account, if at all, due to the large difference with respect to space requirements ($Si^{4+}-0.39$). The calculation of a *ti*-value, in analogy to the *si*-value, dates back to a period prior to the development of modern crystallochemistry, when it was still believed that the mentioned substitution of Ti^{4+} for Si^{4+} took place.

At present it may be taken for granted that Ti should be added to FeO + + MgO. This will also be the main way of handling titanium in the later mentioned QLM-projection based on the so-called basis values. The titanium incorporated in complex anions, e. g., in titanite, with titanium having a cationic character, as in augites and hornblendes and having an oxide-type bond (rutile, ilmenite, perovskite) does have much influence on the Ti-content of normal rocks, which is never very high. Al is treated in the same manner, since the Al_2O_3 of the chemical analysis includes the Al combined in the aluminosilicate anions as well as the Al having a cationic character, e. g., in the augites, hornblendes and micas. The relationships here are, however, more convenient, since the part of Al which is contained in the aluminosilicate anions can be readily estimated due to its ratio of 1 : 1 with respect to (Na, K)$_2$O as well as CaO, to which the Al is bound. Such simple considerations cannot be applied to Ti. It is for purely practical reasons that the previously adopted procedure should be adhered to, especially as this calculation method pays more attention to the role of Ti. According to this method the *ti*-value is calculated in analogy to the *si*-value. This method also has the advantage of providing simple data needed for the comparison of the Ti-contents characteristic of rock series or petrographic provinces.

The equivalent numbers included within groups I-IV are recalculated to the sum of 100, and designated as *al, fm, c,* and *alk.*

The hitherto unconsidered equivalent numbers of SiO_2, TiO_2, P_2O_5 and ZrO_2, and also of CO_2, SO_3, Cl_2, F_2, S and H_2O, are also recalculated on a similar basis to that used for deriving $al + fm + c + alk = 100$. These values are deisgnated as *si, ti, p, zr, co$_2$, so$_3$, cl$_2$, f$_2$, s, h*.

The values *si, al, fm, c,* and *alk* in most cases suffice for the chemical characterization of rocks. In numerous other cases, however, it is desirable to obtain more detailed information on the two constituent components *alk* and *fm*. This is information obtained by the construction of the two ratios

$$k = \frac{K_2O}{K_2O + Na_2O} = \frac{K^+}{K^+ + Na^+}$$

and

$$mg = \frac{MgO}{FeO + MgO + MnO} = \frac{Mg^{2+}}{Fe^{2+} + Fe^{3+} + Mg^{2+} + Mn^{2+}},$$

whereby it is understood that the total iron is recalculated to the divalent form and given as "FeO".

---

* Unfortunately, a minor inconsistency exists with respect to these designations. Although the expressions which were based on Cl_2 and F_2 and defined for the content of chlorine and fluorine, were formulated as as cl_2 and f_2, this is, however, not the case with respect to the term p, corresponding to phosphorus-pentoxide, although it is calculated on the basis of P_2O_5. According to logic, this term should be designated as p_2 with the index 2, in contrast to the values $si, ti, co_2, s,$ and h, which were defined on a monoatomic basis or a simpler formula unit. It is, however, inadvisable to change the old mode of indexing.

If, in special cases, one is interested in the degree of oxidation of the Fe, the following ratio is used:

$$w^* = \frac{2\,Fe_2O_3}{2\,Fe_2O_3 + FeO} = \frac{Fe^{3+}}{Fe^{3+} + Fe^{2+}}.$$

The k-, mg- and w-values can vary from 0 to 1.

The values al, fm, c, and alk are often rounded-off or given to the accuracy of 0.5 in order to bring the sum to 100.0. This is completely adequate, if they are to be used primarily for graphical representation, e.g., in variation diagrams. If, however, it is intended to use the Niggli-values as a basis for the further calculation of values derived therefrom, they should be given to an accuracy of 0.1, as will always be the case hereafter.

Examples**:

1. Quartz-diorite, Spanish Peak, California

	Weight %	Formula weight	Molecular equivalent numbers multiplied by 1000			
SiO_2	59.68	60.06	994			
Al_2O_3	17.09	101.94	168	168	$al = \dfrac{168 \cdot 100}{524} = 32.1$	
Fe_2O_3	2.85	159.68	18, as FeO 36			
FeO	2.75	71.84	38	162	$fm = \dfrac{162 \cdot 100}{524} = 30.9$	
MgO	3.54	40.32	88			
CaO	6.62	56.08	118	118	$c = \dfrac{118 \cdot 100}{524} = 22.5$	
Na_2O	3.87	61.994	62			
K_2O	1.31	94.192	14	76	$alk = \dfrac{76 \cdot 100}{524} = 14.5$	
TiO_2	0.65	79.90	8	524		100.0
P_2O_5	0.25	141.96	2	994 : 524 = si : 100		
H_2O+	1.00	18.016	56	$si = \dfrac{994 \cdot 100}{524} = 190$	$k = \dfrac{14}{14+62} = 0.18$	
H_2O-	0.15					
incl.	0.27	and likewise:	$ti = \dfrac{8 \cdot 100}{524} = 1.5$	$mg = \dfrac{88}{36+38+88} = 0.54$		
	100.03					

---

* In the past, the ratio $o = \dfrac{Fe^{3+}}{Fe^{3+} + Fe^{2+} + Mn^{2+} + Mg^{2+}}$ was used for the same purpose. The new w-value has the advantage of being related only to the element iron with its different valencies.

** The same three examples will always be referred to for illustrating the different calculation methods, as far as igneous rocks are considered, thus allowing for interesting possibilities of comparison and control. The examples are as follows:

   1. A rock with a "normal" chemism, i.e. having neither alumina nor alkali excess: hornblende-quartz diorite, Spanish Peak, Pulmas Co. California, USA. Anal. H.N. Stockes, in H.W. Turner, US Geol. Surv. Ann. Rep. 17. I (1896) 724. Also in H.S. Washington, US Geol. Surv. Prof. Pap. 99 (1917) 381, No. 82.

   2. A rock having a pronounced excess of alumina: garnet-bearing cordierite andesite, Hoayzo near Nijar, Prov. Almeria, Spain. Anal. I. Parga-Pondal; in C. Burri and I. Parga-Pondal, Schweiz. Min. Petr. Mitt. 16 (1936) 245.

   3. A rock having a pronounced excess of alkali: arfvedsonite-lujaurite, Lille Elv, Kangerdluarsuk, Greenland. Anal. C. Winter; in V. Ussing, Medd. om Gronland, 38 (1911) 175. Also in H.S. Washington, loc. cit. (1917) 555, No. 9.

MnO	Tr.
CO$_2$	0.20
SO$_3$	Sp.
Cl	0.03
BaO	0.04
SrO	Tr.
Li$_2$O	Tr.
	0.27

$$p = \frac{2 \cdot 100}{524} = 0.4$$

$$h = \frac{56 \cdot 100}{524} = 10.7$$

$$w = \frac{2 \cdot 18}{2 \cdot 18 + 38} = 0.49$$

2. **Cordierite-andesite**, Hoyazo near Nijar, Spain.

Weight percentages:

SiO$_2$	Al$_2$O$_3$	Fe$_2$O$_3$	FeO	MnO	MgO	CaO
61.08	17.25	3.16	3.10	0.06	2.82	3.14

Na$_2$O	K$_2$O	TiO$_2$	P$_2$O$_5$	H$_2$O+	H$_2$O—	C	Total
1.75	3.33	0.79	0.12	2.63	0.05	0.80	100.08

Molecular equivalent numbers:

SiO$_2$	Al$_2$O$_3$	Fe$_2$O$_3$	FeO	MnO	MgO	CaO	Na$_2$O	K$_2$O	TiO	P$_2$O$_5$
1017	169	20	43	1	70	56	28	35	10	1

Niggli-values:

si	al	fm	c	alk	k	mg	ti	p	w	h
229	38.2	34.9	12.7	14.2	0.55	0.45	2.3	0.2	0.48	33.0

3. **Arfvedsonite-lujaurite** from Lille Elv, Kangerdluarsuk, Greenland.

Weight percentages:

SiO$_2$	Al$_2$O$_3$	Fe$_2$O$_3$	FeO	MnO	MgO	CaO
53.01	15.33	9.14	4.44	0.13	0.10	0.67

Na$_2$O	K$_2$O	TiO$_2$	P$_2$O$_5$	ZrO$_2$	Cl	H$_2$O+	H$_2$O—	Total
11.86	2.60	0.33	Sp.	0.65	0.23	1.88	0.20	100.57

Molecular equivalent numbers:

SiO$_2$	Al$_2$O$_3$	Fe$_2$O$_3$	FeO	MnO	MgO	CaO
882	150	57	61	2	2	12

Na$_2$O	K$_2$O	TiO$_2$	P$_2$O$_5$	ZrO$_2$	Cl$_2$
191	28	4	Sp.	5	3

Niggli-values:

si	al	fm	c	alk	k	mg	ti	p	zr	cl$_2$	w
157	26.8	32.0	2.1	39.1	0.13	0.01	0.7	Tr.	0.9	0.5	0.65

A graphical method for calculating the Niggli-values was proposed by G. Deicha (1955).

β) *Calculation of the weight percentages from given Niggli-values.* The opposite task, namely, of obtaining the weight-percentage composition of a rock from given Niggli-values, is less often called for. However, under certain circumstances, it is desirable to obtain a picture of the weight-percentage composition, expressed as oxides, from given Niggli-values, e. g., they may be arrived at by graphic interpolation, as mean values, from a variation diagram. The following procedure can be applied.

If the number of the formula units $SiO_2 + Al_2O_3 + FeO$ (total iron including MnO) $+ MgO + CaO + Na_2O + K_2O + TiO_2 + P_2O_5$, etc, is indicated as Z, expressed as Niggli-values:

$$Z = si + al + fm + c + alk + ti + p, \text{ etc.} = si + 100 + ti + p, \text{ etc.}$$

If Z is calculated to the sum of 100, the so-called molecular percentages or oxide-equivalent percentages will be obtained. From these the required weight percentages can be obtained by multiplication by the formula weights of the corresponding oxides ("molecular weights") and recalculation to the sum of 100. The weight percentages may be obtained directly, provided that no importance is attached to the molecular percentages, by the calculation, to the sum of 100, of the relative numbers of the oxide formula units, expressed in Niggli-values, multiplied by their corresponding formula weights.

If, for example, the degree of oxidation of the iron is already known from the value

$$w = \frac{2\,Fe_2O_3}{2\,Fe_2O_3 + FeO} = \frac{Fe^{3+}}{Fe^{3+} + Fe^{2+}}$$

the calculation of Fe_2O_3 and FeO may also be carried out separately. In this case the number of trivalent Fe-atoms is given by $w(1-mg)/m$ and that of the Fe_2O_3 formula units — by $^1/_2 w(1-mg)/m$. The sum Z' of the oxide formula units, as expressed in Niggli-values, will be:

$$Z' = \underset{SiO_2}{si} + \underset{Al_2O_3}{al} + \underset{Fe_2O_3}{^1/_2 w(1-mg)/m} + \underset{FeO+MnO}{(1-w)(1-mg)/m} + \underset{MgO}{mg \cdot fm} + \underset{CaO}{c}$$

$$+ \underset{Na_2O}{(1-k)alk} + \underset{K_2O}{k \cdot alk} + \underset{TiO_2}{ti} + \underset{P_2O_5}{p}, \text{ etc.}$$

$$Z' = si + 100 - {^1/_2} w(1-mg)/m + ti + p, \text{ etc.}$$

The molecular and weight percentages can be obtained herefrom, as previously described.

Examples:

1. For the Quartz-diorite from Spanish Peak, California, the following Niggli-values were originally obtained:

si	al	fm	c	alk	k	mg	ti	p	w
190	32.1	30.9	22.5	14.5	0.18	0.54	1.5	0.4	0.49

The derived relative number of the oxide formula units is as follows:

$$Z' = \begin{array}{c} SiO_2 \\ si \\ 190 \end{array} + \begin{array}{c} Al_2O_3 \\ al \\ 32.1 \end{array} + \begin{array}{c} Fe_2O_3 \\ {}^1/_2 w(1-mg)/m \\ 3.5 \end{array} + \begin{array}{c} FeO+MnO \\ (1-w)(1-mg)/m \\ 7.2 \end{array} + \begin{array}{c} MgO \\ mg \cdot fm \\ 16.7 \end{array}$$

$$+ \begin{array}{c} CaO \\ c \\ 22.5 \end{array} + \begin{array}{c} Na_2O \\ (1-k)alk \\ 11.9 \end{array} + \begin{array}{c} K_2O \\ k \cdot alk \\ 2.6 \end{array} + \begin{array}{c} TiO_2 \\ ti \\ 1.5 \end{array} + \begin{array}{c} P_2O_5 \\ p \\ 0.4 \end{array}$$

$$Z' = 288.4 = si + 100 - {}^1/_2 w(1-mg)/m + ti + p.$$

The molecular-equivalent percentages are obtained by recalculation to the sum of 100:

SiO_2	Al_2O_3	Fe_2O_3	$FeO+MnO$	MgO	CaO	Na_2O	K_2O	TiO_2	P_2O_5	Total
65.9	11.1	1.2	2.5	5.8	7.8	4.1	0.9	0.5	0.1	99.9

The following numbers are obtained after multiplication by the corresponding formula weights:

3957 1132 192 180 234 437 254 85 40 14 6525.

The recalculation to the sum of 100 provides the weight percentages sought after (in brackets are given the analysis values calculated on a H_2O-free basis, for the sake of comparison).

SiO_2	Al_2O_3	Fe_2O_3	$FeO+MnO$	MgO	CaO	Na_2O	K_2O	TiO_2	P_2O_5	Total
60.6	17.3	2.9	2.8	3.6	6.7	3.9	1.3	0.6	0.2	99.9
(60.5)	(17.3)	(2.9)	(2.8)	(3.6)	(6.7)	(3.9)	(1.3)	(0.7)	(0.3)	(100.0)

2. The following Niggli-values were obtained for the **cordierite-andesite** from Hoyazo near Nijar, Spain:

si	al	fm	c	alk	k	mg	ti	p	w
229	38.2	34.9	12.7	14.2	0.55	0.45	2.3	0.2	0.48

The following relative number of oxide formula units are obtained:

SiO_2	Al_2O_3	Fe_2O_3	$FeO+MnO$	MgO	CaO	Na_2O	K_2O	TiO_2	P_2O_5
229	38.2	4.6	10.0	15.7	12.7	6.4	7.8	2.3	0.2

$$Z' = 326.9 = si + 100 - {}^1/_2 w(1-mg)/m + ti + p.$$

If it is desired to calculate the weight percentages directly, the aforementioned numbers should be multiplied by the corresponding formula weights. After rounding off the values to the nearest ten, these will yield:

1375 389 74 72 63 71 40 73 18 3   Total: 2178

The weight percentages sought after will be obtained after recalculation to the sum of 100:

SiO$_2$	Al$_2$O$_3$	Fe$_2$O$_3$	FeO+MnO	MgO	CaO	Na$_2$O	K$_2$O	TiO$_2$	P$_2$O$_5$	Total
63.1	17.9	3.4	3.3	2.9	3.3	1.8	3.4	0.8	0.1	100.0
(63.2)	(17.9)	(3.3)	(3.2)	(2.9)	(3.3)	(1.8)	(3.4)	(0.8)	(0.1)	(99.9)

The H$_2$O- and C-free original values are given in brackets for comparison.

3. For the Arfvedsonite-lujaurite from Lille Elv, Kangerdluarsuk, the previously calculated Niggli-values are as follows:

si	al	fm	c	alk	k	mg	ti	p	zr	cl$_2$	w
157	26.8	32.0	2.1	39.1	0.13	0.01	0.7	tr.	0.9	0.5	0.65

From these the relative number of oxide formula units is obtained:

SiO$_2$	Al$_2$O$_3$	Fe$_2$O$_3$	FeO+MnO	MgO	CaO	Na$_2$O	K$_2$O	TiO$_2$	P$_2$O$_5$	ZrO$_2$	Cl$_2$
157	26.8	10.3	11.1	0.3	2.1	34.0	5.1	0.7	tr.	0.9	0.5

$$Z' = 248.8 = si + 100 + ti + p + zr + cl_2 - \tfrac{1}{2} w(1 - mg)/m.$$

After multiplicaztion by the corresponding formula weights and rounding off to the nearest 10, it is found that:

943 273 165 80 1 12 211 48 6 — 11 2

Total : 1754

The weight percentages are obtained by recalculation to the sum of 100 (the original analysis, calculated on a H$_2$O-free basis, is again provided in brackets):

SiO$_2$	Al$_2$O$_3$	Fe$_2$O$_3$	FeO+MnO	MgO	CaO	Na$_2$O	K$_2$O	TiO$_2$	ZrO$_2$	Cl	Total
53.8	15.6	9.4	4.6	0.1	0.7	12.0	2.7	0.3	0.6	0.2	100.0
(53.8)	(15.6)	(9.3)	(4.6)	(0.1)	(0.7)	(12.0)	(2.6)	(0.3)	(0.7)	(0.2)	(99.9)

The three examples show that the recalculation of the weight or molecular-equivalent percentage (molecular percentage) composition of rocks from given Niggli-values, can indeed be accomplished by the above-mentioned procedure with a sufficient degree of accuracy.

b) **Niggli-Values and Cation Percentages**

α) Definition and calculation of the cation percentages. In the introduction it was mentioned that although the representation of a rock analysis in oxide weight percentages SiO$_2$, Al$_2$O$_3$, Fe$_2$O$_3$, FeO, MnO, etc, is obviously based on geochemical and crystallochemical rules, other possibilities may also be considered. One of these is the calculation of the so-called cation (metal atom) percentages, which has recently been used to an increased extent. Although these values were already in use in the early stages of development of petrochemistry (used by H. Rosenbusch, among others), they were only rarely used subsequently. The cation percentages provide the percentile number of the electropositive elements (metal atoms, cations), oxygen and the other anions being neglected.

Their calculation is carried out by dividing the weight percentages by the formula unit with reference to 1 cation of the corresponding oxide. Thus, the weight percentages of SiO_2, FeO, MnO, MgO, CaO, TiO_2 are divided by their corresponding formula weights. However, the weight percentages of Al_2O_3, Fe_2O_3, Na_2O, K_2O, P_2O_5, are only divided by half of their corresponding formula weights. Thus are obtained the so-called atomic equivalent numbers, which are proportional to the number of the metal atoms or cations. The required cation percentages are obtained by recalculation to the sum of 100. The atomic-equivalent numbers are determined by means of the tables provided at the end of this book. The atomic-equivalent number of oxides having two metal atoms, e.g., Al_2O_3, can be found by multiplying the molecular-equivalent number by a factor of 2.

Examples:

1. Quartz-diorite, from Spanish Peak, California.

	Weight %	Molecular equivalent numbers × 1000	Atomic equivalent numbers × 1000	Metal atom or cation %	
SiO_2	59.68	994	994	Si^{4+}	56.0
Al_2O_3	17.09	168	336	Al^{3+}	18.9
Fe_2O_3	2.85	18	36	Fe^{3+}	2.0
FeO	2.75	38	38	Fe^{2+}	2.1
MgO	3.54	88	88	Mg^{2+}	5.0
CaO	6.62	118	118	Ca^{2+}	6.7
Na_2O	3.87	62	124	Na^+	7.0
K_2O	1.31	14	28	K^+	1.6
TiO_2	0.65	8	8	Ti^{4+}	0.5
P_2O_5	0.25	2	4	P^{5+}	0.2
			1774		100.0

2. Garnet-bearing cordierite-andesite from Hoyazo near Nijar, Almeria province, Spain

	Weight %	Molecular equivalent numbers × 1000	Atomic equivalent numbers × 1000	Metal atom or cation %	
SiO_2	61.08	1017	1017	Si^{4+}	59.7
Al_2O_3	17.25	169	338	Al^{3+}	19.8
Fe_2O_3	3.16	20	40	Fe^{3+}	2.4
FeO	3.10 }	44	44	Fe^{2+}	2.6
MnO	0.06				
MgO	2.82	70	70	Mg^{2+}	4.1
CaO	3.14	56	56	Ca^{2+}	3.3
Na_2O	1.75	28	56	Na^+	3.3
K_2O	3.33	35	70	K^+	4.1
TiO_2	0.79	10	10	Ti^{4+}	0.6
P_2O_5	0.12	1	2	P^{5+}	0.1
incl.	3.48				
	100.08		1703		100.0

3. Lujaurite from Lille Elv, Kangerdluarsuk, Greenland

	Weight %	Molecular equivalent numbers × 1000	Atomic equivalent numbers × 1000		Metal atom or cation %
SiO_2	53.01	882	882	Si^{4+}	48.4
Al_2O_3	15.33	150	300	Al^{3+}	16.5
Fe_2O_3	9.14	57	114	Fe^{3+}	6.3
FeO	4.44	61 ⎫	63	Fe^{2+}	3.5
MnO	0.13	2 ⎭			
MgO	0.10	2	2	Mg^{2+}	0.1
CaO	0.67	12	12	Ca^{2+}	0.7
Na_2O	11.86	191.5	383	Na^+	21.0
K_2O	2.60	27.5	55	K^+	3.0
TiO_2	0.33	4	4	Ti^{4+}	0.2
ZrO_2	0.65	5	5	Zr^{4+}	0.3
incl.	2.31		1820		100.0
	100.57				

β) Relationships between cation percentages and Niggli-values. Bearing in mind the given definitions, the following basic relationships are found to exist between cation percentages and Niggli-values:

$$Si^{4+} = si$$
$$Al^{3+} = 2\,al$$
$$Fe^{3+} + Fe^{2+} + Mn^{2+} = (1 - mg)fm$$
$$Mg^{2+} = mg \cdot fm$$
$$Ca^{2+} = c$$
$$Na^+ = 2(1 - k)alk$$
$$K^+ = 2\,k \cdot alk$$
$$Ti^{4+} = ti$$
$$P^{5+} = 2\,p$$
$$\Sigma = si + 2\,al + fm + c + 2\,alk + ti + 2\,p$$
$$= si + 100 + al + alk + ti + 2\,p.$$

If the degree of oxidation of the iron is known, the following should be substituted for the third equation:

$$Fe^{3+} = w(1 - mg)fm$$
$$Fe^{2+} + Mn^{2+} = (1 - w)(1 - mg)fm.$$

These relationships allow for the recalculation of cation percentages to Niggli-values and vice-versa.

γ) Calculation of the cation percentages from given Niggli-values. The required cation percentages can be derived from the relations detailed in section β) by recalculation to the sum of 100.

Examples:

1. Quartz-diorite of Spanish Peak

si	al	fm	c	alk	k	mg	ti	p	w
190	32.1	30.9	22.5	14.5	0.18	0.54	1.5	0.4	0.49

Si^{4+}	Al^{3+}	Fe^{3+}	$Fe^{2+}+Mn^{2+}$	Mg^{2+}	Ca^{2+}	Na^+	K^+	Ti^{4+}	P^{5+}	Total
190	64.2	6.95	7.25	16.7	22.5	23.8	5.2	1.5	0.8	338.9

Recalculated to the sum of 100:

56.0	18.9	2.1	2.1	5.0	6.7	7.0	1.5	0.5	0.2	100.0
(56.0)	(18.9)	(2.0)	(2.1)	(5.0)	(6.7)	(7.0)	(1.6)	(0.5)	(0.2)	(100.0)

2. Cordierite andesite from Hoyazo near Nijar, Spain

si	al	fm	c	alk	k	mg	ti	p	w
229	38.2	34.9	12.7	14.2	0.55	0.45	2.3	0.2	0.48

Si^{4+}	Al^{3+}	Fe^{3+}	$Fe^{2+}+Mn^{2+}$	Mg^{2+}	Ca^{2+}	Na^+	K^+	Ti^{4+}	P^{5+}	Total
229	76.4	9.2	9.95	15.75	12.7	12.8	15.6	2.3	0.4	384.1

Recalculated to the sum of 100:

59.6	19.9	2.4	2.6	4.1	3.3	3.3	4.1	0.6	0.1	100.0
(59.7)	(19.8)	(2.4)	(2.6)	(4.1)	(3.3)	(3.3)	(4.1)	(0.6)	(0.1)	(100.0)

3. Lujaurite from Lille Elv, Kangerdluarsuk, Greenland

si	al	fm	c	alk	k	mg	ti	zr	w
157	26.8	32.0	2.1	39.1	0.13	0.01	0.7	0.9	0.65

Si^{4+}	Al^{3+}	Fe^{3+}	$Fe^{2+}+Mn^{2+}$	Mg^{2+}	Ca^{2+}	Na^+	K^+	Ti^{4+}	Zr^{4+}	Total
157	53.6	20.6	11.1	0.3	2.1	68.0	10.2	0.7	0.9	324.5

Recalculated to the sum of 100:

48.4	16.5	6.3	3.4	0.1	0.7	21.0	3.1	0.2	0.3	100.0
(48.4)	(16.5)	(6.3)	(3.5)	(0.1)	(0.7)	(21.0)	(3.0)	(0.2)	(0.3)	(100.0)

An excellent agreement is found to exist between the results thus obtained and the values in brackets, which were calculated directly from the weight percentages.

δ) Calculation of the Niggli-values from given cation percentages. By reversing the procedure outlined under β) it follows that:

$$si = Si^{4+}$$
$$ti = Ti^{4+}$$
$$p = 1/2\, P^{5+}$$
$$al = 1/2\, Al^{3+}$$
$$fm = Fe^{3+} + Fe^{2+} + Mn^{2+} + Mg^{2+}$$

$$c = Ca^{2+}$$
$$alk = {}^1/_2(Na^+ + K^+)$$

where, according to their definition, **al + fm + c + alk** should be calculated to the sum of 100 and **si, ti** and **p** should be reduced on the same basis. In addition:

$$k = \frac{K^+}{K^+ + Na^+} \quad \text{and} \quad mg = \frac{Mg^{2+}}{Mg^{2+} + Fe^{3+} + Fe^{2+} + Mn^{2+}}.$$

Examples:

1. Quartz-diorite of Spanish Peak

Si^{4+}	Al^{3+}	Fe^{3+}	$Fe^{2+}+Mn^{2+}$	Mg^{2+}	Ca^{2+}	Na^+	K^+	Ti^{4+}	P^{5+}	Total
56.0	18.9	2.0	2.1	5.0	6.7	7.0	1.6	0.5	0.2	100.0

$$\begin{aligned}
{}^1/_2 Al^{3+} = {}^1/_2\, 18.9 = &\quad 9.45 \quad al = 32.0 \quad (32.1)\\
Fe^{3+} + Fe^{2+} + Mg^{2+} = 2.0 + 2.1 + 5.0 = &\quad 9.1 \quad fm = 30.8 \quad (30.9)\\
Ca^{2+} = &\quad 6.7 \quad c = 22.7 \quad (22.5)\\
{}^1/_2(Na^+ + K^+) = {}^1/_2(7.0 + 1.6) = &\quad 4.3 \quad alk = 14.5 \quad (14.5)\\
\hline
&\quad 29.55 \quad\quad 100.0 \quad (100.0)
\end{aligned}$$

$$si = \frac{56 \cdot 100}{29.55} = 189.5\ (190) \quad ti = \frac{0.5 \cdot 100}{29.55} = 1.7\ (1.5) \quad p = \frac{0.1 \cdot 100}{29.55} = 0.3\ (0.4)$$

$$k = \frac{1.6}{1.6 + 7.0} = 0.19\ (0.18) \quad mg = \frac{5.0}{2.0 + 2.1 + 5.0} = 0.55\ (0.54)$$

$$w = \frac{2.0}{2.0 + 2.1} = 0.49\ (0.49).$$

2. For the cordierite-andesite from Hoyazo one similarity obtains:

Si^{4+}	Al^{3+}	Fe^{3+}	Fe^{2+}	Mg^{2+}	Ca^{2+}	Na^+	K^+	Ti^{4+}	P^{5+}	Total
59.7	19.8	2.4	2.6	4.1	3.3	3.3	4.1	0.6	0.1	100.0

$$\begin{aligned}
{}^1/_2 Al^{3+} &= 9.9 \quad al = 38.1\ (38.2) \quad si = 230\ (229)\\
Fe^{3+} + Fe^{2+} + Mg^{2+} &= 9.1 \quad fm = 35.0\ (34.9) \quad ti = 2.3\ (2.3)\\
Ca^{2+} &= 3.3 \quad c = 12.7\ (12.7) \quad p = 0.2\ (0.2)\\
{}^1/_2 Na^+ + {}^1/_2 K^+ &= 3.7 \quad alk = 14.2\ (14.2) \quad k = 0.55\ (0.55)\\
\hline
&\ 26.0 \quad\quad 100.0\ (100.0) \quad mg = 0.45\ (0.45)
\end{aligned}$$

3. For the Arfvedsonite-lujaurite from Lille Elv one obtains:

Si^{4+}	Al^{3+}	Fe^{3+}	Fe^{2+}	Mg^{2+}	Ca^{2+}	Na^+	K^+	Ti^{4+}	Zr^{4+}	Total
48.4	16.5	6.3	3.5	0.1	0.7	21.0	3.0	0.2	0.3	100.0

$$\begin{aligned}
{}^1/_2 Al^{3+} &= 8.25 \quad al = 26.7\ (26.8) \quad si = 157\ (157)\\
Fe^{3+} + Fe^{2+} + Mg^{2+} &= 9.9 \quad fm = 32.1\ (32.0) \quad ti = 0.6\ (0.7)\\
Ca^{2+} &= 0.7 \quad c = 2.3\ (2.1) \quad zr = 1.0\ (1.1)\\
{}^1/_2 Na^+ + {}^1/_2 K^+ &= 12.0 \quad alk = 38.9\ (39.1) \quad k = 0.12\ (0.13)\\
\hline
&\ 30.85 \quad\quad 100.0\ (100.0) \quad mg = 0.01\ (0.01)
\end{aligned}$$

By comparison with the values shown in the brackets, which were directly calculated from the analyses, it may be noted that the agreement is satisfactory.

c) Niggli Values and Molecular Equivalent Percentages (Molecular Percentages)

In earlier works rock analyses were occasionally expressed in molecular equivalent percentages (molecular percentages). These were used also as a basis for calculating Osann-values which were often used at that time. These values are obtained by recalculation of the molecular equivalent numbers to the sum of 100. Their relations to the Niggli values will be discussed in brief.

a) Calculation of the Niggli-values from the molecular equivalent percentages. Since the molecular equivalent percentages, as well as the molecular equivalent numbers from which the Niggli-values were previously calculated, are proportional to the number of the molecular formula units, the Niggli-values can be readily obtained from them.

Example:

Quartz diorite, Spanish Peak

	Weight %	Molecular equivalent numbers × 1000	Molecular equivalent %	
SiO_2	59.68	994	65.9	
Al_2O_3	17.09	168	11.1	11.1
Fe_2O_3	2.85	18	1.2 as FeO 2.4	
FeO	2.75	38	2.5	10.7
MgO	3.54	88	5.8	
CaO	6.62	118	7.8	7.8
Na_2O	3.87	62	4.1	5.0
K_2O	1.31	14	0.9	
TiO_2	0.65	8	0.5	34.6
P_2O_5	0.25	2	0.1	
		1510	99.9	

$$al = \frac{11.1 \cdot 100}{34.6} = 32.1 \qquad si = \frac{65.9 \cdot 100}{34.6} = 190$$

$$fm = \frac{10.7 \cdot 100}{34.6} = 30.9 \qquad ti = \frac{0.53 \cdot 100}{34.6} = 1.5$$

$$c = \frac{7.8 \cdot 100}{34.6} = 22.5 \qquad p = \frac{0.13 \cdot 100}{34.6} = 0.4$$

$$alk = \frac{5.0 \cdot 100}{34.6} = 14.5 \qquad w = \frac{2 \cdot 1.2}{2 \cdot 1.2 + 2.5} = 0.49$$

$$\overline{100.0}$$

$$k = \frac{0.9}{0.9 + 4.1} = 0.18 \qquad mg = \frac{5.8}{5.8 + 2.5 + 2 \cdot 1.2} = 0.54$$

β) Calculation of the molecular-equivalent percentages from given Niggli-values. There is seldom need to carry out the reverse of the process outlined in **a**). This may be readily performed according to the instructions already provided in section BI 1 β and will, therefore, not be described in further detail.

2. The Representation of Fundamental Petrochemical Relationships by Means of the Niggli-Values

a) The Quartz Number *qz* as an Expression of Saturation with Respect to SiO_2

A number of interesting concepts on idealized normative mineral compositions can be obtained immediately from the Niggli-values. Since the amount of SiO_2 required for the formation of the highly silicated normative ingredients (feldspars, augite-type compounds) can be calculated and expressed by Niggli-values, it is possible to compare this quantity with the quantity of silica which is actually present. Immediately one can discern whether it is possible for free quartz to occur along with the highly silicated constituents, whether the latter are present without free quartz, or whether the available amount of silica is insufficient for their formation. In other words, if the rock is oversaturated, saturated or undersaturated with respect to silica.

Two different cases should be distinguished in connection with the calculation of the amount of SiO_2 required for the formation of the highly silicated components.

1a) The case when *al* > *alk* and, at the same time *al* < (*alk* + *c*), i.e., no alumina excess is present. After combining the total $Na_2O + K_2O$ with Al_2O_3 in the ratio 1:1 there remains an excess of Al_2O_3, *T*, which is combined with CaO in a ratio of 1:1 to form anorthite. Every unit of *alk* is combined with a 6-fold amount of SiO_2, i.e., 6 *alk*, corresponding to the composition of the alkali feldspars $6 SiO_2 \cdot Al_2O_3 \cdot (Na, K)_2O$. A quantity of Al_2O_3 equal to *al* — *alk* is still available for the formation of anorthite. An amount of 2 (*al* — *alk*) (with respect to *si*), corresponding to the formula $2SiO_2 \cdot Al_2O_3 \cdot CaO$, should be combined with it. The amount of *c* available for wollastonite $SiO_2 \cdot CaO$ (in diopside) is given by *c* — (*al* — *alk*). The total *fm* is still available for the formation of enstatite or hypersthene. Thus, the amount of *si* required for the formation of the highly silicated components is given by:

si' =   6 *alk*   + 2 (*al* — *alk*) + [*c* — (*al* — *alk*)] +   *fm*
     Alkali feldspar   Anorthite   Diopside   Ortho-augite

si' = 100 + 4 *alk* (since, according to the definition, *al* + *fm* + *c* + *alk* = 100).

1b) Once again *al* > *alk*, but at the same time *al* > (*alk* + *c*), i.e., alumina is present in excess over the sum of the alkalis and calcium. This situation only occurs rarely in the case of fresh, non-endomorphic igneous rocks, but may be the rule in the case of weathered igneous rocks and argillaceous sediments or their metamorphic derivatives. In this instance, the equation si' = 100 + 4 *alk* will be valid only for the case when the Al_2O_3 excess $al - (alk + c)$ is assumed to be combined with si in the ratio 1 : 1, in sillimanite, andalusite or kyanite.

An amount of si equal to 6 alk is again required in order to combine with the available alk for the formation of alkali feldspar. The available c, and not the al excess, determines the formation of anorthite. Therefore an amount of si equal to 2 c is used. To the remainder of al, which is equal to $al - (alk + c)$, is added an equal amount of si, for the formation of sillimanite. Wollastonite does not occur, and the total fm is incorporated in ortho-augite. Hence one obtains:

$$si' = \underset{\text{Alkali feldspar}}{6\ alk} + \underset{\text{Anorthite}}{2\ c} + \underset{\text{Sillimanite}}{al - (alk + c)} + \underset{\text{Ortho-augite}}{fm} = 100 + 4\ alk\ .$$

Should the alumina excess be calculated as **corundum** and not as **sillimanite**, an amount of si smaller by $al - (alk + c)$ is required and one obtains:

$$si' = 6\ alk + 2\ c + fm = 100 + 5\ alk - al\ .$$

2. The case when $al < alk$, i.e., after combining the total alumina with $(K, Na)_2O$ in a ratio of 1:1, there remains an **excess of alkali**, designated as alk' which is used for normative aegirine formation. In this case the formation of anorthite is impossible. Since the composition of aegirine corresponds to $4SiO_2 \cdot Fe_2O_3 \cdot Na_2O$, the necessary amount of si will be:

$$si' = \underset{\text{Alkali feldspar}}{6\ al} + \underset{\text{Aegirine}}{4\ (alk - al)} + \underset{\text{Diopside + Ortho-Augite}}{c + [fm - 2\ (alk - al)]}$$

$$si' = 100 + 3\ al + alk\ .$$

The "**quartz number**" is now defined as $qz = si - si'$. Were only the most simple, idealized mineral compounds present in the rock under consideration, as was assumed here for the sake of simplicity, and if all the iron was bivalent (with respect to aegirine, thereby excluding such minerals as magnetite or hematite) and all the al always combined to alk in the ratio 1:1 (i.e., no mica minerals are present), the following associations could be anticipated:

$qz > 0$ free SiO_2 as oxide, e.g., quartz

$qz = 0$ quartz-free parageneses of highly silicated crystalline compounds (feldspars, pyroxenes)

$qz < 0$ occurrence of poorly silicated silicates, e.g., olivine, melilite, feldspathoids.

Other minerals which were not considered in the schematic, simplified derivation of the si'-value may, however, be present. These may be silica-poor (Al-bearing) augites and hornblendes, biotite, ore minerals, as well as accessory minerals such as titanite, apatite and ilmenite. In such cases quartz may even occur at slightly negative qz-values. In the case of hemicrystalline rocks, on the other hand, quartz may often be absent, despite the fact that $qz > 0$. The SiO_2 excess is in the form glass (cryptomorphic according to A. Lacroix, 1924, 531).

The following qz-numbers are obtained for the three aforementioned examples:

1. Quartz-diorite
from Spanish Peak, California. $qz = 190 - (100 + 4 \cdot 14.5) = +32$

2. Cordierite-andesite
from Hoyazo near Nijar $qz = 229 - (100 + 4 \cdot 14.2) = +72.2$
(Al-excess taken as $SiO_2 \cdot Al_2O_3$)
$qz = 229 - (100 + 5 \cdot 14.2 - 38.4) = 132.6$
(Al-excess taken as Al_2O_3)

3. Arfvedsonite-lujaurite
from Lille Elv $qz = 157 - (100 + 3 \cdot 26.8 + 39.1) = -62.5$

1. Quartz-bearing, as already shown by the designation.
2. Very rich in glass without occurrence of quartz.
3. Nepheline-and sodalite-bearing in accordance with the negative qz value.

b) The Alumina Excess

The following points are significant for the derivation of a simplified normative mineral composition: $2\,alk^*$ is a measure of the quantity of the alkali feldspars, $k\,2\,alk$ being specific for potassium feldspar and $(1-k)\,2\,alk$ — for sodium feldspar.

$(al - alk) = T$, i.e., the excess of alumina over the alkalis, is a measure of the anorthite quantity, since the amount of alumina not combined with the alkalis is combined with calcium in the ratio $Al_2O_3 : CaO = 1:1$. The total amount of feldspar is therefore $F = 2\,alk + (al - alk) = al + alk$.

The following $T = (al - alk)$ values are obtained for the three examples mentioned previously:

1. Quartz-diorite from Spanish Peak $T = 32.1 - 14.5 = 17.6$
2. Cordierite-andesite from Hoyazo $T = 38.2 - 14.2 = 24.0$
3. Arfvedsonite-lujaurite from Lille Elv $T = 26.8 - 39.1 = -12.3$

According to the above, 1 and 2 bear plagioclase whereas 3 does not.

The value $(al - (c + alk)) = t$ is useful for arriving at the anticipated mineral composition. If $t > 0$, i.e., $al > (c + alk)$, an excess of alumina will remain after the combination of the alumina with alkalis and calcium in the ratio of $1:1$. This "alumina-excess proper", designated

---

* It is $2\,alk$, and not alk, which serves as a measure in the case of the formulae of the following isomorphic relations: consider $[(SiO_2)_3 AlO_2]K$ for potassium feldspar or $[(SiO_2)_3 AlO_2]Na$ for sodium feldspar, and $[(SiO_2)_2(AlO_2)_2]Ca$ for anorthite. In earlier presentations the simple alk-value was used by P. Niggli (for example P. Niggli, 1923) as a measure of the quantity of alkali feldspars, corresponding to the formula unit $6SiO_2 \cdot Al_2O_3 \cdot K_2O$ or $6SiO_2 \cdot Al_2O_3 \cdot Na_2O$.

by t *, has very rarely been applied to endomorphic igneous rocks. If present in a large amount, it may serve as an important criterion in distinguishing the sedimentary origin of metamorphic rocks.

For the three mentioned examples t is calculated as follows:
1. Quartz-diorite from Spanish Peak $t = 32.1 - (22.5 + 14.5) = -4.9$
2. Cordierite-andesite of Hoyazo $t = 38.2 - (12.7 + 14.2) = +11.3$
3. Arfvedsonite-lujaurite from Lille Elv $t = 26.8 - (2.1 + 39.1) = -14.4$

The markedly positive t-value of the (completely fresh) cordierite-andesite 2. immediately indicates its endomorphic character, which, however, is also expressed by the abundance of garnet and cordierite.

If $al - (c + alk) = 0$, i.e., $al = (c + alk)$, the total quantity of the calcium and the alkalis is combined with alumina in the ratio of 1:1 (feldspars and feldspathoids). Therefore no calcium is left for the formation of calcium-bearing augites or hornblendes. If, however,

$$al - (alk + c) < 0, \quad \text{i.e.} \quad c > (al - alk),$$

this indicates that, after the formation of alkali feldspars (and perhaps their substitutes) and anorthite, some calcium will remain, which cannot be combined with alumina and must, therefore, be incorporated in the dark constituents. This part of the calcium, corresponding to the "chaux non feldspatisable" of the French authors, is designated by c'.

The following relations hold for the three examples under consideration:
1. Quartz-diorite from Spanish Peak $c' = c - (al - alk) = 22.5 - (32.1 - 14.5) = +4.9$
2. Cordierite-andesite from Hoyazo $t > 0$, as mentioned previously, therefore $c' < 0$
3. Arfvedsonite-lujaurite from Lille Elv $T = -12.3$, as mentioned previously, i.e., $(alk - al) = 12.3$.

In accordance with these figures, 1. is hornblende-bearing, whereas in case 3. the alumina does not even suffice for the saturation of the alkalies, and normative anorthite formation is therefore out of the question. The alkali excess $(alk - al)$, together with Fe^{3+}, must provide alkali-augites or hornblendes, which is actually the case. In 2. the alumina excess is expressed by the presence of abundant cordierite and garnet.

c) Derivation of a Simplified Normative Mineral Composition from the Niggli-Values

One is not often required to derive a normative mineral composition from the Niggli-values. In order to do this it is preferable to start the calculation directly from the weight percentages, as will be shown further on in section B II when dealing with the equivalent norm. Under certain circumstances it may, however, be desirable to recalculate a rock's

---

* The designations T and t were not always used consistently by P. Niggli. The last presentation (P. Niggli, 1948, 336) will be here followed, as defined: $t = al - (alk + c)$ and $T = (al - alk)$, when the value is positive. This is in accordance with the publication by P. Niggli (1936a) where the definition of $t = al - (alk + c)$, is provided, as opposed to other publications.

chemistry, as defined by Niggli-values, to a normative mineral composition. This may arise, for example, when the Niggli-values are obtained by interpolation from a variation diagram. Thus, this problem merits a brief discussion. The example of the quartz-diorite from Spanish Peak, California, will serve as an illustration of the procedure. The following Niggli-values were calculated for this rock:

si	al	fm	c	alk	k	mg
190	32.1	30.9	22.5	14.5	0.18	0.54

In accordance with the considerations made at the time of introduction of the quartz-number qz, simply constructed, silica-saturated, feldspar- and pyroxene-type compositions will be selected as the normative components which should be arrived at by calculation. Once again we will make use of the following formula units dictated by the isomorphic relationships: $[(SiO_2)_3 AlO_2]K$ for Or, $[(SiO_2)_3 AlO_2]Na$ for Ab; and $[(SiO_2)_2(AlO_2)_2]Ca$ for An. This corresponds to the substitution of 1 Na or 1 Ca for 1 K.

If the number of K-atoms present is taken as a measure of Or-formation, then this number directly multiplied by 5 will provide the quantity of Or present, as the sum Si + Al + K in the selected formula is equal to 5. Since the Niggli-values are taken as a starting point, 1 alk must be considered as being equivalent to 2 (Na + K), because its calculation was based on $Na_2O + K_2O$. On the other hand, the number of the K- or Na-atoms is given by $k \cdot 2\,alk$ or $(1-k)\,2\,alk$, respectively. The An-quantity may be obtained from the Al by multiplying by 5/2, in accordance with the formula $[(SiO_2)_2(AlO_2)_2]Ca$; in this case, too, the sum Si + Al + Ca equals 5. As the calculation of al is based on the quantity of Al_2O_3, i.e., 1 al is equivalent to 2 Al, only the alumina $(al - alk)$ remaining after the formation of alkali feldspars remains available for An. Therefore the quantity $5\,(al-alk)$ will represent An. Thus, the ratio $Or:Ab:An$ is given by $2\,k \cdot alk : 2\,(1-k)\,alk : (al-alk)$.

The following simplified normative mineral compositions are thus obtained:

			Summed to 100	Simplified equivalent norm
Or:	$5\,k \cdot 2\,alk = 10\,k \cdot alk =$	26.1	7.8	(8.0)
Ab:	$5\,(1-k)\,2\,alk = 10\,(1-k)\,alk =$	119.0	35.4	(35.3)
An:	$5\,(al-alk) =$	88.0	26.1	(26.3)
Wo:	$2\,[c-(al-alk)] =$	9.8	2.9	(2.3)
En:	$2\,mg \cdot fm =$	33.4	9.9	(9.9)
Hy:	$2\,(1-mg)\,fm =$	28.4	8.4	(8.4)
Q:	$si - (100 + 4\,alk) = qz =$	32.0	9.5	(9.8)
		336.7	100.0	(100.0)

The corresponding values of a similarly simplified "equivalent norm" are given in brackets for comparison. They show satisfactory agreement. The simplification employed here, as compared with the usual calculation procedure explained in detail in section B II, mainly entails the fact that the total Fe is taken as bivalent, as in the case of the Niggli-values. Thus, magnetite formation is precluded, as well as the representation of Ti and P. If insufficient SiO_2 is present for the formation of highly silicated compounds, olivine or felspathoids will be formed.

3. Application Limits of the Niggli-Values

The Niggli values have found wide application in the literature. They have certainly been put to great use in the comparative study of the chemistry of rocks. The **variation diagrams** (previously termed differentiation diagrams) based thereon, in which the values *al, fm, c, alk* are plotted as ordinates against the corresponding *si*-value on the abscissa, have the nature of correlation diagrams. They have proved especially valuable for the characterization and comparative study of rock series and petrographic provinces of different classes. This is due to their ability of showing clearly that certain kinds of associations or differentiation types repeat themselves and occur under analogous tectonic conditions in different geological epochs and regions. Thus is illustrated one of the essential facts of petrography, i.e., that the mode of association of different igneous rocks of the same age is neither singular nor arbitrary, and principles of universal application must have been active during the origin of magmatic kindreds.

During geological history, at least in post-Cambrian times, these events must have repeated themselves under analogous, large-scale tectonic conditions. Thus, a number of more or less sharply-defined province types may be distinguished. It should be noted primarily that these remained essentially independent of the neighboring rocks during their main period of activity. Therefore it should be concluded that any influence exerted upon them, if at all, must have been small. Locally established assimilation phenomena are exceptions. This holds true especially in contact zones as well as in processes of desilication and formation of synthetic melts through carbonate assimilation. These exceptions are well illustrated by means of the Niggli-values.

Despite the fact that in a large number of cases the Niggli-values have proved to be a very useful tool for the development of chemical-petrographical concepts, their theoretical basis has, however, not been paid the attention it has merited for some time already. The possibility does exist that insufficient attention was paid to the fact that these values do not possess the nature of absolute numbers of the corresponding oxides, and that they represent **ratios**, according to definition, of the corresponding oxide quantities calculated to the sum of the basic oxides Al_2O_3 + FeO (total iron) + MnO + MgO + CaO + Na_2O + K_2O. Therefore they are only proportional to these absolute quantities, as is clearly evident from the calculated example. Thus, the Niggli values of a rock provide a means of undertaking considerations on a stoichiometric basis, using the known ideal chemism of the rock-forming minerals as a basis. These considerations lead to the calculation of the amount of free silica or the quartz-number *qz*, and the *c'*-value or the alumina excess, as already mentioned before. The fact that the Niggli-values do not represent absolute numbers, but only their ratios, immediately comes to light when correlation problems are dealt with.

The correlation of two Niggli-values of the same rock, e.g., *si* and *alk*, is not the correlation of two absolute numbers, but the correlation of two ratios having a common denominator. In this connection, reference should be made to what was said in section A II 2. It should be especially emphasized that the existence of a correlation between the ratios does not allow one to assume the same for the absolute values, which are the main point of interest. F. Chayes (1949), who was the first to indicate this fact, provided an instructive example (F. Chayes, 1949). He starts by stating that in rocks of metasomatic origin a negative correlation should be established between the original material and later supplied material, i.e., the addition of the second should correspond to the removal of the first. Since H. G. Backlung has expressed the view that the Swedish and Finnish rapakivi granites were formed by the alkali-assimilation of Jotnian sandstones, F. Chayes examined the correlation conditions between the content of alkali feldspar and free quartz in a number of well-known rapakivi-granite series. This was carried out by using the CPIW-normative content of (*or + ab*) and *Q*, on the one hand, and the Niggli-value *alk* (2 *alk*, would be more correct, but this fact does not bear an essential influence on the result), as a measure of the alkali-feldspar content, and the quartz number *qz*, as a measure of the free quartz, on the other hand. It was shown that the results obtained by means of the Niggli-values were not only not in agreement, but directly contradictory, and therefore misleading, when compared with the results derived from the normative values expressed as percentage values. The difference lies not only in the degree of correlation but also in the misleadingly obtained positive correlation, instead of the actual negative correlation which was derived from the norm values.

Considering the great significance attributed to this statement, with regard to the application of the Niggli-values, the relationships of one of the rapakivi series, which was also used by F. Chayes, should be dealt with in greater detail, i.e., the rapakivi series of the Swedish Loos-Hamra area (H. V. Eckermann, 1936). Figure 21 shows the correlation diagram* for the values 2 *alk*, as a measure of the alkali-feldspars, and *qz*, as a measure for the free SiO_2 not used for the formation of silicates. A weak positive correlation is quite distinct. This could have led to the assumption that the content of alkali feldspars increases slightly with the increment of SiO_2-excess.

If these relations are checked on the basis of the metal atom (cation) percentages, the value $5(K^+ + Na^+)$ will provide a measure of the quantity of the alkali feldspars corresponding to the ratio $Si^{4+} : Al^{3+} : (K^+ + Na^+) = 3 : 1 : 1$. The quantity of Si which is needed for the building of the highly silicated silicates (feldspars and pyroxenes) can be calculated if the calculation of magnetite and apatite is carried out first. Fe^{3+} should be disregarded, bearing in mind considerations analogous to those made during the computation of the quartz number *qz*. This is given by

$$Si' = 3(K + Na) + [Al - (K + Na)] + Ca - \frac{1}{2}[Al - (K + Na)] + Fe^{2+} + Mg$$

$$Si' = \frac{1}{2}[5(K + Na) + Al] + Ca + Fe^{2+} + Mg.$$

If the formation of magnetite and apatite is taken into consideration, only a quantity of $(Fe^{2+} - 1/2\ Fe^{3+})$ and $(Ca - 3/2 P)$ will be available for silicate formation and thus:

$$Si' = \frac{1}{2}[5(K + Na) + Al - Fe^{3+} - 3P] + Ca + Fe^{2+} + Mg.$$

* The analyses used are the same as those taken by F. Chayes. In the original work (H. V. Eckermann, 1936) they were numbered: 74-79, 81-83 and 92-98, pages 271-275 and 294-297.

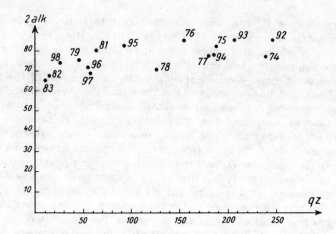

FIGURE 21. Rapakivi rocks from the Loos-Hamra area (Sweden). Weak positive correlation between $2\,alk$ and qz. The designation of the analyses corresponds to the original work by H. V. Eckermann (1936). After C. Burri (1956)

The correlation diagram for the values $5(K + Na)$ and $Q = Si - Si'$ (Figure 22), thus calculated, shows a distinct negative correlation, i. e., decrease in alkali-feldspar content with increment of SiO_2 excess. This is in agreement with F. Chayes's correlation derived from the CPIW—normative content of $(Or + Ab)$ and Q.

It can readily be shown that the reason for the misleading picture on Figure 21 should be attributed to the relative nature of the Niggli-values, i. e., the conclusions were drawn with regard to the correlation of absolute values on the basis of the correlation of relative values, this being inadmissible, as already shown. For this purpose it is merely necessary to multiply the Niggli-values $2\,alk$ and qz by the total oxides Al_2O_3 + FeO (total iron) + MnO + MgO + CaO + K_2O + Na_2O = S. The molecular-equivalent numbers are thus obtained, which again represent relative measures of the alkali oxides, and thus also of the alkali feldspars, as well as of the SiO_2-excess.

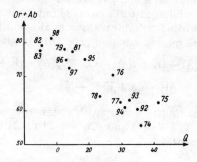

FIGURE 22. Distinct negative correlation between the equivalent-normative content of alkali feldspar and free quartz for the same rocks as in Figure 21. After C. Burri (1956)

$$Or + Ab = 5(Na^+ + K^+) = 2\,alk\,\frac{100}{S}, \qquad Q = Si - Si' = qz\,\frac{100}{S}.$$

A correlation diagram established on the basis of these values will then show the same relations as Figure 22 and will be in agreement with the facts. The falsification of the true correlation between the alkali-feldspar content and SiO_2-excess, from a negative one to a slightly positive one, due to the application of the Niggli-values, is apparently due to the fact that the sum of the basic oxides S, by which value all the numerical values were divided, is not constant. Figure 23 shows, moreover, that a distinct negative correlation exists between S and SiO_2-excess: $SiO_2 - SiO_2' = qz \cdot S$.

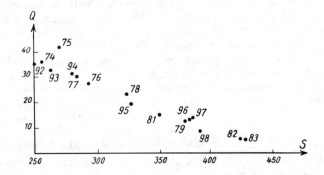

FIGURE 23. Correlation between free quartz and the sum of the basic oxides S for the same rocks as in Figures 21 and 22

This disadvantage of the Niggli-values, i.e., the fact that they are not pure measured values but relative numbers with a common denominator, which may, however, differ from case to case (from rock to rock), becomes irrelevant on eliminating this common denominator. This always occurs, for example, in the case of ratio construction between two Niggli-values, whereby the common denominator is eliminated.

Among others, the method of calculating the composition and quantitative relationships of the normative feldspars from the Niggli-values, as described in the next chapter, as well as Ritmann's values $Si°$ and $Az°$ for the characterization of the silication conditions, may serve as examples. The relative nature of the Niggli-values also does not influence the calculation of a simplified mineral composition, as given under section 2c, since the denominator is the same for all terms.

4. Further Applications of the Niggli-Values

a) Compositional and Quantitative Relations of the Normative Feldspars

The determination of the compositional and quantitative relations of the normative feldspars is of special importance as far as SiO_2-saturated and oversaturated igneous rocks are concerned. By comparing them with the model relationships, as determined by means of optical investigation and thin-section measurements, one is able to draw conclusions about the chemism of the other minerals present, including the possible sesquioxide content of augites and hornblendes. The type and quantitative relationship of the normative feldspars are less important in the case of rocks undersaturated with respect to SiO_2. This is due to the simplifications made during the norm of calculation, whereby only one of several possible combinations is selected, e. g., the frequent heteromorphic relation feldspar + olivine $\rightleftarrows$ feldspar substitute + pyroxene is not taken into consideration.

On considering SiO_2-saturated and oversaturated rocks it is of practical significance to be able to obtain all the interesting data about the type and quantitative relations of the normative feldspars without resource to calculating a complete norm. These data should be obtained in equivalents and not in weight percentage ratios, in as simple a way as possible, from the Niggli-values *al*, *alk* and *k* Niggli, 1927, 118-126).

Initially, the consideration should be confined to the case where $si \geq 100 + 4\,alk$ and $(alk + c) \geq al \geq alk$, i. e., the rock is either saturated or oversaturated with respect to SiO_2 and lacks alumina or alkali excesses. The formula units given by the isomorphic relations are again chosen, namely:

71

$6\,SiO_2 \cdot Al_2O_3 \cdot K_2O$ or $[(SiO_2)_3 AlO_2]\,K$ for potassium feldspar = *Kf*,
$6\,SiO_2 \cdot Al_2O_3 \cdot Na_2O$ or $[(SiO_2)_3 AlO_2]\,Na$ for sodium feldspar = *Ab*,
$4\,SiO_2 \cdot 2\,Al_2O_3 \cdot 2\,CaO$ or $[(SiO_2)_2 (AlO_2)_2]\,Ca$ for calc. feldspar = *An*.

The number of formula units (molecules) $[(SiO_2)_3(AlO_2)](K, Na) = Kf + Ab$ is given by $2\,alk$, as already indicated, and not by *alk*, because the *alk*-value is calculated on the basis of $(Na_2O + K_2O)$. In a more detailed form: $Kf = k\,2\,alk$ and $Ab = (1-k)\,2\,alk$. The remaining alumina is then available for the composition of anorthite, namely $(al - alk)$. Since *al* was calculated on the basis of Al_2O_3, $(al - alk)$ will, at the same time, provide the number of the formula units (molecules) of $[(SiO_2)_2(AlO_2)_2]Ca = An$. Thus the sum of the total feldspars is given by:

$$F = 2\,alk + (al - alk) = (al + alk).$$

The following relations can be immediately expressed therefrom:

$$\frac{\text{Alkali feldspars}}{\text{Total feldspars}} = \frac{Ab + Kf}{Ab + A + Kf} = \frac{2\,alk}{al + alk},$$

$$\frac{\text{Potassium feldspar}}{\text{Total feldspar}} = \frac{Kf}{Ab + An + Kf} = \frac{k\,2\,alk}{al + alk},$$

$$\frac{\text{Potassium feldspar}}{\text{Alkali feldspar}} = \frac{Kf}{Kf + Ab} = \frac{k\,2\,alk}{2\,alk} = k,$$

$$\frac{\text{Sodium feldspar}}{\text{Alkali feldspar}} = \frac{Ab}{Kf + Ab} = \frac{(1-k)\,2\,alk}{2\,alk} = (1-k)$$

$$\frac{\text{Sodium feldspar}}{\text{Calcium feldspar}} = \frac{Ab}{An} = \frac{(1-k)\,2\,alk}{al - alk},$$

$$\frac{\text{Sodium feldspar}}{\text{Plagioclase}} = \frac{Ab}{Ab + An} = \frac{(1-k)\,2\,alk}{(1-k)\,2\,alk + (al - alk)} = \frac{(1-k)\,2\,alk}{al + alk - k\,2\,alk},$$

$$\frac{\text{Potassium feldspar}}{\text{Alkali feldspar}} = \frac{Kf}{Ab+An} = \frac{k\,2\,alk}{(1-k)\,2\,alk + (al-alk)} = \frac{k\,2\,alk}{al + alk - k\,2\,alk}.$$

The common denominator (the sum of the metallic oxides Al_2O_3 + FeO + + MnO + MgO + CaO + Na_2O + K_2O) of the Niggli-values, which varies from rock to rock, is eliminated through the construction of these ratios. Thus we arrive at ratios of pure numbers of the oxide content. These may be compared to each other. These ratios are then also in agreement with those obtained directly from the normative values.

All of these relations may be arrived at directly from known values of *al*, *alk* and *k* by means of the slide rule. They also provide for a simple and lucid graphical representation, if we take:

$$k = x \qquad \frac{2\,alk}{al + alk} = y$$

and these values are plotted against one another within a rectangular system of coordinates. The following may be read immediately from the diagram:

$$\frac{\text{Potassium feldspar}}{\text{Alkali feldspar}} = x \quad \text{and} \quad \frac{\text{Alkali feldspar}}{\text{Total feldspar}} = y.$$

Taking into account the forementioned, it follows that:

$$\frac{\text{Potassium feldspar}}{\text{Total feldspar}} = \frac{k \cdot 2\,alk}{al + alk} = x\,y = C$$

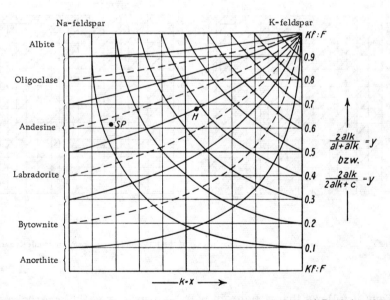

FIGURE 24. Diagram for the determination of the normative feldspar relations $Kf:F$ and $An:Ab$ from the Niggli-values. After P. Niggli (1927)

SP corresponds to the projection point of the quartz-diorite from Spanish Peak, with the coordinates $k = 0.18$ and $2\,alk/(al + alk) = 0.62$. Therefrom it follows that $Kf:F = 0.12$ and the plagioclase composition is An_{43}. *H* corresponds to the projection point of the cordierite-andesite from Hoyazo near Nijar, with the coordinates $k = 0.55$ and $2\,alk/(2\,alk + c) = 0.69$. Therefrom it follows that $Kf:F = 0.38$ and the plagioclase is $Plag\ An_{50}$.

A family of equilateral hyperbolae can be drawn for actual values of these ratios, for which the coordinate axes constitute the asymptotes. Obviously, however, true significance is only attributed to the Ist quadrant. The curves for the ratios $Kf:(Kf+Ab+An)$ from values of 0.1 to 0.9 are illustrated in Figure 24. This diagram can be converted into a rectilinear one by logarithmic transformation and the usage of a double logarithmic paper (Figure 24a), thus facilitating probable interpolations.

If the ratio $Kf:(Ab+An) = Kf:Plag$, is of interest, the following procedure may be employed.

From $\dfrac{Kf}{Kf+Ab+An} = C$, it follows that $Kf(1-C) = C(Ab+An)$

and thus that $\dfrac{Kf}{Ab+An} = \dfrac{C}{1-C} = C'$.

The ratio $Kf:(Ab+An) = C'$ can therefore be obtained immediately either from the calculated x and y values or from the ratio $Kf:(Kf+Ab+An) = C$, which may be read from Figures 24 or 24a. C'-values from 0 to ∞ correspond to C-values from 0 to 1, as follows:

C	0	0.1	0.2	0.3	0.4	0.5	0.6	0.7	0.8	0.9	1.0
C'	0	0.11	0.25	0.43	0.67	1.0	1.50	2.33	4.00	9.00	∞

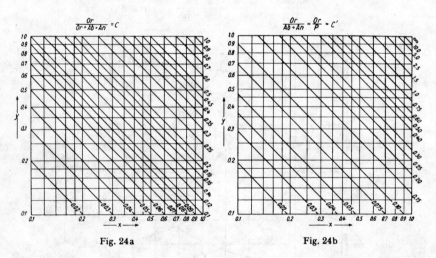

Fig. 24a   Fig. 24b

FIGURES 24a, 24b. Diagrams for the determination of the ratios $Or:F$ and $Or:P$, on double-logarithmic coordinate paper.

Thus it follows that a diagram may be drawn or valid hyperbolae may be used for the ratio $Kf:(Kf+Ab+An)$; only in this case other numerical values should be applied. This follows from the already-stated fact that the ratio $Kf:(Kf+Ab+An) = C$ varies from 0 to 1 and that $Kf:(Ab+An) = C'$ varies from 0 to ∞. The difference is relatively insignificant for small values up to about 0.25, but increases rapidly for larger values.

Here again the diagram may be converted into a rectilinear form by using double-logarithmic paper (Figure 24b).

Of greater importance is the ratio $Ab:An$, i.e., the plagioclase composition. It is given by:

$$\frac{Ab}{An} = \frac{(1-k)\,2\,alk}{al-alk} = \frac{(1-k)\,2\,alk}{al+alk-2\,alk} = \frac{(1-k)\dfrac{2\,alk}{al+alk}}{1-\dfrac{2\,alk}{al+alk}} = (1-x)\frac{y}{(1-y)} = C''.$$

By transformation it follows that

$$y\,[x-(1+C'')] = -C'',$$

i.e., we again obtain an equation of an equilateral hyperbola which, however, has its mid-point on the X-axis at an interval of $(1+C'')$ from the origin, and is located in the negative quadrant, as C' is negative. Only the branch located in quadrant II is of real significance. Different values of $C'' = Ab/An$, representing different plagioclase compositions, correspond to different hyperbolae, all of which pass through point (1, 1). Figure 24 provides the appropriate picture—the so-called **feldspar diagram**. The ratios Kf/F and Ab/An can be read directly from the diagram, using the plotted projection points based on the corresponding values of k and $2\,alk/(al+alk)$. If it is desired to express the An-content in the convenient form of $An\%$ instead of by the ratio of $Ab:An$, the following relation will hold: $Ab = C''An$. By addition of An to both sides of this equation it follows that $Ab + An = An(1 + C'')$ and

$$An\% = \frac{An}{Ab+An}\,100 = \frac{100}{1+C''}.$$

If cation percentages and not Niggli-values are available, the diagram is equally suitable, since,

$$\frac{2\,alk}{al+alk} = \frac{K+Na}{\tfrac{1}{2}(Al+K+Na)} = \frac{2(K+Na)}{Al+K+Na} \quad \text{and} \quad k = \frac{K}{K+Na}.$$

For the quartz-diorite from Spanish Peak the ratio $Kf:F = 0.11$ is obtained, as based on the given values of $k = 0.18$ and $2\,alk/(al+alk) = 0.62$. Therefrom the ratio $Kf:Plag$ is calculated as 0.12. A composition of approximately An_{43} is obtained from the plagioclase diagram. The calculation results in $Ab/An = 1.35$, corresponding to $An_{42.5}$.

This representation is also very suitable for providing an overall picture of the composition and relationship of the normative feldspars of rock series or entire petrographic provinces.

In addition to the main case just discussed, where no alkali-or alumina-excess was present, two other special cases still remain to be considered. These are, namely:

a) $si \geq (100 + 3\,al + alk)$ and $(al < alk)$, i.e., the case of alkali-excess and a positive qz-number. In the presence of alkali-excess, normative aegirine will appear, thus precluding the formation of anorthite. Thus, the ratios

Ab : *An* and *Kf* : *Plag* are no longer of interest. Only that part of Na = = 2 *alk* (1 — *k*) which is not bound within aegirine, and not its total quantity, is available for albite formation. An extention of the feldspar diagram (Figure 25) allows for the immediate determination of the ratio *Kf* : *Ab* in the presence of aegirine. If *al* < *alk*, 2 *alk*/(*al* + *alk*) >1. If no alkali feldspar is present, but only aegirine, then *al* = O and 2 *alk*/(*al* + *alk*) = = 2 *alk*/*alk* = 2. Thus, the aegirine pole can immediately be designated by the coordinates *x* = 0 and *y* = 2 (Figure 25). The ratio *Kf* : *Ab*, or the *k'*-number, which is only decisive for the feldspars, can now be derived for rocks with (*alk* ≥ *al*) by displacing the point of projection from the aegirine pole toward the abscissa. Thus one may read off the required alkali feldspar ratio. According to this method the arfvedsonite-lujaurite from Lille Elv will yield a *k'*-value of 0.16 from *k* = 0.13 and 2 *alk*/(*al* + *alk*) = 1.19. In a rock with sufficient silica, this will correspond to an alkali feldspar of composition $Kf_{19} Ab_{81}$.

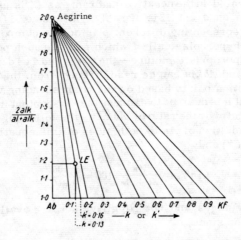

FIGURE 25. Extension of the diagram shown in Figure 24 for rocks with *al* < *alk*. According to P. Niggli (1927) *LE* corresponds to the projection point of the arfvedsonite-lujaurite from Lille Elv, with the coordinates *k* = 0.13 and 2 *alk*/(*al* + *alk*) = 1.19. The *k'*-number, which is decisive for the alkali-feldspar ratio (a sufficient SiO_2-content is presumed), is shown to be 0.16. This can be read from the point of intersection of the line originating from the aegirine pole, drawn through *LE*, with the abscissa.

b) *si* ≥ (100 + 4 *alk*) and *c* < (*al* — *alk*) or *al* > (*alk* + *c*), i. e., an alumina-excess *t* is present. If an excess of alumina is present, the value of *c*, and not of (*al* — *alk*), will be decisive for the formation of anorthite. The *c*-value should therefore be substituted for (*al* — *alk*) in the previous expressions, thus giving

$$y = \frac{2\,alk}{2\,alk + c} = \frac{Na + K}{Na + K + Ca}$$

For the cordierite-andesite from Hoyazo this method gives *k* = 0.55 and *y* = 0.69. Using these values as a basis, the diagram yields a ratio of

$Kf : F = 0.38$ and a normative plagioclase An_{50}, which is in agreement with the calculation.

b) The State of Silication

On studying rock series or petrographic provinces in which changes of the state of silication took place during the course of their development, we deal with processes of silication or desilication. The problem arises as to how these processes are to be expressed quantitatively. On the occasion of his investigation of the Vesbic volcano (Somma-Vesuvius), A. Rittmann (1933) indicated that this can be accomplished by means of the Niggli-values. Therefore he defined the following:

"Degree of silication" $Si° = \dfrac{\text{Actual } SiO_2 \text{ present}}{SiO_2 \text{ required for high silication}}$

"Degree of acidity" $Az° = \dfrac{\text{Number of } SiO_2 \text{ equivalents present}}{\text{Number of total oxide equivalents present}}$

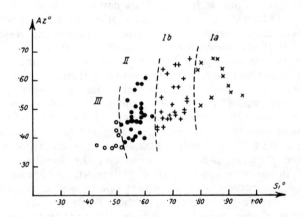

FIGURE 26. $Si°/Az°$-diagram for the young volcanic rocks of Cape Verde and Dakar

Expressed in Niggli-values we obtain:

$$Si° = \frac{si}{100 + 4\,alk} = \frac{si}{si - qz} \quad \text{for the case when } al > alk,$$

$$Si° = \frac{si}{100 + 3\,al + alk} = \frac{si}{si - qz} \quad \text{for the case when } al < alk.$$

$$Az° = \frac{si}{si + al + fm + c + alk} = \frac{si}{si + 100}.$$

Here, too, the common denominator (sum of the metallic-oxide equivalents) is eliminated.

For rocks with $qz \lesseqgtr 0$, $Si° \lesseqgtr 1$. As shown by Rittmann with regard to the Vesbic volcano (Somma-Vesuvius)(Rittmann, 1933), the plotting of $Si°$ against $Az°$ in a rectangular diagram allows for the drawing of certain interesting conclusions. This pertains to the occurrence of rock series of various silica contents within a single petrographic province. As far as the Vesbic volcano is concerned, it can be shown that the change in the degree of silication with the time, trended persistently toward a decrease in silication. In Figure 26 such a diagram is given for the young volcanic rocks of Cape Verde and the neighboring mainland of Dakar. The analyses were taken from works by J. Bacelar Bebiano, J. Chautard, H. Ermert, A. Lacroix, A. Mario de Jesus, G. M. Part, R. Reinisch, and C. F. Torre de Assuncao, and were compiled by C. Burri (1959). For all of the rocks $Si° < 1$, i.e., they are undersaturated with respect to SiO_2, possessing values down to 0.45. Four groups are readily distinguished with regard to the silication. These will be dealt with in greater detail in section B II 10a.

c) The *al-fm-c-alk* -Tetrahedron and the Representation of the Igneous Field

The four Niggli-values *al, fm, c* and *alk* can be regarded as the corners of a concentration tetrahedron, within which any rock chemism can be represented by a single point. The corresponding *si*-values may be indicated near these points for the sake of completion. In order to provide a true-scale representation in a plane, one should either have recourse to one of the projections mentioned in section A II 3b, or one should regard sectional planes appropriately located within the tetrahedron, as suggested by P. Niggli (1920, pp. 476-477, 1923, pp. 60-62, 1924, pp. 35-39 and 70-73, 1948, pp. 24-25). For this purpose the edge *c-fm* of the tetrahedron should be divided into 10 equal parts, corresponding to the *c : fm* ratios of 9:1, 8:2, 7:3, 6:4, etc. Section planes are now constructed through these points and the edge *al-alk*, thus resolving the tetrahedron into 10 wedge-shaped, three-dimensional parts. Points which happen to fall within these parts are projected onto the nearest plane, which in turn is unfolded into the plane of drawing. Strictly speaking, these sectional planes (Figure 27) deviate increasingly from an equilateral triangle as they become more distantly located from the *c* or *fm* corners. For the purposes of graphical representation it is, however, sufficient to handle them as if they were equilateral. This has the advantage in that commercial triangular coordinate paper may be used.

According to Niggli's suggestion, the representation is carried out in such a way that sections of the tetrahedron (taken in pairs), located symmetrically in relation to the central plane, are opened up, bookwise, about the common edge *al-alk* and are drawn as double triangles in the plane of the drawing. The insertion of a point corresponding to the given *al, fm, c* and *alk* values is then carried out, as in the case of a concentration triangle, using the values *al, alk* and (*c* + *fm*), on the section having the appropriate *c/fm* ratio. This type of representation is especially suitable for the representation of the so-called igneous field.

This forms a disc-like section within the tetrahedron, and appears in the section planes in the form of a belt (Figure 27). Since the total possible variation range of the **al-fm-alk-c** ratios is covered by the tetrahedron, the chemical relations in sediments can also be studied by this method. The igneous field bisects the tetrahedron into the T-space located on its concave side, opposite the **al**-corner, and the P-space situated on its convex side, opposite the **alk**- and **c**-corners. The igneous rocks and their metamorphic derivatives comprise the primary source material for the formation of sediments. During weathering, a more or less complete separation of soluble from insoluble material takes place. The most important soluble

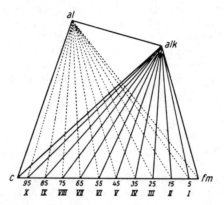

Concentration tetrahedron *al-fm-c-alk* with indicated section planes

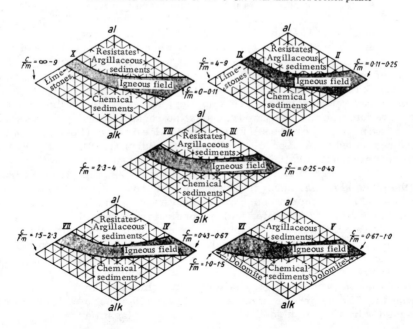

FIGURE 27. Concentration tetrahedron *al-fm-c-alk* with indicated sectional planes through the edge *al-alk* for illustration of the igneous field. After P. Niggli (1920)

components are the alkalis — CaO and MgO. They are precipitated from the solutions, partly by the agency of organisms and partly in purely inorganic form as carbonates, sulfates and chlorides. Al_2O_3, Fe_2O_3 and SiO_2 are enriched in the insoluble portion, and are either introduced into the residual clays or are transported in suspension and redeposited. Thus, chemical compositions lying in the igneous field within the tetrahedron undergo splitting. The soluble portion is displaced towards the P-(precipitate) space, and the insoluble portion — toward the T-(argillaceous) space. Only sediments with a very limited primary separation of the soluble portion from the insoluble, fall within the igneous field (soil-breccias, arkoses for example), as well as secondary mixtures of argillaceous residues and calcitic-dolomitic precipitates. This is of great importance, e. g., when considering calcitic-dolomitic marl, calcitic-sandstones, saline clays, etc. Tetrahedral representation is very suitable for considering whether a given metamorphic rock is — with respect to its chemism — either of igneous or sedimentary origin. If its point of projection clearly falls beyond the igneous field, a sedimentary origin of the rock is generally probable. If the point falls within the igneous field, an igneous origin is possible, but a sedimentary origin is not, however, entirely precluded. By considering the *si*-value at the same time, the decision is made easier, since in an igneous rock the variation range of *si*, corresponding to a given *al-fm-c-alk* ratio, is always narrow, whereas for sediments no such restrictions exist.

The *al-fm-c-alk* tetrahedron is less suitable for the representation of rock series or petrographic provinces, owing to the difficulties involved in projecting the three-dimensional illustration on the plane of drawing. In all events, recourse should be made to the already-mentioned experiments of F. Becke (1925, pp. 27-56) and his pupils Ch. Bacon (1926, pp. 126-172, especially p. 164) and A. Marchet (1931, pp. 490-537).

Sometimes, however, in special cases when the content of one of the four components is very small throughout the entire rock series, so that the points of projection fall very close to one particular tetrahedron plane and can therefore be projected thereon, situations may arise which allow for a very clear representation. This applies, for example, in the case of very calcium-poor to calcium-free alkaline rocks, including such rocks as the osannite-syenites from Alter Pedroso (Portugal), which were described by A. Lacroix (1916) and C. Burri (1928), and their basic differentiates, the alkaline rocks from Evisa (Corsica) and similar occurrences. Figure 28 shows the projection of the rocks from Alter Pedroso on the plane *al-fm-alk*, according to the procedure described in section A II 3cβ. All of them are more or less saturated with SiO_2 and contain alkali feldspars and osannite, with subordinate amounts of aegirine-augite and analcite or quartz as accessories. With minor deviations, the points of projection of several analyses fall along the line joining the pole of the theoretical alkali feldspar or analcite composition (*al = alk*) with the field of the osannite analyses. The rock series can thus be clearly expressed by a differentiation process of gravitational enrichment of the melanocratic constituents. The occurrence of melanocratic varieties such as schlieren or late gangue (pedrosite) is in good accord with this explanation.

d) Further Representations of the Igneous Field

Further expressions may be derived from the Niggli-values *al*, *fm*, *c* and *alk*, the sum of which should also equal 100. By this procedure they can also be adapted for representation in the concentration tetrahedron.

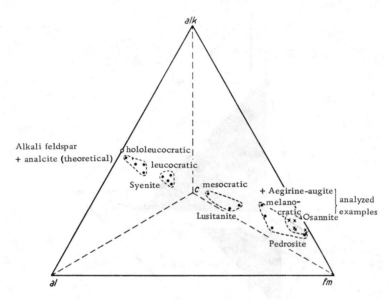

FIGURE 28. The alkali syenites from Alter Pedroso (Alemtejo Province, Portugal) and their basic differentiates in an *al-fm-c-alk*-tetrahedron, projected on the *al-fm-alk*-plane. This type of projection was recommended because of the low content of *c* throughout the series. The arrangement of the points of projection of the rocks, which are mainly composed from alkali feldspar ± some analcite and osannite (riebeckite) and subordinate amounts of aegirine-augite, along a straight line, indicates a simple differentiation process involving separation and gravitational enrichment of the melanocratic constituents. The holomelanocratic pedrosites appear as gangue. After C. Burri (1928).

If $T = (al - alk) > 0$, then

$$(al - alk) + fm + c + 2\,alk = T + fm + c + 2\,alk = 100,$$

whereas if $(al < alk)$ or $(alk - al) = -T = alk'$, it holds that

$$(alk - al) + fm + c + 2\,al = alk' + fm + c + 2\,al = 100.$$

Since in some rock series or petrographic provinces rocks having $al \lessgtr alk$ occur together, a means of representation including both cases is recommended. P. Niggli (1948, p. 25) used two tetrahedra with either the plane *fm-c-2 alk* or *fm-c-2 al* as the common base and $T = (al - alk)$ or $-T = alk' = (alk - al)$ as the opposite corner. In order to be able to use ordinary rectangular coordinate paper for the important section planes normal to the edge *fm-2 alk* or *fm-2 al*, the two tetrahedra are deformed in a rectangular tendency. Thereby the planes $2\,alk$-T-c or $2\,al$-$(-T)$-c become equilateral, whereas the other triangles become rectangular isosceles (Figure 29). It is now possible to construct sectional planes for specific values of $2\,alk$, $2\,al$ or fm, parallel to the plane $T\text{-}c\text{-}fm\text{-}(-T)$. On these planes the points of projection may be inserted using rectangular coordinates.

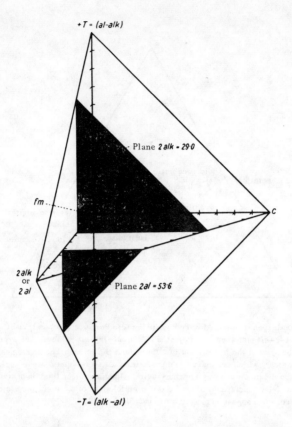

FIGURE 29. Rectangularly deformed double-tetrahedron $(al-alk)\text{-}2\,alk\text{-}fm\text{-}c$ or $(alk-al)\text{-}2\,al\text{-}fm\text{-}c$. After P. Niggli (1948)

SP corresponds to the quartz-diorite from Spanish Peak. Its point of projection is located in the plane $2\,alk = 29.0$ and may be plotted according to the rectangular coordinates $c = 22.5$ and $(al-alk) = T = 17.6$, since these are illustrated undistorted.

LE corresponds to the point of projection of the arfvedsonite-lujaurite from Lille Elv. It is located on the plane $2\,al = 53.6$ with the coordinates $c = 2.1$ and $(alk-al) = -T = -12.3$. The cordierite-andesite from Hoyazo near Nijar would be situated on the plane $2\,alk = 28.4$ having the coordinates $c = 12.6$ and $(al-alk) = T = 24.2$. In order not to complicate the figure it has, however, not been inserted.

If the double-tetrahedron's boundary plane $T\text{-}(-T)\text{-}c$ is treated on its own, an illustration of the igneous field is obtained as a special case. This was already suggested by Niggli (P. Niggli, F. de Quervain, R. U. Winterhalter 1930, pp. 40, 67, 70-71, 340-341) and is of special value for the consideration of the chemism of metamorphic rocks of sedimentary origin (Niggli, 1929). It is especially well adapted for the representation of the chemism of sedimentary rocks of the type clay-calcareous marl-limestone, or clay-dolomitic marl-dolomite, i. e., for sediments the chemical composition of which falls partly within the igneous field (Figure 30). Using this representation, it can be clearly shown, among others, that the para-rocks at present exhibiting a high-metamorphic facies with kyanite, staurolite, garnet, hornblende, mica, etc, on the southern boundary of the Gotthard Massif, originally constituted a sedimentary series similar to the so-called Tremola series of Paleozoic age, which are likewise located on the southern margin of the same massif (P. Niggli, 1929).

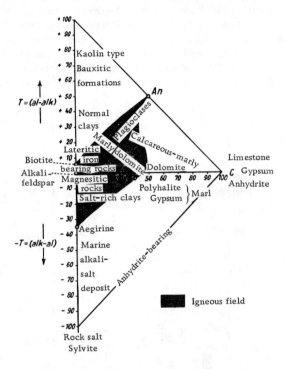

FIGURE 30. Location of the igneous field and the main sedimentary rocks in the triangle $+T\text{-}(-T)\text{-}c$, as achieved by the projection of the double-tetrahedron of Figure 29 on the plane $+T\text{-}(-T)\text{-}c\text{-}fm$. After P. Niggli, F. De Quervain and R. U. Winterhalter (1930)

Additional representations of the igneous field for special requirements can be obtained immediately. Only the representation of si versus $(c + alk)$, in which the projection points of the important rock-forming minerals are also indicated, will be cited here in Figure 31 (C. Burri and P. Niggli, 1945, p. 71). These relationships determine the boundaries of the field. Only certain rare, exceptional rocks fall beyond its limits.

e) The Variation Diagram

Not only the visuality is improved by reducing the 12 or more oxide components, which are generally determined by the rock analysis, to the five values si, al, fm, c and alk, but the possibility of graphical illustration in a plane also arises. Therefore, the values al, fm, c and alk are plotted as ordinates against *si*, which is taken as the abscissa. The values of *al* or *fm*, etc, are then connected eventually by drawing a smooth curve through the points if a high point-density is obtained. The procedure thus bears a close similarity to that used for the diagrams introduced by J. P. Iddings and A. Harker, where the metallic oxides, taken as molecular percentages, are plotted against SiO_2. The diagrams constructed by P. Niggli on the basis of the Niggli-values were originally "differentiation diagrams", since they were primarily intended for representing the chemical variability of differentiation series. Later, preference was given to the term "variation diagram" which is a neutral expression, free of assumptions, a priori, on the genesis of the considered association. This change in terminology serves to emphasize that this mode of representation is aimed at the clarification of the chemical variation within a rock series or petrographic province without introducing a bias as to its formation.

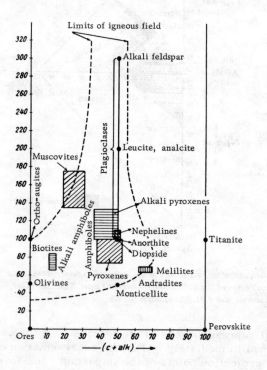

FIGURE 31. Illustration of the igneous field and the important rock-forming minerals in the *si*-(*c* + *alk*)-diagram. After C. Burri and P. Niggli (1945)

F. Chaye's (1949) objection to the variation diagrams is quite valid, namely, that they deal with correlations of ratios having a common denominator. Therefore, a correlation may be assumed to exist even if the absolute values from which the ratios were derived, are entirely uncorrelated. However, if a comparison is made with Harker's variation diagrams, in which the oxide values are directly plotted against one another, it may be recognized that a distinct correlation indeed exists between the SiO_2-content and the content of Al_2O_3, Fe_2O_3, FeO, MgO, MnO, CaO, Na_2O, K_2O, TiO_2, etc. In all events, this holds true for cases of rock series which are products of a single magma source. Simple volcanoes of the type of Lassen Peak, California, may serve as example, if only to mention a well-known and investigated example. The correlation degree of the variation diagrams of the Niggli-values need not necessarily be identical to that of the oxide values. However, from experience it is known that the Niggli variation diagrams are very suitable for the comparative study of petrographic provinces. The characterization of these provinces, based on the configuration of the curves, provide most valuable assistance, among others, during the study of the relationships between the differentiation type and the tectonic setting. They clearly illustrate the chemical variation caused by differentiation. On the other hand, they are not suitable for expressing the exact nature of the differentiation processes which took place. From these Niggli variation diagrams no conclusions may be drawn as to which quantitative ratios of the oxide species were enriched or depleted, since the correlation of oxide ratios does not provide a basis for conclusions on the absolute values. Other methods, based on the Niggli equivalent norms, should be used for these purposes.

It should also be mentioned that Niggli's variation diagrams enable one to arrive at a number of important conclusions, based to a certain degree on purely empirical data, despite the theoretical objections raised against them from the point of view of mathematical statistics. Provided that a certain density of analyses are at hand, it may be shown, for example, that a good to very good correlation will exist in rock series or petrographic provinces for which a uniform magma source of relatively restricted size is assumed. In these cases the point fields will allow for the drawing of smooth middle curves. Bearing this in mind, the variation diagram of Lassen Peak, Calif., shown on Figure 33, provides an instructive example. The analyses were performed by H. Williams (1932) and E.S. Shepherd (1938). If, on the contrary, the correlation is less complete to poor, i.e., a marked scattering of the points occurs, it is almost always noted that the province under consideration does not fall within a clear tectonic setting, e.g., the province is situated in the border region between an orogenic zone and a craton, i.e., in the foreland, in the "Rückland"* or in the Zwischengebirge, where different trends of differentiation of different ages take place. The possible active role of assimilation may constitute a second reason. This may be the case in individual volcanoes or volcanic groups of a limited scale. Indeed, a uniform differentiation process and, in consequence, a good correlation in the variation diagram may be anticipated, because of the relatively restricted space available for the underlying magma. The following occurrences are typical in this respect: the Vesbic volcano (Somma-Vesuvius), the Albanian Mountains (Volcano Laziale), Monte Vulture, as well as Cape Verde. In this connection it should be mentioned that the poor correlation in the variation diagram of the Vesbic Volcano (Figure 33) served as the starting point for A. Rittmann's (1933 and 1944) fruitful investigations of the magmatic development of the Somma-Vesuvius magmas and the significance of the carbonate assimilation there.

* [Literally translated as rear land].

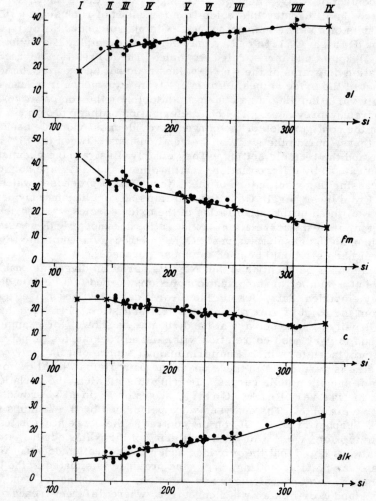

FIGURE 32a. Variation of *al, fm, c* and *alk* relative to *si* for the young lavas from Lassen Peak, California, with interpolated average characteristic values

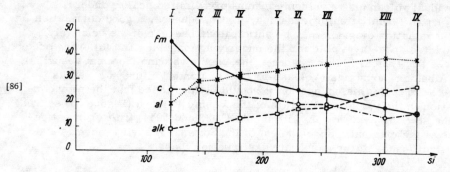

FIGURE 32b. Variation diagram of the young lavas of Lassen Peak, based on the average characteristic values from Figure 32a

In order to make the variation diagram more complete, the values which shed light on the details of the composite components, *alk* and *fm*, are used,

$$k = \frac{K_2O}{K_2O + Na_2O} = \frac{K^+}{K^+ + Na^+},$$

$$mg = \frac{MgO}{MgO + FeO + MnO} = \frac{Mg^{2+}}{Mg^{2+} + Fe^{2+} + Fe^{3+} + Mn^{+2}},$$

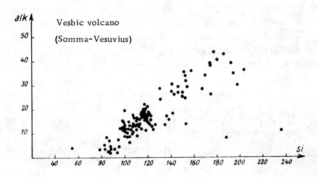

FIGURE 33. Variation of *alk* as dependent on *si* for the rocks of the Vesbic volcano (Somma-Vesuvius)

whereby it is understood that the total iron, recalculated to FeO, is included under "FeO" in this case. These values are plotted against each other in a rectangular, so-called *k-mg* diagram, as already suggested in 1915 by H. S. Washington (1915). Since, according to their definition, both *k* and *mg* may vary from 0 to 1, the whole correlation diagram is situated within a quadrangle with an edge length equal to unity.

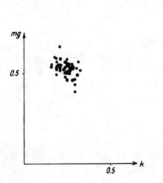 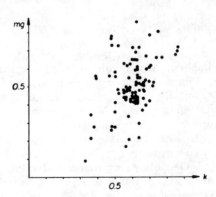

FIGURE 34a. *k-mg* diagram for the lavas of Lassen Peak, California

FIGURE 34b. *k-mg* diagram for the Vesbic volcano (Somma-Vesuvius)

In order to provide an example of such, Figure 34a illustrates the *k-mg*-diagram for the rocks of Lassen Peak. If the number of analyses is very large, the procedure outlined previously under section A II 2 may be applied. Accordingly, the total variation range is divided into classes. Thereby the entire field is broken down into squares in each of which is indicated the number of the points falling therein. In this way the absolute number or the percentage frequency may be noted. Figure 35 shows such a representation for the three main provinces already mentioned previously: the young igneous rocks of the North American Cordilliera (878 analyses), young volcanic rocks from Germany and North Bohemia (508 analyses) and young igneous rocks from Italy between Toscana and Campania (338 analyses). Distinct differences are here confirmed, characterizing the three provincial types "Pacific", "Atlantic" and "Mediterranean". From experience it is known that the *k-mg*-diagrams show characteristic fields of points of different, sometimes high, correlation. It should, however, also be noted that a typical case of ratio-correlation is considered, which is the general case already mentioned, for which none of the absolute

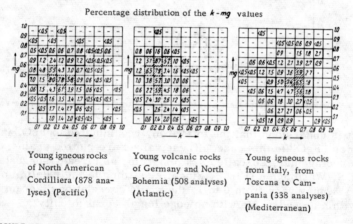

Young igneous rocks of North American Cordilliera (878 analyses) (Pacific)

Young volcanic rocks of Germany and North Bohemia (508 analyses) (Atlantic)

Young igneous rocks from Italy, from Toscana to Campania (338 analyses) (Mediterranean)

FIGURE 35. *k-mg* diagram for three selected main provinces (compare with Figures 5-7). After C. Burri and P. Niggli (1945)

values x_1, x_2, x_3, x_4 are equal. Only the following conclusions may be arrived at from a diagram showing a clear correlation between k and *mg*, e.g., Figure 34b: small values of the ratio $k = K_2O/(K_2O + Na_2O)$ will preferentially be associated with similar small values of the ratio *mg* = = $MgO/(MgO + FeO + MgO)$; large values of k will preferentially be associated with large *mg* values; and an increase in the value of the first generally heralds a parallel increase of the second. However, it cannot be concluded, for example, that a high FeO content will occur with a similar high Na_2O-content, or that the absolute content of Na_2O will vary parallel to that of FeO, or that of K_2O — parallel to MgO.

It is worth mentioning that, somewhat in contrast to H.S. Washington, P. Niggli never made such inferences from his *k-mg*-diagrams. Instead, he always concluded, from the correlation between the k and *mg*-values, i.e., from the pattern and location of the fields of points, only on the mutual dependence between the ratios k and *mg*. He also used them primarily as comparison characteristics in the study of rock series or petrographic provinces.

f) **Schematic Review of the Mutual Dependence Between si and alk for Certain Idealized Normative Mineral Compositions**

When al-, fm-, c-, alk- values are given, the mineral composition of igneous rocks mainly depends on the silication. Bearing this in mind, a means of graphic representation may be used to illustrate the occurrences of a number of common mineral combinations as determined by their dependence on si and alk (P. Niggli, 1927, pp. 125-126).

The following cases may be distinguished:

   I. In the normal case of oversaturation or saturation with respect to SiO_2 the following is valid:

$$si \geq (100 + 4\,alk) \text{ bzw. } alk \leq \frac{si - 100}{4}, \text{ provided that } al > alk.$$

   II. If olivine-type compounds (forsterite, fayalite, Ca-orthosilicate or monticellite) are taken instead of augite-type compounds (enstatite, hypersthene, wollastonite or diopside), the required si quantity is calculated to be

$$si = \underset{\text{Alkali feldspar}}{6\,alk} + \underset{\text{Anorthite}}{2\,(al - alk)} + \underset{\text{Ca-orthosilicate}}{\tfrac{1}{2}[c - (al - alk)]} + \underset{\text{Olivine}}{\tfrac{1}{2}fm}$$

$$= 6\,alk + 2\,al - 2\,alk + \tfrac{1}{2}c - \tfrac{1}{2}al + \tfrac{1}{2}alk + \tfrac{1}{2}fm$$

$$= 4\,alk + al + 50, \text{ whence it follows that } alk = \frac{si - al - 50}{4}$$

$$\text{or if } (al = alk),\quad si = 5\,alk + 50 \quad\text{or}\quad alk = \frac{si - 50}{5}.$$

   III. If the aluminosilicates are considered to be of the type $4\,SiO_2 \cdot Al_2O_3 \cdot (K, Na)_2O$ or $(SiO_2)_2 AlO_2(K, Na)$ (leucite, analcite), and in addition anorthite and augite, then for this case the following is valid:

$$si = \underset{\text{Leucite}}{4\,alk} + \underset{\text{Anorthite}}{2\,(al - alk)} + \underset{\text{Wollastonite}}{[c - (al - alk)]} + \underset{\text{Enstatite, hypersthene}}{fm}$$

$$= 4\,alk + 2\,al - 2\,alk + c - al + alk + fm,$$

$$si = 2\,alk + 100, \text{ whence it follows that } alk = \frac{si - 100}{2}.$$

   IV. If the aluminosilicates are considered to be of type $2\,SiO_2 \cdot Al_2O_3 \cdot (Na, K)_2O$ or $[SiO_2 \cdot AlO_2](Na, K)$ (nepheline), and in addition anorthite and augite, it then follows that:

$$si = \underset{\text{Nepheline}}{2\,alk} + \underset{\text{Anorthite}}{2\,(al - alk)} + \underset{\text{Wollastonite}}{[c - (al - alk)]} + \underset{\text{Enstatite, hypersthene}}{fm}$$

$$= al + fm + c + alk = 100.$$

   V. Lastly, if the combination nepheline, anorthite and olivine is taken, it follows that:

$$si = 2\,alk + 2\,(al - alk) + \tfrac{1}{2}\,[c - (al - alk)] + \tfrac{1}{2}\,fm,$$
$$\text{Nepheline}\quad\ \ \text{Anorthite}\quad\ \ \text{Ca-orthosilicate}\quad\ \ \text{Olivine}$$

$$si = al + 50,\quad\text{or when}\quad al = alk\ \text{ist},\ si = alk + 50\quad\text{or}\quad alk = si - 50.$$

During all of these derivations, it was presumed that alk and al are bound to each other in the ratio of $1:1$, as is indeed the case in alkali-feldspars, leucite and nepheline. It hence follows that alk can never be greater than 50. Thus, si_{max} is a function of alk and al with regard to the various considered mineral combinations, al occurring only if olivine-type compounds are presumed.

At a first glance, si may be considered to be merely a function of alk, if (as is indeed highly probable) we set $al = alk$. The following curves should be drawn:

$$alk = \frac{si - 100}{4};\quad alk = \frac{si - 50}{5};\quad alk = \frac{si - 100}{2};\quad alk = si - 50.$$

If the alk-values are plotted as a function of si, or the alk-curves are drawn, for a rock series or a petrographic province, these operations are already sufficient to allow one to determine which of the considered idealized mineral combinations is at all possible, or which may be precluded as impossible. It should be recalled that a number of other possibilities, e.g., a high ore content or the occurrence of aegirine, were not taken into account. A general picture of certain heteromorphic possibilities may also be obtained, as illustrated by the example of the partial overlapping of the fields of olivine-metasilicate-feldspars and nepheline-metasilicate-feldspars (Figure 36).

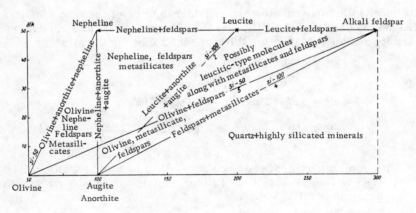

FIGURE 36. Schematic review of the relationship between si and alk for certain idealized normative mineral compositions. After P. Niggli (1937)

g) Distribution of SiO_2 Among Normative Leucocratic and Melanocratic Components

In certain cases it may be of interest to know what quantity of the available SiO_2 is introduced into the light or dark normative components, and what part of it is available as free SiO_2 for the formation of quartz after the calculation of the norm (P. Niggli, 1927, pp. 120-133). If it is present in such a quantity that oversaturation with respect to SiO_2 exists, the only light normative components which bear consideration are the feldspars. They require a quantity of si given by

$$si' = \underbrace{6\,alk}_{\text{Alkali feldspars}} + \underbrace{2\,(al - alk)}_{\text{Anorthite}} \quad \text{when } al > alk,$$

$$si' = \underbrace{6\,al}_{\text{Alkali feldspars}} \quad \text{when } al < alk.$$

The total iron, i.e., bivalent and trivalent (the latter taken as divalent), as well as the total Mg and the c which is not bound in anorthite $c - (al - alk)$, should be combined with SiO_2 in the ratio of 1:1. If however, $al < alk$, the formation of aegirine will require $4\,si$ and $2\,fm$ for each unit of $(alk - al)$, corresponding to the formula $4\,SiO_2 \cdot Fe_2O_3 \cdot Na_2O$. Thus, the si-requirement of the dark constituents is given by:

$$si'' = \underbrace{fm}_{\text{Enstatite + hypersthene}} + \underbrace{c - (al - alk)}_{\text{Wollastonite}} = 100 - 2\,al, \quad \text{when } al > alk,$$

$$si'' = \underbrace{4\,(alk - al)}_{\text{Aegirine}} + \underbrace{fm - 2\,(alk - al)}_{\text{Enstatite + hypersthene}} + \underbrace{c}_{\text{Wollastonite}} = 100 - 3\,al + alk, \quad \text{when } al < alk.$$

The remaining unused si is given by:

$$si''' = si - (100 + 4\,alk) \quad \text{when } al > alk,$$

$$si''' = si - (100 + 3\,al + alk) \quad \text{when } al < alk.$$

If si is set as unity, and the amount of si bound to the normative light constituents is designated as $si'/si = L_s$, the amount of the same bound to the normative dark constituents is designated as $si''/si = F_s$, and the remainder is designated as $si'''/si = Q_s$, it follows that:

for the case $al > alk$

$$L_s = \frac{si'}{si} = \frac{4\,alk + 2\,al}{si},$$

for the case $al < alk$

$$L_s = \frac{si'}{si} = \frac{6\,al}{si},$$

$$F_s = \frac{si''}{si} = \frac{100 - 2\,al}{si}, \qquad F_s = \frac{si''}{si} = \frac{100 - 3\,al + alk}{si},$$

$$Q_s = \frac{si'''}{si} = \frac{si - (100 + 4\,alk)}{si}, \qquad Q_s = \frac{si'''}{si} = \frac{si - (100 + 3\,al + alk)}{si},$$

whence it always follows that $L_s + F_s + Q_s = 1$.

Throughout the calculation it has been presumed that highly silicated compounds may be formed, corresponding to $qz \geq 0$. If, however, $qz < 0$, and thus the available SiO_2 is insufficient for the formation of highly silicated compounds, two possibilities may arise: according to the first, compounds low in silica are formed in a quantity corresponding to the available amount of SiO_2. The second possibility is to form the highly silicated components, disregarding the actual available amount of SiO_2, and to state the SiO_2 deficit as a negative value, so that the equation $L_s + F_s + Q_s = 1$ still holds true. The second procedure is often preferred, since no assumptions are made as to the mode of distribution of the SiO_2-deficit, i.e. whether olivine, leucite, or nepheline are formed.

The values L_s, F_s and Q_s were represented by P. Niggli in a triangular projection which provides a clear picture of possible modes of distribution of the SiO_2 among the various compounds poor in silica. Since this illustration procedure has nowadays been replaced by the QLM-triangle, which will be described in more detail further on, it will not be discussed in greater detail here.

5. Magma-Types

a) The Concept of a Magma-Type

P. Niggli also used the values si, al, fm, c, alk, k and mg for the formulation of a system of magma-types in order to characterize the rock's chemism. This refers exclusively to the chemical relations and therefore does not represent any form of rock classification. It was first introduced by P. Niggli in 1920, partly in conjunction with A. Osann, and later repeatedly extended. The final representations followed in 1923 (P. Niggli, 1923) and 1936 (P. Niggli, 1936), as well as in 1945 (C. Burri and P. Niggli, 1945). A principally acceptable draft on the concept of magma-types was provided by E. B. Bailey and H. H. Thomas in 1924 (E. B. Bailey, C. T. Clough, W. B. Wright, J. E. Richey, G. V. Wilson and others, 1924, especially in 13-28). In the context used here, as well as in that used by E. B. Bailey and H. H. Thomas, a "magma-type" is defined merely as an abbreviated designation for the characterization of the rock's chemism. The actual mineral composition is not taken into consideration. The presentation of the magma-type is thus insufficient for the complete classification of an igneous rock. Such a classification calls for the consideration of the actual mineral composition (modus), which is not unequivocally determined by the chemism. This is also proved by the existence of different heteromorphic facies types. The nomenclature must therefore be twofold, including the possibility of expressing, separately, the chemical and mineralogical characteristics. "Quartz-dioritic andesite" or "quartz-dioritic dacite" or "aplite-granitic" or "alkali-granitic liparite" serve as examples of such a means of twofold indication.

The average composition of the igneous rocks of the upper lithosphere may be used as a point of reference for the systematics of the magma-types.

Thus, the following average is valid:

si	al	fm	c	alk	k	mg
200	30	32	18	20	0.35	0.50

Magma-types for which $al \sim fm$ (approximately between 26 and 32), as in the previously mentioned example, are generally termed isofalic. Using this as a reference, the following designations may be introduced. Here $(+)$, $(-)$, and $(\sim)$ designate values larger than, smaller than, and similar to those existing for isofalic types respectively.

al	fm	
$(+)$	$(-)$	salic magmas
$(-)$	$(+)$	femic magmas
$(\sim)$	$(-)$	subfemic magmas
$(-)$	$(\sim)$	subalic magmas
$(-)$	$(-)$	subalfemic magmas
$(+)$	$(+)$	peralfemic magmas
$(+)$	$(\sim)$	semialic magmas
$(\sim)$	$(+)$	semifemic magmas.

Figure 37 clarifies these relationships graphically.

Further definitions are:

$al < alk$	peralkalic magmas
$al \sim alk$ to $alk > 2/3\, al$	relatively alkali-rich magmas
$alk \sim 2/3\, al$ to $1/2\, al$	intermediate alkali magmas
$alk < 1/2\, al$	relatively alkali-poor magmas

and with regard to the c-value

$c > 25$	c-rich magmas
c 15–25	c-normal magmas
$c \leq 15$	c-poor magmas.

These limits are shown diagramatically in Figure 38.

Using these criteria, a tabulated review of the magma-types will be presented in the following sections (C. Burri and P. Niggli, 1945). The upper division consists of the three magma series: the calc-alkali series, the sodic series, and the potassic series. This arrangement is justified by the fact that in nature a large number of igneous provinces are found, which are characterized by the common occurrence of magmas of one of these three series. Such associations undoubtedly merit a corresponding geological significance. The occurrence of transitional and mixed provinces can immediately be anticipated, and their regional distribution, as dependent on the major tectonics, readily explains these occurrences.

b)

TABLE III
Review of magma-types (C. Burri and P. Niggli, 1945)

Magma-type	si	al	fm	c	alk	k	mg

A. Magmas of the Calc-Alkali Series

a) Leucogranitic magmas (markedly salic, acid, mostly rich in alkali and poor in c)

	si	al	fm	c	alk	k	mg
1. Aplite-granitic	460	47	8	5	40	0.45	0.25
2. Yosemitite-aplitic	350	45	6	13	36	0.4	0.3
3. Engadinite-granitic	380	43	13	8	36	0.5	0.25
4. Yosemitite-granitic	350	43	14	13	30	0.45	0.3

b) Granitic magmas (isofalic to weakly salic or subfemic, generally acid, intermediate to alkali-rich, calc-poor to c-normal)

	si	al	fm	c	alk	k	mg
1. Adamellitic	300	37.5	22.5	13.5	26.5	0.45	0.3
2. Tasna-granitic	300	36	28	9	27	0.45	0.35
3. Moyitic	380	33	32	15	20	0.45	0.3
4. Granitic	270	34	29	13	24	0.45	0.35
5. Opdalitic	225	32	32	18	18	0.45	0.45

c) Granodioritic magmas (salic, lower in alkali than leucogranitic, mostly still distinctly acid)

	si	al	fm	c	alk	k	mg
1. Granodioritic	280	39	22	17	22	0.45	0.4
2. Leucotonalitic	220	39	24	21	16	0.5	0.3
3. Farsunditic	300	42	20	15	23	0.25	0.4
4. Leucopeleeitic	200	38	21	24	17	0.2	0.4

d) Trondjemitic magmas (salic, acid, alkali-rich to intermediate, c-poor to c-normal)

	si	al	fm	c	alk	k	mg
1. Soda granite-aplitic	450	47	7,5	3,5	42	0.2	0.2
2. Trondjemite-aplitic	400	44	5	13	38	0.2	0.3
3. Quartz diorite-aplitic	420	46.5	4	15	34.5	0.2	0.3
4. Soda-engadinitic	400	43.5	15	3,5	38	0.25	0.25
5. Trondjemitic	370	42	12	11	35	0.25	0.3
6. Soda-rapakivitic	340	42	20	8	30	0.25	0.3
7. Leucoquartz-dioritic	300	42	17.5	13	27.5	0.25	0.4
8. si-oligoclasitic	380	44	8	20	28	0.15	0.4

e) Quartz-dioritic magmas (isofalic to weakly femic, neutral to weakly acid, intermediate to alkali-poor, c-variable)

	si	al	fm	c	alk	k	mg
1. Quartz-dioritic	225	32	31	19	18	0.25	0.45
2. Melaquartz-dioritic	200	30	40	10	20	0.3	0.5
3. Peleeitic	180	33	32	23	12	0.2	0.45
4. Tonalitic	180	33	33	22	12	0.4	0.4

f) Dioritic magmas (isofalic to weakly femic, neutral to weakly basic, mostly alkali-poor to intermediate and c-normal)

	si	al	fm	c	alk	k	mg
1. Dioritic	155	30	35	21	14	0.3	0.5
2. Lamprodioritic	150	25	40	21.5	13.5	0.25	0.5
3. Orbitic	135	27	42	21.5	9.5	0.25	0.5

Magma-type	si	al	fm	c	alk	k	mg

g) **Gabbro-dioritic magmas** (femic, neutral to basic, mostly alkali-poor and c-normal)

Magma-type	si	al	fm	c	alk	k	mg
1. si-gabbroidal	150	25	43	22	10	0.2	0.5
2. Normal gabbro-dioritic	130	23	44	22.5	10.5	0.2	0.5
3. Melagabbro-dioritic	130	19	51	21	9	0.25	0.5

h) **Gabbroidal magmas** (femic, basic, alkali-poor and c-rich to c-normal)

Magma-type	si	al	fm	c	alk	k	mg
1. Gabbroidal	108	21	51	22	6	0.2	0.5
2. fm-gabbroidal	80	24	54	17	5	0.2	var.
3. Eucritic	100	18	55	24	3	0.1	0.7
4. c-gabbroidal	100	25	46	25	4	0.1	0.7
5. Miharaitic	130	23	42	27.5	7.5	0.2	0.5
6. Pyroxene-gabbroidal	100	23.5	40.5	31.5	4.5	0.2	0.7
7. Hawaiitic	120	18.5	48	27.5	6	0.1	0.5

i) **Leucogabbroidal magmas** (isofalic to weakly femic or subalic, basic to neutral, mostly c-rich, alkali-poor)

Magma-type	si	al	fm	c	alk	k	mg
1. Leucomiharaitic	140	26.5	38	26.5	9	0.25	0.45
2. Achnahaitic	100	29	40	27	4	0.2	0.5
3. Cumbraitic	200	27	31	27	15	0.3	0.3
4. Belugitic	130	29	32	29	10	0.15	0.5
5. Ossipitic	110	30	35	30	5	0.15	0.6

j) **Plagioclase magmas** (salic, mostly neutral to basic, in part c-rich and alkali-poor)

Magma-type	si	al	fm	c	alk	k	mg
1. Oligoclastic	190	43	11	22	24	0.2	0.3
2. Andesinitic	190	46.5	5	28	20.5	0.2	0.3
3. Labradorite-felsitic	150	46	4	35	15	0.1	0.45
4. Anorthositic	120	43	10	40	7	0.1	0.45
5. si-melaplagioclastic (in case of low si, only melaplagioclastic)	220	35	15	35	15	0.1	0.6
6. Anorthosite-gabbroidal	130	37	22	33	8	0.1	0.55

k) **Hornblenditic magmas** (strongly femic, basic, alkali-poor, c-normal to c-poor)

Magma-type	si	al	fm	c	alk	k	mg
1. al-hornblenditic	120	19	61	15	5		0.7
2. Hornblenditic	80	15	60	20	5		0.6
3. Hornblende-peridotitic	80	10	74	14	2		0.7
4. Diallagitic	90	7	68	23	2		0.7
5. Websteritic	95	1	80	18	1		0.8

l) **Pyroxenitic magmas** (femic to strongly femic, basic, alk-variable, c-rich)

Magma-type	si	al	fm	c	alk	k	mg
1. si-pyroxenitic	125	13	50	30	7		0.7
2. Ariegitic	80	19	49	31	1		0.8
3. Ostraitic	60	18	53	27	2		0.6
4. Issitic	65	14	53	30	3		0.5
5. Jacupirangitic	70	7	56	35	2		0.6
6. Pyroxenitic	90	9	59	28	4		0.7
7. Batukitic	80	5	60	30	5		0.6
8. Koswitic	63	3	66	30.5	0.5		0.7

Magma-types	si	al	fm	c	alk	k	mg

m) **Orthoaugitic-perioditic magmas** (very strongly femic, basic, *c*-poor)

	si	al	fm	c	alk	k	mg
1. Orthoaugitic	95	4	90	5	1	mostly high	
2. Peridotitic	60	5	90	4	1	,, ,,	
3. Ore-peridotitic	15	6	91	2	1	mostly low	

B. Magmas of the Sodic Series

a) **Alkali granitic magmas** (salic, acid, alkali-rich, *c*-poor)

	si	al	fm	c	alk	k	mg
1. Alkali granite-aplitic	450	46	6	3	45	0.35	0.15
2. Alkali-granitic	400	41	15	3	41	0.35	0.2
3. Alkali syenite-aplitic (albititic)	280	43	8	7	42	0.33	0.2
4. Nordmarkitic	280	41	15	5	39	0.3	0.15
5. Gibelitic	260	35	21	9	35	0.3	0.15

b) **Evisitic magmas** (subfemic, isofalic to weakly femic, acid, peralkalic, *c*-poor)

	si	al	fm	c	alk	k	mg
1. Leucoevisitic	300	30	20	7	43	0.3	0.2
2. Evisitic-groruditic	300	30	30	2	38	0.35	0.1
3. Evisitic-pantelleritic	320	23	35	2	40	0.4	0.1
4. Normal evisitic	320	22.5	42	3	32.5	0.4	0.1

c) **Foyaitic magmas** (salic to subfemic, neutral to basic, alkali-rich, *c*-poor to *c*-normal)

	si	al	fm	c	alk	k	mg
1. Umptekitic	220	37	18	9	36	0.3	0.25
2. Foyaitic	180	40	15	5	40	0.3	0.2
3. Lardalitic	170	35	20	13	32	0.3	0.4
4. Tahititic	150	32	22	18	28	0.25	0.3
5. Urtitic	115	41	12	6	41	0.2	0.2

d) **Lujauritic magmas** (subfemic, isofalic to subalic, neutral to basic, peralkalic, *c*-poor)

	si	al	fm	c	alk	k	mg
1. Tinguaitic	170	36	15	7	42	0.2	0.15
2. Urtite-tinguaitic	100	36	15	7	42	0.2	0.15
3. Melatinguaitic	160	30	23	7	40	0.15	0.15
4. Greenlanditic	135	29	23	6	42	0.1	0.1
5. Lujauritic	160	29	28	5	38	0.2	0.2
6. Tawitic	130	20	32	8	40	0.1	0.2

e) **Subplagiofoyaitic magmas** (salic, mostly neutral to basic, alkali-rich to intermediate, *c*-poor to *c*-normal)

	si	al	fm	c	alk	k	mg
1. Bostonitic	230	46	12.5	2	39.5	0.3	0.3
2. Pulaskitic	210	40	18	10	32	0.3	0.3
3. Essexite-foyaitic	175	39	18	12	31	0.3	0.3
4. Monmouthitic	100	36	17	17	30	0.2	0.25

f) **Essexite-diorite magmas** (salic to subfemic)

	si	al	fm	c	alk	k	mg
1. Larvikitic	170	37	21	17	25	0.3	0.35
2. Kassaitic	170	33	23	19	25	0.25	0.35
3. Rouvillitic	140	37	20	23	20	0.25	0.4

Magma-type	si	al	fm	c	alk	k	mg

g) **Sodic-syenitic magmas** (mostly isofalic to weakly femic, mostly acid to intermediate, mostly alkali-rich and c-poor)

Magma-type	si	al	fm	c	alk	k	mg
1. si-sodic syenitic	300	35	28	5	32	0.25	0.2
2. si-maenaitic	300	34	29	9	28	0.2	0.5
3. sodic-syenitic	160	32	28	12	28	0.25	0.3
4. Maenaitic	210	33	28	14	25	0.3	0.4
5. Nosykombitic	150	33	28	14	25	0.3	0.4
6. Mela-sodic syenitic	140	26.5	39	11.5	23	0.25	0.3

h) **Ijolitic magmas** (subalic to subfemic, mostly basic, often still alkali-rich and -poor)

Magma-type	si	al	fm	c	alk	k	mg
1. Syenite-ijolitic	213	29	25	19	27	0.25	0.7
2. Ijolitic	100	25	25	25	25	0.25	0.4
3. Melteigitic	100	20	32.5	30	17.5	0.25	0.4
4. c-melteigitic	120	17	27	37.5	18.5	0.25	0.3
5. Turjaitic	70	15	33	41	11	0.25	0.5
6. Okaitic	55	14	30	49	7	0.15	0.5
7. Muritic	120	25	25	20	30	0.25	0.25

i) **Essexitic magmas** (isofalic, mostly neutral, intermediate and c-normal)

Magma-type	si	al	fm	c	alk	k	mg
1. Essexite-akeritic	175	30	30	20	20	0.3	0.4
2. Essexitic	130	30	30	20	20	0.3	0.4

j) **Theralitic magmas** (femic to subalic, basic, alkali-rich, mostly c-normal)

Magma-type	si	al	fm	c	alk	k	mg
1. Theralitic	110	21	38	23	18	0.25	0.45
2. Melatheralitic	85	17.5	47	21	14.5	0.2	0.55

k) **Sodic-gabbroidal magmas** (weakly femic to femic, basic, alkali-intermediate to alkali-poor, mostly c-normal to c-poor)

Magma-type	si	al	fm	c	alk	k	mg
1. Beringitic	125	23.5	39	22.5	15	0.25	0.45
2. Mugearitic	135	24	42	18	16	0.25	0.6
3. Sodic-lamprosyenitic	135	22	50	13.5	14.5	0.25	0.6
4. Essexite gabbro-dioritic	105	23	43	24	10	0.25	0.45
5. Essexite gabbroidal	95	20	49	21.5	9.5	0.25	0.5

l) **Theralite-gabbroidal magmas** (isofalic, subalic to femic, basic, more often alkali-poor and c-rich)

Magma-type	si	al	fm	c	alk	k	mg
1. Gabbromelteigitic	110	27	31	27	15	0.25	0.4
2. Theralite-gabbroidal	105	24	38	25	13	0.25	0.5
3. Berondritic	90	20	40	32	8	0.25	0.5
4. Turjaite-gabbroidal	100	24	33	32	11	0.25	0.5

m) **Gabbro-theralitic magmas** (femic, basic, intermediate to alkali-rich, c-rich)

Magma-type	si	al	fm	c	alk	k	mg
1. Gabbro-theralitic	100	17	43	27.5	12.5	0.2	0.45
2. c-gabbro-theralitic	100	15	40	35	10	0.25	0.5

n) **Mela-sodic gabbroidal magmas** (femic to strongly femic, basic, relatively alkali-rich, c-normal to high)

Magma-type	si	al	fm	c	alk	k	mg
1. Alkali jacupirangitic	95	8.5	52	32.5	7	0.3	0.4
2. Ankaratritic	75	13	52	27	8	0.25	0.6

Magma-type	si	al	fm	c	alk	k	mg
3. *alk*-issitic	75	17	49	28	6	0.3	0.6
4. Vesecite-polzenitic	50	9	54.5	30	6.5	0.3	0.7
5. Kaulaitic	100	13	55	22	10	0.15	0.6
6. Modlibovite-polzenitic	50	13	56	23	8	0.2	0.65
7. Alkali hornblende-peridotitic	65	8	68.5	16	7.5	0.3	0.8

o) Alkalipyrobolic magmas (femic, acid to basic, peralkalic, *c*-variable)

	si	al	fm	c	alk	k	mg
1. Lusitanitic	160	19	52	3	26	0.2	0.1
2. Rockallitic	250	13	50	3	34	0.1	0.1
3. Alkali-mafitic	140	5	60	3	32	0.1	0.1
4. Sodic-hornblenditic	120	2	78	5	15	0.15	0.10
5. Pienaaritic	130	16	29	35	20	0.2	0.25
6. Salitritic	100	2	38	50	10	0.25	0.6

C. Magmas of the Potassic Series

a) Leucosyenite granitic magmas (salic, acid, alkali-rich, *c*-poor)

	si	al	fm	c	alk	k	mg
1. Rapakivitic	350	41	18	9	32	0.45	0.3
2. Granosyenitic	260	39	18	11	32	0.45	0.3

b) Juvitic magmas (salic, acid to neutral, alkali-rich, mostly *c*-poor)

	si	al	fm	c	alk	k	mg
1. Potassic-nordmarkitic	270	40	15	5	40	0.4	0.25
2. Leucosyenitic	190	39	18	11	32	0.5	0.3
3. Potassic-gibelitic	260	35	21	9	35	0.4	0.2
4. Potassic-foyaitic	170	39	14	11	36	0.5	0.25
5. Monzonite-syenitic	180	36	23	15	26	0.45	0.35

c) Arkitic magmas (mostly subfemic to subalfemic, neutral to basic, alkali-rich, *c*-normal to high)

	si	al	fm	c	alk	k	mg
1. Sviatonossitic	190	30	20	24	26	0.4	0.3
2. Leucosommaitic	160	32.5	21	19	27.5	0.3	0.4
3. Arkitic	100	28	23	22	27	0.4	0.3
4. Borolanitic	120	27	22	31	20	0.6	0.35
5. Vesbitic	100	23	18	39,5	19.5	0.8	0.8

d) Syenite-granitic magmas (isofalic to weakly femic, acid to neutral, alkali-rich, *c*-poor to normal)

	si	al	fm	c	alk	k	mg
1. *si*-syenite-granitic	330	30	28	16	26	0,4	0.2
2. Syenite-granitic	250	30	29	13	28	0.5	0.4
3. Kammgranitic	225	26	39	12	23	0.6	0.6

e) Syenitic magmas (isofalic to weakly femic, intermediate to weakly basic, rather alkali-rich, *c*-poor)

	si	al	fm	c	alk	k	mg
1. Syenitic	180	30	30	12,5	27.5	0.5	0.4
2. *si*-Kamperitic	185	28	37	12,5	22.5	0.5	0.4
3. Kamperitic	150	29	37	11,5	22.5	0.6	0.5

Magma-type	si	al	fm	c	alk	k	mg

f) **Monzonitic magmas** (isofalic to subfemic or weakly femic, intermediate to basic, mostly normal to poor with regard to *alk* and *c*)

Magma-type	si	al	fm	c	alk	k	mg
1. Leucomonzonitic	180	37.5	25	17	20.5	0.45	0.5
2. *si*-monzonitic	170	30	30	20	20	0.45	0.45
3. Monzonitic	140	29	31	21	19	0.5	0.45
4. Sommaite-monzonitic	140	28	33.5	24.5	14	0.6	0.55
5. Sommaite-tonalitic	105	33	33	22	12	0.4	0.4

g) **Sommaitic magmas** (subalic, basic to neutral, alkali-rich to intermediate, *c*-normal to high)

Magma-type	si	al	fm	c	alk	k	mg
1. Melarkitic	150	22	34	22	22	0.6	0.4
2. Sommaitic	115	24.5	34	24.5	17	0.55	0.5
3. Sommaite-ossipitic	125	26	33	27	14	0.55	0.5

h) **Potassic-dioritic magmas** (weakly femic to femic, acid, neutral to basic, generally relatively alkali-poor to intermediate, *c* somewhat variable)

Magma-type	si	al	fm	c	alk	k	mg
1. Vredefortitic	250	28	42	17.5	12.5	0.5	0.4
2. Sommaite-dioritic	135	23.5	42	23.5	11	0.45	0.55
3. Monzonite-dioritic	135	27	38	21.5	13.5	0.4	0.5
4. Lampro-sommaitic	135	22.5	46.5	18	13	0.5	0.6

i) **Lamproitic magmas** (femic to strongly femic, intermediate to basic, alkali-rich to peralkalic, *c*-poor to normal)

Magma-type	si	al	fm	c	alk	k	mg
1. Lamprosyenitic	150	23	46	13	18	0.6	0.6
2. Wyoming-lamproitic	165	18	41	14	27	0.85	0.75
3. Murcialamproitic	140	17	52	13	18	0.6	0.75
4. Yogoitlamproitic	140	19	41	19	21	0.8	0.7
5. Jumillitic	110	13	60	14	13	0.7	0.8
6. Biotititic	70	14	74	1	11	0.8	0.9

j) **Shonkinitic magmas** (femic, mostly basic, relatively alkali-rich)

Magma-type	si	al	fm	c	alk	k	mg
1. Yogoitic	145	22	40	20	18	0.5	0.55
2. Shonkinitic	100	17.5	47.5	23	12	0.55	0.65

k) **Melashonkinitic magmas** (strongly femic, basic, *alk* intermediate to relatively high, *c* mostly normal)

Magma-type	si	al	fm	c	alk	k	mg
1. Kajanitic	90	13	55	23	9	0.6	0.7
2. Potassic-hornblenditic	100	14	60	17	9	0.4	0.8

l) **Missouritic-allnöitic magmas** (weakly femic, basic, *alk*-variable, *c*-high)

Magma-type	si	al	fm	c	alk	k	mg
1. Shonkinite-missouritic	110	18	34	34	14	0.6	0.55
2. Normal missouritic	95	14.5	44	31.5	10	0.6	0.6
3. Antsohitic	105	19	42	30	9	0.6	0.55
4. Potassic-polzenitic	80	12.5	47	34	6.5	0.6	0.6
5. Pyroxenolitic	80	13	40	43	4	0.6	0.6
6. Alnöitic	50	10	43	40	7	0.6	0.6

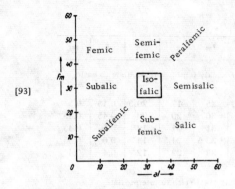
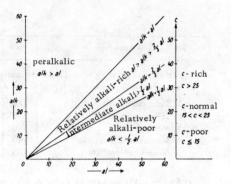

FIGURE 37. Classification principles of magmas. After C. Burri and P. Niggli (1945)

Figure 38. Classification principles of magmas. After C. Burri and P. Niggli (1945)

II. The Equivalent Norm

1. General Considerations of Experiments for the Calculation of the Normative Mineral Composition

The mutual relationships between the chemism and the mineral composition of igneous and metamorphic rocks constitute one of the most fundamental problems of petrography. Already in early days numerous petrographers attempted to calculate the mineral composition of such rocks from their chemical analysis. Mention will only be made of W. C. Brögger whose works on the rocks from the Kristiania (now Oslo) area include numerous excellent examples pertaining to this problem. The most important step in this development was, however, made with the introduction of a norm calculation procedure — the so-called CPIW-norm (Ch. W. Cross, J. P. Iddings, L. V. Pirsson, H. S. Washington, 1902, and 1903, H. S. Washington, 1917, A. Johannsen, 1931). Indeed, this procedure was not intended primarily for the calculation of a mineral composition which would agree with the observed mineral composition (so-called modus). The calculation is carried out according to exact norm stipulations in order to attain an imaginary mineral composition expressed by simply composed components. This corresponds to only one of several possible heteromorphic facies, and no attention is paid to the actual minerals present. Complex and water-bearing minerals, e. g. sesquioxide-bearing augites, hornblendes, and micas, are purposely excluded. However, conclusions can be drawn on the sesquioxide content of augites and hornblendes, by comparing the calculated mineral composition, the so-called norm, with the true mineral composition present, the so-called modus. If the normative An content is considerably higher than the modal content (in the case of a holocrystalline structure), a part of the alumina which was bound normatively in An is in fact included in augites or hornblendes. The calculation of the normative mineral composition is thus not mainly carried out for the purpose of clarifying the relationships between the chemism and the mineral composition of the considered rock. It serves as a basis for the so-called "quantitative classification of igneous rocks" of the four mentioned investigators. In abbreviated form it is also referred to as the "American" or "CPIW-classification" which is based on relations between the so-called "norm-minerals", calculated in accordance with the standard procedure already mentioned.

The influence of the CIPW-norm on the development of chemical petrography has, nevertheless, been extremely deep and long-lasting, and should not be underestimated, even in the majority of countries outside America, where the CIPW-classification, as such, has never found wide application. Without exaggeration it may be stated that the entire petrographic research was given a considerable and long-lasting stimulus with the introduction of the CIPW-norm. Chemical petrography, in particular, took advantage of this development, since chemical analyses of rocks now became necessary for the classification of rocks. Therefore the number of useful published rock analyses rapidly increased. Since more and more support was thus given to the comparative chemical studies, the problem of the petrographic provinces, for example, was able to be dealt with on an entirely new basis, and could now be handled quantitatively. In addition, the fundamental problem of the limits of the chemical variation range of igneous rocks was brought closer to a definite solution by the speedy accumulation of data.

The CIPW-norm was entirely valid to serve as a foundation for the American classification of the igneous rock, i.e. the purpose for which it was originally intended. Certain weighty limitations, however, have restricted its use as a tool for the study of the completely generally-formulated relationships between the gross-chemism and the mineral composition. In order to clearly illustrate such relationships, a petrographical calculation method must satisfy the following conditions:

1. The results of the calculation should be directly comparable with the observation, i.e. with the actual mineral composition present (modus).
2. It should be possible to transform a calculated mineral composition into other heteromorphic facies corresponding to it, without resort to elaborate calculations. This is particularly important for the study of metamorphic rocks. Thus, it is essential that the reaction relationships should be simply formulated.
3. It should also be possible, if required, to introduce analysed examples of naturally occurring, rock-forming minerals into the calculation, in addition to the idealized standard mineral compositions.

These demands were only partly met with, if at all, by the CIPW-norm. This norm provides the mineral composition in weight percentages. These must be recalculated to volume percentages, utilizing the knowledge of the density of the considered components, in order to compare them with the results obtained from thin-section measurements. Alternatively, the volume percentages of the latter should be recalculated to weight percentages, for comparison.

The normative mineral composition, which is expressed in weight percentages, cannot be transformed into another composition, hetermorphic to it, without making resort to elaborate calculations. Disregarding the fact that this method was not intended for this purpose and that, consequently, many normative minerals are missing from the list, the complex procedure mentioned hereafter should be applied: in each case the equivalent ratios should be obtained by dividing the weight-percentage values by their corresponding formula weights, in order to formulate the reaction equations. The results should then be reconverted into weight percentages. The list of normative minerals accounted for in the CPIW-system is relatively restricted, as shown by the following table (the conventional symbols of the corresponding compounds are provided in brackets).

TABLE IV

Compounds used for the calculation of the CPIW-norm

Salic group

Mineral	Formula
Quartz (Q)	SiO_2
Corundum (C)	Al_2O_3
Zircon (Z)	$ZrO_2 \cdot SiO_2$
Orthoclase (or)	$K_2O \cdot Al_2O_3 \cdot 6\,SiO_2$
Albite (ab)	$Na_2O \cdot Al_2O_3 \cdot 6\,SiO_2$
Anorthite (an)	$CaO \cdot Al_2O_3 \cdot 2\,SiO_2$
Leucite (lc)	$K_2O \cdot Al_2O_3 \cdot 4\,SiO_2$
Nepheline (ne)	$Na_2O \cdot Al_2O_3 \cdot 2\,SiO_2$
Kaliophylite (kp)	$K_2O \cdot Al_2O_3 \cdot 2\,SiO_2$
Halite (hl)	$NaCl$
Thenardite (th)	$Na_2O \cdot SO_3$
Na-carbonate (nc)	$Na_2O \cdot CO_2$

Femic group

Mineral	Formula
Acmite (ac)	$Na_2O \cdot Fe_2O_3 \cdot 4\,SiO_2$
Na-metasilicate (ns)	$Na_2O \cdot SiO_2$
K-metasilicate (ks)	$K_2O \cdot SiO_2$
Diopside (di)	$CaO \cdot (Mg, Fe)O \cdot 2\,SiO_2$
Enstatite (en)	$MgO \cdot SiO_2$
Hypersthene (hy)	$FeO \cdot SiO_2$
Wollastonite (wo)	$CaO \cdot SiO_2$
Olivine (ol)	$2(Mg, Fe)O \cdot SiO_2$
Ca-orthosilicate (cs)	$2\,CaO \cdot SiO_2$
Magnetite (mt)	$FeO \cdot Fe_2O_3$
Chromite (cm)	$FeO \cdot Cr_2O_3$
Ilmenite (il)	$FeO \cdot TiO_2$
Hematite (hm)	Fe_2O_3
Titanite (tn)	$CaO \cdot TiO_2 \cdot SiO_2$
Perovskite (pf)	$CaO \cdot TiO_2$
Rutile (ru)	TiO_2
Apatite (ap)	$2(3\,CaO \cdot P_2O_5) \cdot CaF_2$
Fluorite (fr)	CaF_2
Pyrite (pr)	FeS_2
Calcite (Cc)	$CaO \cdot CO_2$

Thus, all mineral compounds of complex structure, such as sesquioxide-bearing pyroxenes, hornblende and micas, are not included in the list. This primarily pertains to the species which occur only in metamorphic rocks, among others — garnets, staurolite, kyanite, etc.

In order to comply with the conditions Nos. 2 and 3 mentioned above, equivalent ratios should be introduced instead of weight percentage ratios. However, it is not practical to simply replace the weight percentage data by molecular- or equivalent-percentage values. Thereby the conditions enlisted under 2 and 3 would be satisfied, since in this way the required reaction equation may be formulated without difficulty, allowing for the transformation of a calculated norm to that of a heteromorphic facies, although the calculations must in all cases be rereckoned to the sum of 100. However, condition 1 would no longer be satisfied, the possibility of comparison with the modus being sacrificed, since the volume- and molecular-equivalent percentages partly differ to a considerable degree.

2. The Principle of Niggli's Equivalent Norm

P. Niggli provided the solution to the dilemma outlined in the previous section by introducing his equivalent-norm (initially termed "molecular-norm"). (P. Niggli, 1936, 1938, 1950, 1950a, 1954; C. Burri and P. Niggli, 1945, especially 72-100 and 579-623; T. N. Muthuswami, 1952; M. J. Fuster, 1954. A first outline is found in P. Niggli and B. V. Lombaard, 1933; numerous works including practical examples, mostly Zürich dissertations, are included in the Schweiz. Mineralog.-Petrograph. Mitt. from 1936 on). Niggli has shown that the most important principle lies in the **choice of the correct formula units**, i.e. the equivalent weights, upon which the calculation is based. The size of those units is, to a certain degree, a matter of free choice, since the molecular concept of classical chemistry is not valid for inorganic crystal compounds. The one and only deciding factor in the chemical composition of orthoclase, for example, is that the ratio Si : K : Al : O is equal to 3 : 1 : 1 : 8. The question whether it is understood that "1 orthoclase" ("1 Or" in short) corresponds to $1[(SiO_2)_3AlO_2]K$, to $1(K_2O \cdot Al_2O_3 \cdot 6SiO_2)$, to $1/3(Si_3O_8AlK)$ or to 1/5 of this formula is in itself rather insignificant. This question should be answered by a convention once and for all. The four formulations presented here, or their equivalent or formula weights, are related to each other in the ratio 1 : 2 : 1/3 : 1/5 or 15 : 30 : 5 : 3. If, however, we deal with more than one mineral species and wish to make comparisons or to formulate reaction relations, it becomes necessary to arrive at a common and agreeable basis. In crystal chemistry, the formula units are frequently calculated, for this purpose, to an equal number of O-atoms, corresponding to an equal content of the unit cell. In petrographic calculations it is of advantage to select the equivalent weights of the important rock-forming minerals in such a manner that they will not differ greatly from one another. If they could be made equal, the difference between weight- and equivalent-percentages data would be completely eliminated. As may be foreseen, this ideal situation cannot be realized. Niggli has shown, however, that this situation may be very closely approached and that for **the great majority of important rock-forming silicates, very similar equivalent weights may be provided.** This is achieved if the sums of the important electropositive elements (cations), excluding H, are made equally large. The electronegative elements O, C, N, F, and Cl are not counted. As far as S is concerned, it should be noted that initially (P. Niggli 1936, 297-298) the sulfate sulfur (analogous to P in the phosphates), and not the sulfide sulfur, was included in the calculation. Later, however (P. Niggli 1954, 80-81) a uniform treatment of sulfur was suggested. Accordingly, this element is counted along with the cations when the formula size is fixed. This procedure is also justified from the point of view of analytical chemistry, since during rock analyses sulfide and sulfate sulfur cannot be distinguished from one another.

According to Niggli's suggestion, formula units are constructed, which include equal numbers of atoms of the elements Si, Al, Fe, Mn, Mg, Ca, Na, K, Ti, P, Zr, S, etc. These are then designated as "conformable formula units". For practical reasons the sum of these elements should be calculated to unity. The formula units thus obtained are designated as "analytical", and the corresponding equivalent weights likewise as "analytical-equivalent weights". In the same manner the corresponding volumes are termed "analytical-equivalent volumes". In similarity to the CPIW-system, these units are indicated by symbols which are abbreviations of the names of the corresponding mineral. However, the first letter is always given in capitals in order to distinguish the Niggli-system from the CIPW-system.

The fact that it is at all possible to obtain formula units the equivalent weights of which differ only slightly from one another, can be explained by geochemical laws as yet unclarified. It is known that all electropositive rock-forming elements are relatively close to one another and have low atomic numbers, and that they do not differ greatly in their atomic weights. If H is overlooked, as, in all events, it does not play any significant role, the extremely light and extremely heavy elements are not of great importance in the upper lithosphere. The reasons for this are found in the nature of the great differentiation processes which took place during the pre-geological epochs when the hard crust of the Earth was formed. The majority of the rock-forming minerals are silicates. Here the relatively small differences in the atomic weights of the other elements are even further equalized or lessened through their combination with Si and O. Some examples will serve to illustrate this fact more clearly. The sum of the electropositive elements of nepheline, taken as $Na_2O \cdot Al_2O_3 \cdot 2SiO_2$, is given by $2Na + 2Al + 2Si = 6$. The conformable formula unit for forsterite, for example, should be $4MgO \cdot 2SiO_2$, in order to obtain an equal number of electropositive elements. If computable formula units the sum of electropositive elements of which equal unity, are chosen, and if the symbols are introduced, we obtain the following:

$$Na_2O \cdot Al_2O_3 \cdot 2SiO_2 = 6Ne \quad \text{or} \quad 1Ne = 1/6 (Na_2O \cdot Al_2O_3 \cdot 2SiO_2) \quad \text{eq. wt. } 47.34$$
$$2MgO \cdot SiO_2 = 3Fo \quad \text{or} \quad 1Fo = 1/3 (2MgO \cdot SiO_2) \quad \text{eq. wt. } 46.90$$

or, as additional examples:

$$1\,En = 1/2\,(MgO \cdot SiO_2) \qquad 50.19$$
$$1\,Or = 1/10\,(K_2O \cdot Al_2O_3 \cdot 6\,SiO_2) \qquad 55.65$$
$$1\,Ab = 1/10\,(Na_2O \cdot Al_2O_3 \cdot 6\,SiO_2) \qquad 52.43$$
$$1\,An = 1/5\,(CaO \cdot Al_2O_3 \cdot 2\,SiO_2) \qquad 55.63$$
$$1\,Q\ = 1\,SiO_2 \qquad 60.06.$$

Using this system, reaction equations attain a very simple form:

$$Na_2O \cdot Al_2O_3 \cdot 2\,SiO_2 + 4\,SiO_2 = Na_2O \cdot Al_2O_3 \cdot 6\,SiO_2$$
$$6\,Ne \qquad\qquad 4\,Q \qquad\qquad 10\,Ab$$
$$\text{or}\quad 3\,Ne \qquad\qquad 2\,Q \qquad\qquad 5\,Ab$$

$$K_2O \cdot Al_2O_3 \cdot 4\,SiO_2 + 2\,SiO_2 = K_2O \cdot Al_2O_3 \cdot 6\,SiO_2$$
$$8\,Lc \qquad\qquad 2\,Q \qquad\qquad 10\,Or$$
$$\text{or}\quad 4\,Lc \qquad\qquad 1\,Q \qquad\qquad 5\,Or$$

$$2\,MgO \cdot SiO_2 \;+\; SiO_2 \;=\; 2\,(MgO \cdot SiO_2)$$
$$3\,Fo \qquad\qquad 1\,Q \qquad\qquad 4\,En$$

i.e. using this formulation, the sum of the reaction coefficients is identical on both sides of the equation, and the sum of the equivalent units remains equal. An important consequence follows from this with regard to calculation practice. If only the percentage distribution is assumed or calculated, it is permissible to carry out all other required recalculations to other mineral species belonging to heteromorphic facies without recalculation to the basis of 100. For example, the above-mentioned relationships yield accordingly 3% *Ne* + 2% *Q* = 5% *Ab*, or 60% *Ne* + 40% *Q* = 100% *Ab*; as other examples 18.6% may be substituted for 14.9% *Lc* + 3.7% *Q* or 11.2% *En* may be replaced by 8.4% *Fo* + 2.8% *Q*. In all these examples the sum of the equivalents is always maintained. Since the equivalent weights of the minerals were always related to the sum of the electropositive elements = 1, computable equivalent weights should also be determined for the constituent oxides in an analogous way. It is seen that these also do not differ very much from one another, and, at any rate, the differences are far smaller than in the case of the formula weights as written in the usual manner. This is shown in Table V (see below).

H_2O and CO_2 should be neglected and placed within brackets in reaction equations. For example,

$$CaO \cdot CO_2 + SiO_2 = CaO \cdot SiO_2 + CO_2$$
$$1\,Cc \qquad 1\,Q \qquad 2\,Wo \qquad (CO_2).$$

This procedure is very advantageous for the study of metamorphic rocks, as in the course of their change H_2O and CO_2 are often added or driven off.

TABLE V

Computable equivalent weights of the important oxides

SiO_2	60.1	MgO	40.3	$\tfrac{1}{2}P_2O_5$	71
$\tfrac{1}{2}Al_2O_3$	51	CaO	56.1		
$\tfrac{1}{2}Fe_2O_3$	79.8	$\tfrac{1}{2}Na_2O$	31		
FeO	71.8	$\tfrac{1}{2}K_2O$	47.1		
MnO	70.9	TiO_2	79.9		

3. The Basis and the Calculation of the Basis Components

A list of simple compounds exists, among which the fluctuations occurring in the computable equivalent weights are much smaller than in the rock-forming oxides. Some of these compounds are hypothetical, others are, however, significant, because the important rock-forming minerals may be derived from them with the aid of simple reactions — mainly addition of silica. These are termed **basis bonds** or **basis components** (former "basis molecules"). Table VI presents these compounds.

TABLE VI

Review of the basis bonds

Designation	Symbol	Chemical composition	Equivalent weight
Kaliophylite	1 Kp	$= 1/6 (K_2O \cdot Al_2O_3 \cdot 2 SiO_2)$	52.7
Nepheline	1 Ne	$= 1/6 (Na_2O \cdot Al_2O_3 \cdot 2 SiO_2)$	47.3
* Ca-aluminate	1 Cal	$= 1/3 (CaO \cdot Al_2O_3)$	52.7
Spinel	1 Sp	$= 1/3 (MgO \cdot Al_2O_3)$	47.4
Hercynite	1 Hz	$= 1/3 (FeO \cdot Al_2O_3)$	57.9
* K-metasilicate	1 Ks	$= 1/3 (K_2O \cdot SiO_2)$	51.4
* Na-metasilicate	1 Ns	$= 1/3 (Na_2O \cdot SiO_2)$	40.7
Ca-orthosilicate	1 Cs	$= 1/3 (2 CaO \cdot SiO_2)$	57.4
Forsterite	1 Fo	$= 1/3 (2 MgO \cdot SiO_2)$	46.9
Fayalite	1 Fa	$= 1/3 (2 FeO \cdot SiO_2)$	67.9
Chromite	1 Cm	$= 1/3 (FeO \cdot Cr_2O_3)$	74.6
Rock salt (halite)	1 Hl	$= 1$ NaCl	58.5
Corundum	1 C	$= 1/2 (Al_2O_3)$	51.0
Hematite	1 Hm	$= 1/2 (Fe_2O_3)$	79.8
Zircon	1 Z	$= 1/2 (ZrO_2 \cdot SiO_2)$	91.6
Quartz	1 Q	$= 1$ SiO_2	60.1
Rutile	1 Ru	$= 1$ TiO_2	79.9
Pyrite	1 Pr	$= 1/3$ FeS_2	40.0[1]
Fluorite	1 Fr	$= 1$ CaF_2	78.1
Ca-phosphate	1 Cp	$= 1/5 (3 CaO \cdot P_2O_5)$	62.0
Calcite	1 Cc	$= 1 (CaO \cdot CO_2)$	100.1
Anhydrite	1 A	$= 1/2 (CaO \cdot SO_3)$	68.1
Thenardite	1 Th	$= 1/3 (Na_2O \cdot SO_3)$	47.3
Na-carbonate	1 Nc	$= 1/2 (Na_2O \cdot CO_2)$	53.0
* Ferrisilicate	1 Fs	$= 1/3 (Fe_2O_3 \cdot SiO_2)$	73.2
* Na-Ferrisilicate	1 Fns	$= 1/6 (Na_2O \cdot Fe_2O_3 \cdot 2 SiO_2)$	57.0
* K-ferrisilicate	1 Fks	$= 1/6 (K_2O \cdot Fe_2O_3 \cdot 2 SiO_2)$	62.3

The basis components indicated by * are hypothetical and are not known as minerals.

[1]) In earlier representations, for example P. Niggli, 1936, 299; C. Burri and P. Niggli, 1945, 74 and 622, the equivalent weight of 1 Pr = 120, corresponding to 1 FeS_2. In accordance with the previous reference to the equal treatment of sulfate and sulfide sulphur, it is here set at 1 Pr = 1/3 FeS_2 = 40.0, as also appears in P. Niggli, 1954, 80-81 and 136.

In order to calculate a rock analysis as basis components the weight percentages obtained from analysis should initially be divided by their formula weights. Thus are obtained equivalent numbers. In contrast to the procedure according to which the Niggli-values were calculated, this calculation is carried out in such a way that the resulting equivalent numbers are proportional to the **metal-atom numbers**. This is achieved by **multiplying the numbers of** Al_2O_3, Fe_2O_3, Na_2O, K_2O, and P_2O_5 by the factor 2, if tables of the complete formula weights of these oxides are used. Instead, tables showing **half the formula weights** may be used, i.e. $1/2\ Al_2O_3$, $1/2\ Fe_2O_3$, $1/2\ Na_2O$, etc; these are provided in the appendix. Small amounts of MnO and NiO should be added in the usual way to FeO; SrO, as well as BaO — to K_2O; Li_2O — to MgO; Cr_2O_3 — to Fe_2O_3, etc. Very small amounts of P, Cl, F, and S should either be neglected, or used for the following combination: $Cp = 3CaO \cdot P_2O_5$; $Hl = NaCl$; $Fr = CaF_2$; $Pr = FeS_2$. If microscopic observation proves the presence of primary calcite, $Cc = CaO \cdot CO_2$ should be formed from CO_2.

The following procedure is applied for the formation of these compounds: 2P should be combined with 3Ca to give Cp, i.e. $xP + 3/2\,xCa = 5/2\,xCp$. In the case of NaCl, x Cl should be combined with an equal amount of Na, resulting in xNaCl (not $2x$, since Cl should not be counted). Likewise, the combination of xF with $x/2$Ca yields $x/2$Fr, whereas xS with $x/2$Fe yields $3/2\ xPr$. After dealing with the accessory constituents, the formation of the main basis compounds is carried out according to the following scheme:

1. $xK + xAl + xSi = 3xKp$, corresponding to $3x\left(\frac{KAlSiO_4}{3}\right)$ (kaliophylite, K-nepheline)

    a) If K > Al, a remainder $yK = (K-Al)$ will be left and is used to form

    $yK + y/2\,Si = 3/2\,yKs$, corresponding to $3/2\ y\left(\frac{K_2SiO_3}{3}\right)$ (potassium metasilicate).

    b) If K < Al, then:

2. $xNa + xAl + xSi = 3xNe$, corresponding to $3x\left(\frac{NaAlSiO_4}{3}\right)$ (nepheline)

    a) If the amount of available Na is larger than the residual Al with which it should combine, i.e., not all the Na is eliminated by the formation of nepheline, it should be used up in the manner analogous to that mentioned in 1a):

    $yNa + \frac{y}{2}Si = \frac{3}{2}yNs$, corresponding to $\frac{3}{2}y\left(\frac{Na_2SiO_3}{3}\right)$ (sodium metasilicate)

    b) If, in constrast, Al is left over, i.e. (Na+ K) < Al, the following is formed:

3. $\frac{x}{2}Ca + xAl = \frac{3}{2}xCal$, corresponding to $\frac{3}{2}x\left(\frac{CaAl_2O_4}{3}\right)$ (calcium aluminate)

    a) If some Ca is still left after all the Al is used up, the following is formed:

$$y\,Ca + \frac{y}{2}\,Si = \frac{3}{2}\,y\,Cs, \text{corresponding to } \frac{3}{2}\,y\left(\frac{Ca_2SiO_4}{3}\right) \text{ (Ca-orthosilicate)}$$

b) If, on the other hand, a certain amount of Al is left over, the following is formed:

$$y\,Al + \frac{y}{2}\,Mg = \frac{3}{2}\,y\,Sp, \text{corresponding to } \frac{3}{2}\,y\left(\frac{MgAl_2O_4}{3}\right) \text{ (spinel)}.$$

If an excess of Al still remains after the formation of spinel, the following should be formed by analogy:

$$z\,Al + \frac{z}{2}\,Fe^{2+} = \frac{3}{2}\,z\,Hz, \text{corresponding to } \frac{3}{2}\,z\left(\frac{Fe^{2+}Al_2O_4}{3}\right) \text{(Fe-spinel, hercynite)}.$$

If some Al still remains, it should be used to form:

$$w\,Al = w\,C, \text{corresponding to } w\left(\frac{Al_2O_3}{2}\right) \text{(corundum)}.$$

4. Fe^{3+} should be calculated as the hypothetical ferrisilicate $Fe_2O_3 \cdot SiO_2$, as follows:

$$x\,Fe^{3+} + \frac{x}{2}\,Si = \frac{3}{2}\,x\,Fs, \text{corresponding to } \frac{3}{2}\,x\left(\frac{Fe_2^{3+}SiO_5}{3}\right) \text{ (ferrisilicate)}.$$

5. If Fe^{2+} and Mg are present, which are not bound to Al, the following is formed:

$$x\,Mg + \frac{x}{2}\,Si = \frac{3}{2}\,x\,Fo, \text{corresponding to } \frac{3}{2}\,x\left(\frac{Mg_2SiO_4}{3}\right) \text{ (forsterite)}.$$

and in analogy:

$$y\,Fe^{2+} + \frac{y}{2}\,Si = \frac{3}{2}\,y\,Fa, \text{corresponding to } \frac{3}{2}\,y\left(\frac{Fe_2^{2+}SiO_4}{3}\right) \text{ (fayalite)}.$$

6. TiO_2 is introduced into the calculation directly as rutile, Ru.

7. ZrO_2 is combined with SiO_2, as follows:

$$x\,Zr + x\,Si = 2x\,Z, \text{corresponding to } 2x\left(\frac{ZrSiO_4}{2}\right) \text{ (zircon)}.$$

8. The remaining SiO_2 provides an equal amount of Q (quartz).

In the extremely rare cases when there is insufficient SiO_2 available for the formation of the listed basis bonds, one should employ the same procedure as if sufficient SiO_2 were present. The deficit should be designated as $-Q$ in the calculation.

Depending on the chemism present, 12 possible combinations of the basis-components can be obtained from this standardized calculation method (compare with the review provided in section BII 5a, Table VIII).

Examples:

The formation of the basis bonds, in accordance with the instructions given, are carried out as shown in the schematic review. The fact that the sum of the basis components before recalculation to the sum of 100 (horizontal line) must be equal to the sum of the equivalent numbers (vertical column) provides some means of control, because we are merely dealing with a different arrangement of the equivalents, without any change in their actual numbers. However, it should be noted that this control, i.e. equivalence of the two sums, merely confirms that no errors of addition or subtraction were introduced. This does not prove, however, that the distribution of the equivalents among the individual basis components was carried out in a correct stoichiometric way. During the recalculation of the basis components to the sum of 100, i.e., the construction of the so-called "basis", the use of the auxiliary means mentioned in section AIII is advantageous.

The basis does not yet comply with a mineral composition, if only for the reason that *Cal* and *Fs*, which are not found as mineral species, are included among the basis components. In addition, *Ne*, *Fo*, and *Fa* do not normally occur together with *Q*. The construction of a basis merely involves a grouping or arrangement of the chemism according to certain concepts. These are selected in such a way as to allow for the formation of important rock-forming minerals by means of simple reactions, and primarily, by the addition of silica. However, the basis serves not only as a starting point for the calculation of the mineral composition, but is also suitable for providing the foundation for a most instructive graphical representation of the chemism of the rock. This will be shown further on.

Examples:

1. Quartz-diorite from Spanish Peak, California. Calculation of basis.

	Wt %	Molecular equivalent numbers × 1000	Atomic equivalent numbers × 1000	*Cp*	*Ru*	*Kp*	*Ne*	*Cal*	*Cs*	*Fs*	*Fa*	*Fo*	*Q*
SiO_2	59.68	994	994			28	124		10	18	19	44	751
Al_2O_3	17.09	168	336			28	124	184					
Fe_2O_3	2.85	18	36							36			
FeO	2.75	38	38								38		
MgO	3.54	88	88									88	
CaO	6.62	118	118	6					92	20			
Na_2O	3.87	62	124				124						
K_2O	1.31	14	28			28							
TiO_2	0.65	8	8		8								
P_2O_5	0.25	2	4	4									
H_2O+	1.00		1774	10	8	84	372	276	30	54	57	132	751
H_2O-	0.15												
incl.	0.27	Basis:	*Q*	*Kp*	*Ne*	*Cal*	*Cs*	*Fs*	*Fa*	*Fo*	*Ru*	*Cp*	Total
	100.03		42.3	4.7	21.0	15.6	1.7	3.0	3.2	7.4	0.5	0.6	100.0

2. **Garnet-bearing cordierite-andesite from Hoyazo, near Nijar, Almeria Province, Spain.** Calculation of the basis.

	Wt %	Atomic equivalent numbers × 1000	Cp	Ru	Kp	Ne	Cal	Sp	Fs	Fa	Fo	Q
SiO_2	61.08	1017			70	56			20	22	8.5	840.5
Al_2O_3	17.25	338			70	56	106	106				
Fe_2O_3	3.16	40							40			
FeO	3.10	43 ⎫ 44								44		
MnO	0.06	1 ⎭										
MgO	2.82	70						53			17	
CaO	3.14	56	3				53					
Na_2O	1.75	56				56						
K_2O	3.33	70			70							
H_2O+	2.63											
H_2O-	0.05											
TiO_2	0.79	10		10								
P_2O_5	0.12	2	2									
C	0.80											
	100.08	1703	5	10	210	168	159	159	60	66	25.5	840.5

Basis: Q Kp Ne Cal Sp Fs Fa Fo Ru Cp Total
49.4 12.3 9.9 9.3 9.3 3.5 3.9 1.5 0.6 0.3 100.0

3. **Lujaurite from Lille Elv, Kangerdluarsuk, Greenland.** Calculation of the basis.

	Wt %	Atomic equivalent numbers × 1000	Ru	Hl	Z	Kp	Ne	Ns	Cs	Fs	Fa	Fo	Q
SiO_2	53.01	882			5	55	245	66	6	57	31.5	1	415.5
Al_2O_3	15.33	300				55	245						
Fe_2O_3	9.14	114								114			
FeO	4.44	61 ⎫ 63									63		
MnO	0.13	2 ⎭											
MgO	0.10	2										2	
CaO	0.67	12							12				
Na_2O	11.86	383		6			245	132					
K_2O	2.60	55				55							
TiO_2	0.33	4	4										
P_2O_5	Tr.												
H_2O+	1.88												
H_2O-	0.20												
ZrO_2	0.65	5			5								
Cl	0.23	(6)		(6)									
	100.57	1820	4	6	10	165	735	198	18	171	94.5	3	415.5

Basis: Q Kp Ne Ns Cs Fs Fa Fo Ru Z Hl Total
22.8 9.1 40.4 10.9 1.0 9.4 5.2 0.2 0.2 0.5 0.3 100.0

111   Example 1 corresponds to case IV of the possible combinations of basis bonds, listed in Table VII. Along with the already extensively discussed examples of the cordierite-andesite of Hoyazo, near Nijar, Spain, and the lujaurite from Lille Elv, Greenland, cases III (alumina excess) and V (alkali excess) will also be treated in more detail hereafter.

112   ## 4. Basis and Cation Percentages

a) **Relationships between basis bonds and cation percentages.**

The following simple relations exist between cation percentages and basis bonds. They may be derived directly from the chemical composition.

$$1\,Kp = \tfrac{1}{3}Si + \tfrac{1}{3}Al + \tfrac{1}{3}K \qquad 1\,Cm = \tfrac{1}{3}Fe^{2+} + \tfrac{2}{3}Cr$$
$$1\,Ne = \tfrac{1}{3}Si + \tfrac{1}{3}Al + \tfrac{1}{3}Na \qquad 1\,C = 1\,Al$$
$$1\,Cal = \tfrac{1}{3}Ca + \tfrac{2}{3}Al \qquad 1\,Q = 1\,Si$$
$$1\,Sp = \tfrac{1}{3}Mg + \tfrac{2}{3}Al \qquad 1\,Cc = 1\,Ca$$
$$1\,Hz = \tfrac{1}{3}Fe^{2+} + \tfrac{2}{3}Al \qquad 1\,Hl = 1\,Na$$
$$1\,Cs = \tfrac{1}{3}Si + \tfrac{2}{3}Ca \qquad 1\,Ru = 1\,Ti$$
$$1\,Fo = \tfrac{1}{3}Si + \tfrac{2}{3}Mg \qquad 1\,Z = \tfrac{1}{2}Zr + \tfrac{1}{2}Si$$
$$1\,Fa = \tfrac{1}{3}Si + \tfrac{2}{3}Fe^{2+} \qquad 1\,Cp = \tfrac{3}{5}Ca + \tfrac{2}{5}P$$
$$1\,Fs = \tfrac{1}{3}Si + \tfrac{2}{3}Fe^{3+} \qquad 1\,Nc = 1\,Na$$
$$1\,Ns = \tfrac{1}{3}Si + \tfrac{2}{3}Na \qquad 1\,Th = \tfrac{2}{3}Na + \tfrac{1}{3}S$$
$$1\,Ks = \tfrac{1}{3}Si + \tfrac{2}{3}K \qquad 1\,Fr = 1\,Ca$$
$$1\,Fns = \tfrac{1}{3}Si + \tfrac{1}{3}Na + \tfrac{1}{3}Fe^{3+} \qquad 1\,Pr = \tfrac{1}{3}Fe^{2+} + \tfrac{2}{3}S$$

b) **Calculation of the basis from given cation percentages.**

The relationships listed under a) allow for the calculation of the basis from the given cation percentages, as shown by the following examples:

Examples:

Quartz-diorite from Spanish Peak, California. The cation percentages which were calculated previously are:

Si^{4+}	Al^{3+}	Fe^{3+}	Fe^{2+}	Mg^{2+}	Ca^{2+}	Na^+	K^+	Ti^{4+}	P^{5+}	Total
56.0	18.9	2.0	2.1	5.0	6.7	7.0	1.6	0.5	0.2	100.0

The basis is obtained in the following manner:

Starting with K = 1.6, one obtains 1.6K + 1.6Al + 1.6Si = 4.8 *Kp*
    "   " Na = 7.0, "   " 7.0 Na + 7.0 Al + 7.0 Si = 21.0 *Ne*
    "   " 18.9 − (1.6 + 7.0) *Al* = 10.3 Al, it follows that 10.3 Al + 5.1 Ca = 15.4 *Cal*
    "   " 0.2 P, one obtains 0.2 P + 0.3 Ca = 0.5 *Cp*

113   "   " 6.7 − (5.1 + 0.3) *Ca* = 1.3 Ca, it follows that 1.3 Ca + 0.6 Si = 1.9 *Cs*
    "   " 2.0 Fe^{3+}, one obtains 2.0 Fe^{3+} + 1.0 Si = 3.0 *Fs*
    "   " 2.1 Fe^{2+}, "   " 2.1 Fe^{2+} + 1.0 Si = 3.1 *Fa*
    "   " 5.0 Mg, "   " 5.0 Mg + 2.5 Si = 7.5 *Fo*
    "   " 0.5 Ti, "   " 0.5 Ti = 0.5 *Ru*

Therefore, the following amount of Si was used:

$$1.6+7.0+0.6+1.0+1.0+2.5 = 13.7 \text{ Si.}$$

Accordingly, there yet remains 56.0-13.7 = 42.3 Si, corresponding to 42.3 Q. Thus, the required basis will be:

Q	Kp	Ne	Cal	Cs	Fs	Fa	Fo	Ru	Cp	Total
42.3	4.8	21.0	15.4	1.9	3.0	3.1	7.5	0.5	0.5	100.0
(42.3)	(4.7)	(21.0)	(15.6)	(1.7)	(3.0)	(3.2)	(7.4)	(0.5)	(0.6)	(100.0)

The values obtained directly from the analysis via the atomic-equivalent numbers are in brackets, for comparison. The agreement between the results is a very good one, especially in view of the fact that only one decimal was considered.

The following metal-atom percentages were calculated for the garnet-bearing cordierite-andesite from Hoyazo, near Nijar, Spain:

Si^{4+}	Al^{3+}	Fe^{3+}	Fe^{2+}	Mg^{2+}	Ca^{2+}	Na^+	K^+	Ti^{4+}	P^{5+}	Total
59.7	19.8	2.4	2.6	4.1	3.3	3.3	4.1	0.6	0.1	100.0

The calculation is similar to that used for deriving the basis from oxide weight percentages:

		Ru	Cp	Kp	Ne	Cal	Sp	Fs	Fa	Fo	Q
Si^{4+}	59.7			4.1	3.3			1.2	1.3	0.5	49.3
Al^{3+}	19.8			4.1	3.3	6.2	6.2				
Fe^{3+}	2.4							2.4			
Fe^{2+}	2.6								2.6		
Mg^{2+}	4.1						3.1			1.0	
Ca^{2+}	3.3		0.2			3.1					
Na^+	3.3				3.3						
K^+	4.1			4.1							
Ti^{4+}	0.6	0.6									
P^{5+}	0.1		0.1								Total
	100.0	0.6	0.3	12.3	9.9	9.3	9.3	3.6	3.9	1.5	49.3	100.0
		(0.6)	(0.3)	(12.3)	(9.9)	(9.3)	(9.3)	(3.5)	(3.9)	(1.5)	(49.4)	(100.0)

The agreement with the values obtained directly from the analysis (in brackets) is again complete.

For the lujaurite from Lille Elv, Kangerdluarsuk, Greenland, one likewise obtains:

	Hl	Ru	Z	Kp	Ne	Ns	Cs	Fs	Fa	Fo	Q
Si^{4+} 48.4			0.3	3.0	13.5	3.6	0.3	3.1	1.7	0.1	22.8
Al^{3+} 16.5				3.0	13.5						
Fe^{3+} 6.3								6.3			
Fe^{2+} 3.5									3.5		
Mg^{2+} 0.1										0.1	
Ca^{2+} 0.7							0.7				
Na^{+} 21.0	0.3				13.5	7.2					
K^{+} 3.0				3.0							
Ti^{4+} 0.2		0.2									
Zr^{4+} 0.3			0.3								
Cl^{-} (0.3)	(0.3)										

Total
100.0 0.3 0.2 0.6 9.0 40.5 10.8 1.0 9.4 5.2 0.2 22.8 100.0
(0.3) (0.2) (0.5) (9.1) (40.4) (10.9) (1.0) (9.4) (5.2) (0.2) (22.8) (100.0)

Here, also, the agreement is very good.

c) Calculation of the cation percentages from the basis.

Reversing the relationships given in section a), one obtains:

$Si^{4+} = Q + 1/3\,(Kp + Ne + Cs + Fs + Fa + Fo + Ns + Ks + Fns) + 1/2\,Z$
$Al^{3+} = 1/3\,(Kp + Ne) + 2/3\,(Cal + Sp + Hz) + C$
$Fe^{3+} = 2/3\,Fs + 1/3\,Fns$
$Fe^{2+} = 2/3\,Fa + 1/3\,(Cm + Pr)$
$Mg^{2+} = 2/3\,Fo + 1/3\,Sp$
$Ca^{2+} = 2/3\,Cs + 1/3\,Cal + 1\,(Cc + Fr) + 3/5\,Cp$
$Na^{+} = 1/3\,(Ne + Fns) + 2/3\,(Ns + Th) + 1\,(Hl + Nc)$
$K^{+} = 1/3\,Kp + 2/3\,Ks$
$Ti^{4+} = Ru$
$P^{5+} = 2/5\,Cp$
$Cr^{3+} = 2/3\,Cm$
$Zr^{4+} = 1/2\,Z$

All of the basis components under consideration are included here, irrespective of whether or not they may possibly occur together, according to the regulations stated previously.

Examples:

From the basis of the quartz-diorite from Spanish Peak, California, which was calculated previously:

Q	Kp	Ne	Cal	Cs	Fs	Fa	Fo	Ru	Cp
42.3	4.7	21.0	15.6	1.7	3.0	3.2	7.4	0.5	0.6,

the following metal-atom percentages may be calculated:

$$Si^{4+} = 1/3(4.7 + 21.0 + 1.7 + 3.0 + 7.4 + 3.2) + 42.3 = \ldots \ldots 56.0 \quad (56.0)$$
$$Al^{3+} = 1/3(4.7 + 21.0) + 2/3(15.6) = \ldots \ldots \ldots \ldots 19.0 \quad (18.9)$$
$$Fe^{3+} = 2/3(3.0) = \ldots \ldots \ldots \ldots \ldots \ldots \ldots \ldots 2.0 \quad (2.0)$$
$$Fe^{2+} = 2/3(3.2) = \ldots \ldots \ldots \ldots \ldots \ldots \ldots \ldots 2.1 \quad (2.1)$$
$$Mg^{2+} = 2/3(7.4) = \ldots \ldots \ldots \ldots \ldots \ldots \ldots \ldots 4.9 \quad (5.0)$$
$$Ca^{2+} = 2/3(1.7) + 1/3(15.6) + 3/5(0.6) = \ldots \ldots \ldots 6.7 \quad (6.7)$$
$$Na^+ = 1/3(21.0) = \ldots \ldots \ldots \ldots \ldots \ldots \ldots \ldots 7.0 \quad (7.0)$$
$$K^+ = 1/3(4.7) = \ldots \ldots \ldots \ldots \ldots \ldots \ldots \ldots 1.6 \quad (1.6)$$
$$Ti^{4+} = 1(0.5) = \ldots \ldots \ldots \ldots \ldots \ldots \ldots \ldots 0.5 \quad (0.5)$$
$$P^{5+} = 2/5(0.6) = \ldots \ldots \ldots \ldots \ldots \ldots \ldots \ldots 0.2 \quad (0.2)$$
$$\overline{}$$
$$100.0 \quad (100.0)$$

For the **cordierite-andesite** from Hoyazo, near Nijar, Spain, the following basis was calculated:

Q	Kp	Ne	Cal	Sp	Fs	Fa	Fo	Ru	Cp	Total
49.4	12.3	9.9	9.3	9.3	3.5	3.9	1.5	0.6	0.3	100.0

The following metal-atom percentages are calculated therefrom:

$$Si^{4+} = 1/3(12.3 + 9.9 + 3.5 + 3.9 + 1.5) + 49.4 = \ldots \ldots 59.8 \quad (59.7)$$
$$Al^{3+} = 1/3(12.3 + 9.9) + 2/3(9.3 + 9.3) = \ldots \ldots \ldots 19.8 \quad (19.8)$$
$$Fe^{3+} = 2/3(3.5) = \ldots \ldots \ldots \ldots \ldots \ldots \ldots \ldots 2.3 \quad (2.4)$$
$$Fe^{2+} = 2/3(3.9) = \ldots \ldots \ldots \ldots \ldots \ldots \ldots \ldots 2.6 \quad (2.6)$$
$$Mg^{2+} = 2/3(1.5) + 1/3(9.3) = \ldots \ldots \ldots \ldots \ldots 4.1 \quad (4.1)$$
$$Ca^{2+} = 1/3(9.3) + 3/5(0.3) = \ldots \ldots \ldots \ldots \ldots 3.3 \quad (3.3)$$
$$Na^+ = 1/3(9.9) = \ldots \ldots \ldots \ldots \ldots \ldots \ldots \ldots 3.3 \quad (3.3)$$
$$K^+ = 1/3(12.3) = \ldots \ldots \ldots \ldots \ldots \ldots \ldots \ldots 4.1 \quad (4.1)$$
$$Ti^{4+} = 1(0.6) = \ldots \ldots \ldots \ldots \ldots \ldots \ldots \ldots 0.6 \quad (0.6)$$
$$P^{5+} = 2/5(0.3) = \ldots \ldots \ldots \ldots \ldots \ldots \ldots \ldots 0.1 \quad (0.1)$$
$$\overline{}$$
$$100.0 \quad (100.0)$$

For the **lujaurite** from Lille Elv, Kangerdluarsuk, Greenland, the previously calculated basis yields:

Q	Kp	Ne	Ns	Cs	Fs	Fa	Fo	Ru	Z	Hl
22.8	9.1	40.4	10.9	1.0	9.4	5.2	0.2	0.2	0.5	0.3

The following metal-atom percentages are derived therefrom:

$Si^{4+} = 1/3(9.1 + 40.4 + 1.0 + 9.4 + 5.2 + 0.2 + 10.9) +$
$\quad + 1/2(0.5) + 22.8 = \ldots\ldots\ldots\ldots\ldots\ldots\ldots$ 48.4 (48.4)
$Al^{3+} = 1/3(9.1 + 40.4) = \ldots\ldots\ldots\ldots\ldots\ldots\ldots$ 16.5 (16.5)
$Fe^{3+} = 2/3(9.4) = \ldots\ldots\ldots\ldots\ldots\ldots\ldots\ldots\ldots$ 6.3 (6.3)
$Fe^{2+} = 2/3(5.2) = \ldots\ldots\ldots\ldots\ldots\ldots\ldots\ldots\ldots$ 3.5 (3.5)
$Mg^{2+} = 2/3(0.2) = \ldots\ldots\ldots\ldots\ldots\ldots\ldots\ldots\ldots$ 0.1 (0.1)
$Ca^{2+} = 2/3(1.0) = \ldots\ldots\ldots\ldots\ldots\ldots\ldots\ldots\ldots$ 0.7 (0.7)
$Na^{+} = 1/3(40.4) + 2/3(10.9) + 0.3 = \ldots\ldots\ldots\ldots$ 21.1 (21.0)
$K^{+} = 1/3(9.1) = \ldots\ldots\ldots\ldots\ldots\ldots\ldots\ldots\ldots$ 3.0 (3.0)
$Ti^{4+} = 1(0.2) = \ldots\ldots\ldots\ldots\ldots\ldots\ldots\ldots\ldots$ 0.2 (0.2)
$Zr^{4+} = 1/2(0.5) = \ldots\ldots\ldots\ldots\ldots\ldots\ldots\ldots\ldots$ 0.2 (0.3)
$\quad\quad\quad\quad\quad\quad\quad\quad\quad\quad\quad\quad\quad\quad\quad\quad$ 100.0 (100.0)

All 3 examples show very good agreement with the directly calculated values (in brackets).

5. Derivation of Normative Mineral Compositions from the Basis

The following components (in an idealized form) should be considered with regard to the calculation of the norm of **igneous rocks**, which alone will be dealt with in the meanwhile. The corresponding computable equivalent weights are given in brackets. The extension of this review to include metamorphic rocks will follow later.

TABLE VII

Equivalent normative components of igneous rocks

		Equivalent weight
Quartz	$1\ Q = 1\ SiO_2$	(60.1)
Feldspars		
Orthoclase, potassium feldspar	$1\ Or = 1/10(6\ SiO_2 \cdot Al_2O_3 \cdot K_2O)$	(55.6)
Albite, sodium feldspar	$1\ Ab = 1/10(6\ SiO_2 \cdot Al_2O_3 \cdot Na_2O)$	(52.4)
Anorthite, calcium feldspar	$1\ An = 1/5(2\ SiO_2 \cdot Al_2O_3 \cdot CaO)$	(55.6)
Feldspathoids		
Kaliophylite, potassium-nepheline	$1\ Kp = 1/6(2\ SiO_2 \cdot Al_2O_3 \cdot K_2O)$	(52.7)
Nepheline	$1\ Ne = 1/6(2\ SiO_2 \cdot Al_2O_3 \cdot Na_2O)$	(47.3)
Leucite	$1\ Lc = 1/8(4\ SiO_2 \cdot Al_2O_3 \cdot K_2O)$	(54.5)
Analcite	$1\ Anc = 1/8(4\ SiO_2 \cdot Al_2O_3 \cdot Na_2O \cdot 2\ H_2O)$	(55.0)
Gehlenite	$1\ Ge = 1/5(SiO_2 \cdot Al_2O_3 \cdot 2\ CaO)$	(54.8)
Na-gehlenite	$1\ Na\text{-}Ge = 1/5(2\ SiO_2 \cdot 1/2\ Al_2O_3 \cdot CaO \cdot 1/2\ Na_2O)$	(51.6)
Fe-gehlenite	$1\ Fe\text{-}Ge = 1/5(SiO_2 \cdot AlFeO_3 \cdot 2\ CaO)$	(60.6)
Akermanite	$1\ Ak = 1/5(2\ SiO_2 \cdot MgO \cdot 2\ CaO)$	(54.5)
Fe-akermanite	$1\ Fe\text{-}Ak = 1/5(2\ SiO_2 \cdot FeO \cdot 2\ CaO)$	(60.8)
Sodalite	$1\ Sod = 1/20(6\ SiO_2 \cdot 3\ Al_2O_3 \cdot 3\ Na_2O \cdot 2\ NaCl)$	(48.4)
Noselite	$1\ Nos = 1/21(6\ SiO_2 \cdot 3\ Al_2O_3 \cdot 3\ Na_2O \cdot Na_2O \cdot SO_3)$	(47.3)
Hauynite	$1\ Hau = 1/22(6\ SiO_2 \cdot 3\ Al_2O_3 \cdot 3\ Na_2O \cdot 2\ CaO \cdot 2\ SO_3)$	(51.2)
Cancrinite	$1\ Canc = 1/20(6\ SiO_2 \cdot 3\ Al_2O_3 \cdot 3\ Na_2O \cdot 2\ CaO \cdot 2\ CO_2)$	(52.6)

			Equivalent weight
Olivine group			
Forsterite	$1\ Fo = 1/_3(SiO_2 \cdot 2\ MgO)$		(46.9)
Fayalite	$1\ Fa = 1/_3(SiO_2 \cdot 2\ FeO)$		(67.9)
Monticellite	$1\ Mont = 1/_3(SiO_2 \cdot MgO \cdot CaO)$		(52.1)
Tephroite	$1\ Tephr = 1/_3(SiO_2 \cdot 2\ MnO)$		(67.3)
Augites and hornblendes			
Enstatite	$1\ En = 1/_2(SiO_2 \cdot MgO)$		(50.2)
Hypersthene	$1\ Hy = 1/_2(SiO_2 \cdot FeO)$		(65.9)
Wollastonite	$1\ Wo = 1/_2(SiO_2 \cdot CaO)$		(58.1)
Diopside	$1\ Di = 1/_4(2\ SiO_2 \cdot MgO \cdot CaO)$		(54.1)
Hedenbergite	$1\ Hed = 1/_4(2\ SiO_2 \cdot FeO \cdot CaO)$		(62.0)
Tschermaks compound	$1\ Ts = 1/_4(SiO_2 \cdot Al_2O_3 \cdot CaO)$		(54.5)
Aegirine (acmite)	$1\ Ac = 1/_8(4\ SiO_2 \cdot Fe_2O_3 \cdot Na_2O)$		(57.7)
K-aegirine (K-acmite)	$1\ K\text{-}Ac = 1/_8(4\ SiO_2 \cdot Fe_2O_3 \cdot K_2O)$		(61.8)
Jadeite	$1\ Jd = 1/_8(4\ SiO_2 \cdot Al_2O_3 \cdot Na_2O)$		(50.5)
Spodumene	$1\ Spod = 1/_8(4\ SiO_2 \cdot Al_2O_3 \cdot LiO_2)$		(46.5)
Common hornblende	$1\ Ho = 1/_{15}(7\ SiO_2 \cdot Al_2O_3 \cdot (4\ Mg, Fe)O \cdot 2\ CaO \cdot H_2O)$		(54.3–62.7)
Riebeckite$_1$	$1\ Rb_1 = 1/_{15}(8\ SiO_2 \cdot Fe_2O_3 \cdot 3\ FeO \cdot Na_2O \cdot H_2O)$		(62.3)
Riebeckite$_2$	$1\ Rb_2 = 1/_{32}(16\ SiO_2 \cdot Fe_2O_3 \cdot 8\ FeO \cdot 3\ Na_2O \cdot 2\ H_2O)$		(59.9)
Riebeckite$_3$	$1\ Rb_3 = 1/_{32}(16\ SiO_2 \cdot 2\ Fe_2O_3 \cdot 6\ FeO \cdot 3\ Na_2O \cdot H_2O)$		(60.4)
Micas			
Mg-biotite	$1\ Mg\text{-}Bi = 1/_{16}(6\ SiO_2 \cdot Al_2O_3 \cdot 6\ MgO \cdot K_2O \cdot 2\ H_2O)$		(52.1)
Fe-biotite	$1\ Fe\text{-}Bi = 1/_{16}(6\ SiO_2 \cdot Al_2O_3 \cdot 6\ FeO \cdot K_2O \cdot 2\ H_2O)$		(64.0)
Muscovite$_1$	$1\ Ms_1 = 1/_{14}(6\ SiO_2 \cdot 3\ Al_2O_3 \cdot K_2O \cdot 2\ H_2O)$		(56.9)
Muscovite$_2$ (phengitic)	$1\ Ms_2 = 1/_{14}(7\ SiO_2 \cdot 2\ Al_2O_3 \cdot (Mg, Fe)O \cdot K_2O \cdot H_2O)$		(56.8 to 59.0)
Lepidolite	$1\ Lep = 1/_{32}(12\ SiO_2 \cdot 5\ Al_2O_3 \cdot 3\ Li_2O \cdot 4\ KF \cdot 2\ H_2O)$		(49.6)
Zinnwaldite	$1\ Zwd = 1/_{16}(6\ SiO_2 \cdot 2\ Al_2O_3 \cdot 2\ FeO \cdot 2\ LiF \cdot 2\ KF)$		(54.7)
Scapolite group			
Chloride-marialite	$1\ Ma_1 = 1/_{32}(18\ SiO_2 \cdot 3\ Al_2O_3 \cdot 3\ Na_2O \cdot 2\ NaCl)$		(52.8)
Carbonate-marialite	$1\ Ma_2 = 1/_{32}(18\ SiO_2 \cdot 3\ Al_2O_3 \cdot 4\ Na_2O \cdot CO_2)$		(52.4)
Sulfate-marialite	$1\ Ma_3 = 1/_{33}(18\ SiO_2 \cdot 3\ Al_2O_3 \cdot 4\ Na_2O \cdot SO_3)$		(52.0)
Chloride-meionite	$1\ Me_1 = 1/_{16}(6\ SiO_2 \cdot 3\ Al_2O_3 \cdot 3\ CaO \cdot CaCl_2)$		(59.1)
Carbonate-meionite	$1\ Me_2 = 1/_{16}(6\ SiO_2 \cdot 3\ Al_2O_3 \cdot 4\ CaO \cdot CO_2)$		(58.4)
Sulfate-meionite	$1\ Me_3 = 1/_{17}(6\ SiO_2 \cdot 3\ Al_2O_3 \cdot 4\ CaO \cdot SO_3)$		(57.1)
Accessories, oxides, etc.			
Ca-phosphate	$1\ Cp = 1/_5(P_2O_5 \cdot 3\ CaO)$		(62.0)
Apatite	$1\ Ap = 1/_{16}(3\ P_2O_5 \cdot 9\ CaO \cdot CaF_2)$		(63.0)
Titanite	$1\ Tit = 1/_3(SiO_2 \cdot TiO_2 \cdot CaO)$		(65.3)
Rutile	$1\ Ru = 1\ TiO_2$		(79.9)
Perovskite	$1\ Pf = 1/_2(TiO_2 \cdot CaO)$		(68.0)
Magnetite	$1\ Mt = 1/_3(Fe_2O_3 \cdot FeO)$		(77.1)
Ilmenite	$1\ Ilm = 1/_2(TiO_2 \cdot FeO)$		(75.9)
Hematite	$1\ Hm = 1/_2\ Fe_2O_3$		(79.8)
Chromite	$1\ Cm = 1/_3(Cr_2O_3 \cdot FeO)$		(74.6)
Fluorite	$1\ Fr = 1\ CaF_2$		(78.1)
Pyrite	$1\ Pr = 1/_3\ FeS_2$		(40.0)
Corundum	$1\ C = 1/_2\ Al_2O_3$		(51.0)
Zircon	$1\ Z = 1/_2(SiO_2 \cdot ZrO_2)$		(91.6)
Calcite	$1\ Cc = 1(CaO \cdot CO_2)$		(100.1)
Anhydrite	$1\ A = 1/_2(CaO \cdot SO_3)$		(68.1)
Thenardite	$1\ Th = 1/_3(Na_2O \cdot SO_3)$		(47.3)
Na-carbonate	$1\ Nc = 1/_2(Na_2O \cdot CO_2)$		(53.0)
Halite	$1\ Hl = 1(NaCl)$		(58.5)

If an excess of alumina is present, i.e. *Sp* or *Hz* appear in the basis, the following should be formed:

Cordierite	1 *Cord* = $^1/_{11}$ (5 $SiO_2 \cdot$ 2 $Al_2O_3 \cdot$ 2 MgO)	(53,2)
Fe-cordierite	1 *Fe-Cord* = $^1/_{11}$ (5 $SiO_2 \cdot$ 2 $Al_2O_3 \cdot$ 2 FeO)	(58,9)

This review indeed shows to what extent the computable equivalent weights are similar to one another. In practice this is even more pronounced, since, for example, olivines, augites and micas generally represent mixed crystals situated between the Mg- and Fe-bearing end-members. Therefore their equivalent weights have intermediate values between these extremities. The following list contains a selection of the important reactions for the formation of normative rock-forming minerals from the basis compounds. Here again the list is restricted to the compounds of importance in igneous rocks. The relationships for metamorphic rocks will be dealt with separately further on.

Feldspars

$$2\,SiO_2 \cdot Al_2O_3 \cdot K_2O \; + 4\,SiO_2 = 6\,SiO_2 \cdot Al_2O_3 \cdot K_2O$$
$$6\,Kp \qquad\qquad 4\,Q \qquad\qquad 10\,Or$$
$$\text{or} \quad 3\,Kp \qquad\qquad 2\,Q \qquad\qquad 5\,Or$$

$$2\,SiO_2 \cdot Al_2O_3 \cdot Na_2O + 4\,SiO_2 = 6\,SiO_2 \cdot Al_2O_3 \cdot Na_2O$$
$$6\,Ne \qquad\qquad 4\,Q \qquad\qquad 10\,Ab$$
$$\text{or} \quad 3\,Ne \qquad\qquad 2\,Q \qquad\qquad 5\,Ab$$

$$Al_2O_3 \cdot CaO \quad + 2\,SiO_2 = 2\,SiO_2 \cdot Al_2O_3 \cdot CaO$$
$$3\,Cal \qquad\qquad 2\,Q \qquad\qquad 5\,An$$

119 Feldspathoids

$$2\,SiO_2 \cdot Al_2O_3 \cdot K_2O \; + 2\,SiO_2 = 4\,SiO_2 \cdot Al_2O_3 \cdot K_2O$$
$$6\,Kp \qquad\qquad 2\,Q \qquad\qquad 8\,Lc$$

$$2\,SiO_2 \cdot Al_2O_3 \cdot Na_2O + 2\,SiO_2(+\,2\,W) = 4\,SiO_2 \cdot Al_2O_3 \cdot Na_2O \cdot 2\,H_2O$$
$$6\,Ne \qquad\qquad 2\,Q \qquad\qquad 8\,Anc$$

$$2\,(Al_2O_3 \cdot CaO) \; + SiO_2 \cdot 2\,CaO + SiO_2 = 2\,(SiO_2 \cdot Al_2O_3 \cdot 2\,CaO)$$
$$6\,Cal \qquad\qquad 3\,Cs \qquad 1\,Q \qquad\qquad 10\,Ge$$

$$2\,(SiO_2 \cdot 2\,CaO) + SiO_2 \cdot 2\,MgO + SiO_2 = 2\,(2\,SiO_2 \cdot MgO \cdot 2\,CaO)$$
$$6\,Cs \qquad\qquad 3\,Fo \qquad 1\,Q \qquad\qquad 10\,Ak$$

$$3\,(2\,SiO_2 \cdot Al_2O_3 \cdot Na_2O) + \quad 2\,NaCl \; = 6\,SiO_2 \cdot 3\,Al_2O_3 \cdot 3\,Na_2O \cdot 2\,NaCl$$
$$18\,Ne \qquad\qquad 2\,Hl \qquad\qquad 20\,Sod$$

$$3\,(2\,SiO_2 \cdot Al_2O_3 \cdot Na_2O) + Na_2O \cdot SO_3 = 6\,SiO_2 \cdot 3\,Al_2O_3 \cdot 4\,Na_2O \cdot SO_3$$
$$18\,Ne \qquad\qquad 3\,Th \qquad\qquad 21\,Nos$$

$$3\,(2\,SiO_2 \cdot Al_2O_3 \cdot Na_2O) + 2\,(CaO \cdot SO_3) = 6\,SiO_2 \cdot 3\,Al_2O_3 \cdot 3\,Na_2O \cdot 2\,CaO \cdot 2\,SO_3$$
$$18\,Ne \qquad\qquad 4\,A \qquad\qquad 22\,Hau$$

$$3\,(2\,SiO_2 \cdot Al_2O_3 \cdot Na_2O) + 2\,(CaO \cdot CO_2) = 6\,SiO_2 \cdot 3\,Al_2O_3 \cdot 3\,Na_2O \cdot 2\,CaO \cdot 2\,CO_2$$
$$18\,Ne \qquad\qquad 2\,Cc \qquad\qquad 20\,Canc$$

Augites and hornblendes

$$SiO_2 \cdot 2\,MgO + SiO_2 = 2(SiO_2 \cdot MgO)$$
$$3\,Fo \qquad\qquad 1\,Q \qquad\quad 4\,En$$

$$SiO_2 \cdot 2\,FeO + SiO_2 = 2(SiO_2 \cdot FeO)$$
$$3\,Fa \qquad\qquad 1\,Q \qquad\quad 4\,Hy$$

$$SiO_2 \cdot 2\,CaO + SiO_2 = 2(SiO_2 \cdot CaO)$$
$$3\,Cs \qquad\qquad 1\,Q \qquad\quad 4\,Wo$$

$$SiO_2 \cdot 2\,MgO + 2(SiO_2 \cdot CaO) + SiO_2 = 2(2\,SiO_2 \cdot MgO \cdot CaO)$$
$$3\,Fo \qquad\qquad 4\,Wo \qquad\qquad 1\,Q \qquad\quad 8\,Di$$

$$SiO_2 \cdot MgO + SiO_2 \cdot CaO = 2\,SiO_2 \cdot MgO \cdot CaO$$
$$2\,En \qquad\qquad 2\,Wo \qquad\qquad 4\,Di$$

$$SiO_2 \cdot FeO + SiO_2 \cdot CaO = 2\,SiO_2 \cdot FeO \cdot CaO$$
$$2\,Hy \qquad\qquad 2\,Wo \qquad\qquad 4\,Hed$$

$$Al_2O_3 \cdot CaO + SiO_2 = SiO_2 \cdot Al_2O_3 \cdot CaO$$
$$3\,Cal \qquad\qquad 1\,Q \qquad\quad 4\,Ts$$

$$SiO_2 \cdot 2\,CaO + 2\,Al_2O_3 + SiO_2 = 2(SiO_2 \cdot Al_2O_3 \cdot CaO)$$
$$3\,Cs \qquad\qquad 4\,C \qquad\quad 1\,Q \qquad\quad 8\,Ts$$

$$2(SiO_2 \cdot CaO) + 2\,Al_2O_3 = 2(SiO_2 \cdot Al_2O_3 \cdot CaO)$$
$$4\,Wo \qquad\qquad 4\,C \qquad\qquad 8\,Ts$$

$$SiO_2 \cdot Na_2O + Fe_2O_3 + 3\,SiO_2 = 4\,SiO_2 \cdot Fe_2O_3 \cdot Na_2O$$
$$3\,Ns \qquad\quad 2\,Hm \qquad 3\,Q \qquad\qquad 8\,Ac$$

$$2\,SiO_2 \cdot Fe_2O_3 \cdot Na_2O + 2\,SiO_2 = 4\,SiO_2 \cdot Fe_2O_3 \cdot Na_2O$$
$$6\,Fns \qquad\qquad\qquad 2\,Q \qquad\qquad 8\,Ac$$

$$2\,SiO_2 \cdot Al_2O_3 \cdot Na_2O + 2\,SiO_2 = 4\,SiO_2 \cdot Al_2O_3 \cdot Na_2O$$
$$6\,Ne \qquad\qquad\qquad 2\,Q \qquad\qquad 8\,Jd$$

$$2(Al_2O_3 \cdot CaO) + (SiO_2 \cdot 2\,CaO) + 4(SiO_2 \cdot 2\,MgO) + 9\,SiO_2(+\,2\,H_2O)$$
$$6\,Cal \qquad\qquad\quad 3\,Cs \qquad\qquad 12\,Fo \qquad\qquad 9\,Q \quad (2\,W)$$
$$= 2(7\,SiO_2 \cdot Al_2O_3 \cdot 4\,MgO \cdot 2\,CaO \cdot H_2O)$$
$$30\,Ho$$

$$2\,SiO_2 \cdot Al_2O_3 \cdot CaO + 2(SiO_2 \cdot 2\,MgO) + SiO_2 \cdot CaO + 2\,SiO_2(+\,H_2O)$$
$$5\,An \qquad\qquad\qquad 6\,Fo \qquad\qquad 2\,Wo \qquad\quad 2\,Q \quad (1\,W)$$
$$= 7\,SiO_2 \cdot Al_2O_3 \cdot 4\,MgO \cdot 2\,CaO \cdot H_2O$$
$$15\,Ho$$

$$SiO_2 \cdot Al_2O_3 \cdot CaO + SiO_2 \cdot CaO + 2(SiO_2 \cdot 2\,MgO) + 3\,SiO_2(+\,H_2O)$$
$$4\,Ts \qquad\qquad\qquad 2\,Wo \qquad\qquad 6\,Fo \qquad\qquad 3\,Q \quad (1\,W)$$
$$= 7\,SiO_2 \cdot Al_2O_3 \cdot 4\,MgO \cdot 2\,CaO \cdot H_2O$$
$$15\,Ho$$

$$2(SiO_2 \cdot Fe_2O_3) + 2(SiO_2 \cdot Na_2O) + 3(SiO_2 \cdot 2\,FeO) + 9\,SiO_2(+ 2\,H_2O)$$
$$\quad 6\,Fs \qquad\qquad 6\,Ns \qquad\qquad 9\,Fa \qquad\qquad 9\,Q \quad (2\,W)$$
$$= 2(8\,SiO_2 \cdot Fe_2O_3 \cdot 3\,FeO \cdot Na_2O \cdot H_2O)$$
$$30\,Rb_I$$

$$2(4\,SiO_2 \cdot Fe_2O_3 \cdot Na_2O) + 3(SiO_2 \cdot 2\,FeO) + 5\,SiO_2(+ 2\,H_2O)$$
$$16\,Ac \qquad\qquad 9\,Fa \qquad\qquad 5\,Q \quad (2\,W)$$
$$= 2(8\,SiO_2 \cdot Fe_2O_3 \cdot 3\,FeO \cdot Na_2O \cdot H_2O)$$
$$30\,Rb_I$$

Micas

$$2\,SiO_2 \cdot Al_2O_3 \cdot K_2O + 3(SiO_2 \cdot 2\,MgO) + SiO_2(+ 2\,H_2O)$$
$$6\,Kp \qquad\qquad 9\,Fo \qquad\qquad 1\,Q \quad (2\,W)$$
$$= 6\,SiO_2 \cdot Al_2O_3 \cdot 6\,MgO \cdot K_2O \cdot 2\,H_2O$$
$$16\,Mg\text{-}Bi$$

$$2\,SiO_2 \cdot Al_2O_3 \cdot K_2O + 3(SiO_2 \cdot 2\,FeO) + SiO_2(+ 2\,H_2O)$$
$$6\,Kp \qquad\qquad 9\,Fa \qquad\qquad 1\,Q \quad (2\,W)$$
$$= 6\,SiO_2 \cdot Al_2O_3 \cdot 6\,FeO \cdot K_2O \cdot 2\,H_2O$$
$$16\,Fe\text{-}Bi$$

121 $\quad 2\,SiO_2 \cdot Al_2O_3 \cdot K_2O + 4\,SiO_2 + 2\,Al_2O_3(+ 2\,H_2O) = 6\,SiO_2 \cdot 3\,Al_2O_3 \cdot K_2O \cdot 2\,H_2O$
$\qquad\quad 6\,Kp \qquad\qquad\quad 4\,Q \quad\; 4\,C \quad\; (2\,W) \qquad\qquad\qquad 14\,Ms$

Accessories

$$2\,TiO_2 + SiO_2 \cdot 2\,CaO + SiO_2 = 2(SiO_2 \cdot TiO_2 \cdot CaO)$$
$$2\,Ru \qquad 3\,Cs \qquad 1\,Q \qquad\qquad 6\,Tit$$

$$2\,TiO_2 + 2\,CaO \cdot SiO_2 = SiO_2 + 2(TiO_2 \cdot CaO)$$
$$2\,Ru \qquad 3\,Cs \qquad\quad 1\,Q \qquad 4\,Pf$$

$$3(3\,CaO \cdot P_2O_5) + CaF_2 = 3\,P_2O_5 \cdot 9\,CaO \cdot CaF_2$$
$$15\,Cp \qquad\qquad 1\,Fr \qquad\quad 16\,Ap$$

$$2(SiO_2 \cdot Fe_2O_3) + SiO_2 \cdot 2\,FeO = 2(Fe_2O_3 \cdot FeO) + 3\,SiO_2$$
$$6\,Fs \qquad\qquad 3\,Fa \qquad\qquad 6\,Mt \qquad\quad 3\,Q$$
$$2\,Fs \qquad\qquad Fa \qquad\qquad 2\,Mt \qquad\quad Q$$

$$SiO_2 \cdot 2\,FeO + 2\,TiO_2 = SiO_2 + 2(TiO_2 \cdot FeO)$$
$$3\,Fa \qquad\quad 2\,Ru \qquad 1\,Q \qquad 4\,Ilm$$

If an **alumina excess** is present (occurrence of Hz or Sp) cordierite should be formed in the following way:

$$2(Al_2O_3 \cdot MgO) + 5\,SiO_2 = 5\,SiO_2 \cdot 2\,Al_2O_3 \cdot 2\,MgO$$
$$\quad 6\,Sp \qquad\quad 5\,Q \qquad\qquad 11\,Cord$$
$$\text{or} \quad 6\,Hz \qquad 5\,Q \qquad\qquad 11\,Fe\text{-}Cord$$

a) The standard katanorm and its derivation from the basis

One particular normative mineral composition of the many which may be calculated, in principle, from the basis, is designated as the **standard katanorm** or, in short, **katanorm**. This alludes to a mineral composition such as will appear under conditions prevailing in deep-seated rocks or during katametamorphism, i.e. at high temperatures and pressures. Simply composed, water-free components presented in an idealized form, are here considered. Principally, this is in agreement with the modal relationships existing in such rocks. In similarity to the treatment applied in the CPIW-norm, no attention is paid to the complex melanocratic constituents. The calculation of the standard katanorm is carried out mainly by addition of silica to the basis components, i.e. by the standardized distribution of the available SiO_2 among these components.

If the rarer components Cl, F, CO_2, ZrO_2, SO_3, S, and Cr_2O_3 (added to Fe_2O_3) as well as MnO, NiO (calculated as FeO), SrO and BaO (as K_2O), Li_2O (as MgO), Rb_2O and Cs_2O (as K_2O) are not considered, then the 12 **basis-component combinations** already mentioned in section II 3 can be taken as the starting point of the calculation, according to the indicated regulations. These are illustrated in the following:

TABLE VIII

Possible combinations of the basis components

I	Ia	II	IIa	III	IIIa	IV	IVa	V	Va	VI	VIa
Ru	Ru	Ru	Ru	Ru	Ru	Ru	Ru	Ru	Ru	Ru	Ru
Cp	Cp	Cp	Cp	Cp	Cp	Cp	Cp	Cp	Cp	Cp	Cp
Kp	Kp	Kp	Kp	Kp	Kp	Kp	Kp	Kp	Kp	Kp	—
Ne	Ne	Ne	Ne	Ne	Ne	Ne	Ne	Ne	—	Ks	Ks
Cal	Cal	Cal	Cal	Cal	Cal	Cal	—	Ns	Ns	Ns	Ns
Sp	Sp	Sp	Sp	Sp	—	Cs	Cs	Cs	Cs	Cs	Cs
Hz	Hz	Hz	—	Fo	Fo	Fo	Fo	Fo	Fo	Fo	Fo
C	—	Fa	Fa	Fa	Fa	Fa	Fa	Fa	Fa	Fa	Fa
Fs	Fs	Fs	Fs	Fs	Fs	Fs	Fs	Fs	Fs	Fs	Fs
Q	Q	Q	Q	Q	Q	Q	Q	Q	Q	Q	Q

So-called "large" alumina-excess t present
$Al > (K + Na + \frac{1}{2} Ca)$
decreasing to the left

So-called "small" alumina-excess T present, i.e.
$Al > (K + Na)$

Alkali-excess present, i.e. $Al < (K + Na)$

The following procedure is provided for the calculation of the standard katanorm, in accordance with the case in question:

A. Elimination of Fs

Cases I and Ia

According to

$$SiO_2 \cdot Fe_2O_3 = Fe_2O_3 + SiO_2 \quad \text{or} \quad Hm = \tfrac{2}{3} Fs \text{ and } Q' = Q + \tfrac{1}{3} Fs$$
$$3\,Fs 2\,Hm 1\,Q$$ [1]*

Hm is formed from Fs, and the freed SiO_2 should be added to Q

Cases II, IIA, III, IIIa, IV, IVa

Magnetite is formed from Fs and Fa according to

$$2(SiO_2 \cdot Fe_2O_3) + SiO_2 \cdot 2\,FeO = 2(Fe_2O_3 \cdot FeO) + 3\,SiO_2$$
$$6\,Fs   3\,Fa 6\,Mt 3\,Q$$

or $2\,Fs + 1\,Fa = 2\,Mt + 1\,Q$ The freed SiO_2 is added to Q. Three cases should be distinguished, namely:

a) If $Fs > 2\,Fa$, Fs will be left over. This excess should be converted to hematite by removal of silica, according to: $3\,Fs = 2\,Hm + 1\,Q$. Thus, one obtains: $Mt = 2\,Fa$, $Hm = \tfrac{2}{3}(Fs - 2\,Fa)$ and $Q' = Q + \tfrac{1}{3}(Fs - 2\,Fa)$.

b) If $Fs = 2\,Fa$, then $Mt = Fs$ and $Q' = Q + Fa$.

c) If $Fs < 2\,Fa$, Fa will be left over and one obtains $Mt = Fs$, $Fa' = (Fa - \tfrac{1}{2} Fs)$ and $Q' = Q + \tfrac{1}{2} Fs$.

B. Elimination of $Ns + Ks$

Cases V, Va, VI, VIa

Fks and Fns are formed from Ks and Ns according to:

$$SiO_2 \cdot K_2O + SiO_2 \cdot Fe_2O_3 = 2\,SiO_2 \cdot Fe_2O_3 \cdot K_2O$$
$$3\,Ks 3\,Fs 6\,Fks \text{ (Potassium ferrisilicate)}$$

and, by analogy:

$$3\,Ns \quad + \quad 3\,Fs \quad = 6\,Fns \text{ (Sodium ferrisilicate)}$$

Two cases should be distinguished:

a) If $Fs > (Ks + Ns)$, an excess of $Fs' = Fs - (Ks + Ns)$ will remain, which should be converted to Mt and perhaps to Hm by the removal of silica, as mentioned under A.

b) If $Fs < (Ks + Ns)$, Ns', and perhaps also Ks', will remain.

After the completion of these preparatory operations the combinations I-IV will yield:

* Hereafter, the corresponding Q- amounts available for further operations will be designated by Q', Q'', Q''' etc. By analogy, the same will also apply with regard to other components, e.g. Fa', Fs', etc.

Ru, Cp, Kp, Ne, Q' are common to all these groups. In addition to these components the groups also contain the following:

I	II		III		IV	
Hm	*Mt* $\}$ or $\{$ *Mt*		*Mt* $\}$ or $\{$ *Mt*		*Mt* $\}$ or $\{$ *Mt*	
C	*Hm* $}$ $\phantom{\{}$ *Fa'*		*Hm* $}$ $\phantom{\{}$ *Fa'*		*Hm* $}$ $\phantom{\{}$ *Fa'*	
Hz	and in addition		and in addition		(*Cal*)	
Sp	*Cal*		*Cal*		*Fo* (*Cs*)	
	Sp (*Hz*)		*Fo* (*Sp*)			

124 For case V the following combinations will hold:

Ru, Cp, Kp (Ne), Cs, Fo, Q' are common to all these groups. In addition to these components, the following also occur:

$$Fns \pm Ns' + Fa$$

$$\text{or } Fns + Mt + \begin{cases} Hm \\ \text{or} \\ Fa' \end{cases}$$

In VI *Ne* is missing (and possibly also *Kp*). Instead of *Fns*, *Fns + Fks* will be present and instead of *Ns'*, *Ks'* is present, either with or without *Ns'*.

According to the following scheme, the distribution of *Q'* among the remaining basis-components should now be commenced, thereby forming more highly silicated types:

C. 1. If *Cal* is present, anorthite should be formed according to:

$$\underset{3\ Cal}{Al_2O_3 \cdot CaO} + \underset{2\ Q}{2\ SiO_2} = \underset{5\ An}{2\ SiO_2 \cdot Al_2O_3 \cdot CaO}$$

or
$$An = {}^5/_3\ Cal,\ Q'' = Q' - {}^2/_3\ Cal.$$

2. If *Fns* and *Fks* are present, aegirine (*Ac + K-Ac*) should be formed, as far as *Q'* suffices, according to:

$$\underset{6\ Fns}{2\ SiO_2 \cdot Fe_2O_3 \cdot Na_2O} + \underset{2\ Q}{2\ SiO_2} = \underset{8\ Ac}{4\ SiO_2 \cdot Fe_2O_3 \cdot Na_2O}$$

and in analogy:

$$\underset{6\ Fks}{} + \underset{2\ Q}{2\ Q} = \underset{8\ K\text{-}Ac}{8\ K\text{-}Ac}$$

or
$$\left. \begin{array}{l} Ac = {}^4/_3\ Fns \\ K\text{-}Ac = {}^4/_3\ Fks \end{array} \right\} Q'' = Q' - {}^1/_3\ (Fns + Fks).$$

3. Additional *Q*-excess serves to form orthoclase and, provided that sufficient *Q* is available, to form albite:

$$\underset{6\ Kp}{2\ SiO_2 \cdot Al_2O_3 \cdot K_2O} + \underset{4\ Q}{4\ SiO_2} = \underset{10\ Or}{6\ SiO_2 \cdot Al_2O_3 \cdot K_2O}$$

and, in analogy:

$$6\ Ne \qquad + 4\ Q = 10\ Ab$$

or

$$\left.\begin{array}{l} Or = {}^5/_3\ Kp \\ Ab = {}^5/_3\ Ne \end{array}\right\} Q''' = Q'' - {}^2/_3 (Kp + Ne).$$

125 According to earlier instructions (P. Niggli, 1936, 308), and in agreement with the laws of the CPIW-system, wollastonite was formed from Cs prior to the formation of the alkali feldspars. However, in rocks low in silica (alnoites, polzenites, for example), Cs may also be introduced into monticellite, and the melilites which form instead wollastonite can also be formed from anorthite + orthosilicate. For example:

$$2\ (2\ SiO_2 \cdot Al_2O_3 \cdot CaO) + SiO_2 \cdot 2\ CaO = 2\ (SiO_2 \cdot Al_2O_3 \cdot 2\ CaO) + 3\ SiO_2$$
$$10\ An \qquad\qquad 3\ Cs \qquad\qquad 10\ Ge \qquad\qquad 3\ Q$$

Therefore, since 1945 (C. Burri and P. Niggli, 1945, 584) the aluminosilicates are considered prior to the others during silication.

4. Additional available Q should be combined in the basis components Hz, Sp, C or to Cs, Fa, Fo respectively, in the given order, and according to the following relationships:

$$2\ (Al_2O_3 \cdot FeO) + 5\ SiO_2 = 5\ SiO_2 \cdot 2\ Al_2O_3 \cdot 2\ FeO$$
$$6\ Hz \qquad\qquad 5\ Q \qquad\qquad 11\ Fe\text{-}Cord$$

and, in analogy:

$$6\ Sp \quad + 5\ Q = \quad 11\ Cord$$

as well as

$$Al_2O_3 \quad + SiO_2 = \quad SiO_2 \cdot Al_2O_3$$
$$2\ C \qquad\quad 1\ Q \qquad\qquad 3\ Sil$$

or (provided that sufficient Q is available):

$$\left.\begin{array}{l} Fe\text{-}Cord = {}^{11}/_6\ Hz \\ Cord = {}^{11}/_6\ Sp \\ Sil = {}^3/_2\ C \end{array}\right\} Q'''' = Q''' - ({}^5/_6\ Hz + {}^5/_6\ Sp + {}^1/_2\ C).$$

5. For the cases IV, V, and VI, the reactions yield:

$$SiO_2 \cdot 2\ CaO + SiO_2 = 2\ (SiO_2 \cdot CaO)$$
$$3\ Cs \qquad\quad 1\ Q \qquad\quad 4\ Wo$$

and, in analogy:

$$3\ Fa \quad + 1\ Q = \quad 4\ Hy$$
$$3\ Fo \quad + 1\ Q = \quad 4\ En,$$

or (provided that sufficient Q is available):

$$\left.\begin{array}{l} Wo = {}^4/_3\ Cs \\ Hy = {}^4/_3\ Fa \\ En = {}^4/_3\ Fo \end{array}\right\} Q'''' = Q''' - {}^1/_3 (Cs + Fa + Fo).$$

126 If Ks or Ns or $(Ks + Ns)$ still remain after Bb), the remaining amounts should be combined with silica prior to Cs:

$$SiO_2 \cdot Na_2O + SiO_2 = 2\,SiO_2 \cdot Na_2O$$
$$3\,Ns \quad\quad + 1\,Q = \quad 4\,Ns^*, \quad \text{where } 1\,Ns^* = {}^1/_4(2\,SiO_2 \cdot Na_2O)$$

and, in analogy

$$3\,Ks \quad + 1\,Q = \quad 4\,Ks^*, \text{ where } 1\,Ks^* = {}^1/_4(2\,SiO_2 \cdot K_2O).$$

In cases II and III Hz, and Sp should be combined with silica prior to Fa and Fo.

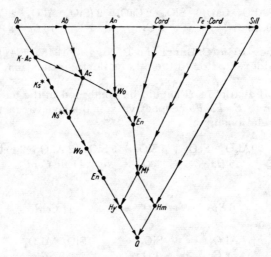

FIGURE 39. Review of the possible mineral combinations of the standard katanorm. After C. Burri and P. Niggli (1945)

Evidently, the number of basis components which can combine with silica, is restricted by the available amount of Q, Q', Q'' etc. In practice, either of two alternative calculation procedures can be applied: the silica balance can be reckoned after each step of the calculation, or the most highly silicated compounds are formed first, without taking into account the actual available amount of Q. Only after completion of the calculation is the Q-balance finally reckoned. This balance compares all the used Q-amounts with the gained amounts (e.g. the Q-amount gained through the formation of Mt), thus giving to the various amounts their appropriate signs. This will show whether an excess or a deficiency is present. If the latter is the case, the last-formed components should be depleted of their silica, in the reverse order of their formation, as far as is necessary.

After carrying out this procedure, the katanorms of each of the 6 different main cases of basis combinations will assume a very specific "paragenesis". This is reviewed in detail schematically (Figure 39). This scheme should
127 be interpreted as follows: only those components which follow one another consecutively, as indicated by the direction of the arrows, may occur

together normatively. In addition, individual components may be assembled together as composite components. This allows for further graphical representations of the possible combinations. The following composite components may serve as an example:

Alkali feldspars	$AF = Or + Ab$
Feldspars	$F = Or + Ab + An$
Aegirine	$Aeg = Ac + K\text{-}Ac$
Alkali augite	$AP = Ac + K\text{-}Ac + Ns^* + Ks^*$
Cordierite	$[Cord] = Fe\text{-}Cord + Cord$
Ortho-augite	$Orthaug = En + Hy$
Pyroxene	$P = En + Hy + Wo$
Ore	$= Mt + Hm$

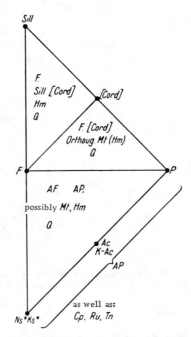

FIGURE 40. Only those components which fall within the limits of one of the three inner triangles can coexist in the katanorm. (Tn in the figure corresponds to Tit in the text). After C. Burri and P. Niggli (1945)

Figure 40 provides a picture of the possible combinations of these composite components. Figures 41-44 illustrate which of the composite components mutually exclude one another. From the diagrams it should be understood that only those components which fall within the limits of the same small triangle are capable of normative coexistence, and that components which do not fall within the same triangle do not represent a possible normative combination, e. g. the following components mutually exclude one another: An and Ac, $K\text{-}Ac$, Ns^*, Ks^* or AP; other examples are: $[Cord]$ and Ac, $[Cord]$ and AP, and Sil and Wo or ortho-augite.

Under certain circumstances, these figures provide one with a desired means of control of the calculation, in that the appearance of discrepancies will indicate a deviation from the calculation rules.

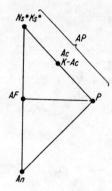

FIGURE 41. Either AP or An occur together with AF and P. After C. Burri and P. Niggli (1945)

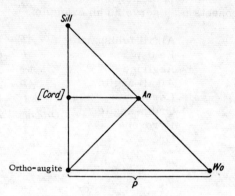

FIGURE 42. In the katanorm Sil does not coexist with Wo or with ortho-augite. After C. Burri and P. Niggli (1945)

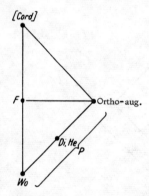

FIGURE 43. $Cord$ does not coexist with Wo in the katazone. After C. Burri and P. Niggli (1945)

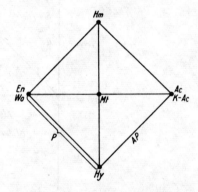

FIGURE 44. Hy and Hm oppose each other in the katazone. After C. Burri and P. Niggli (1945)

Examples:

For the quartz-diorite of Spanish Peak, California, the following basis was calculated:

Q	Kp	Ne	Cal	Cs	Fs	Fa	Fo	Ru	Cp	Total
42.3	4.7	21.0	15.6	1.7	3.0	3.2	7.4	0.5	0.6	100.0

Since this combination of basis components corresponds to case IV, or since $Al > (Na + K)$, no formation of Fks or Fns is attempted, and one should start directly with the formation of magnetite from Fs:

$$3.0\,Fs + 1.5\,Fa = 3.0\,Mt + 1.5\,Q,$$

whereby $3.2 - 1.5 = 1.7$ Fa' remains.

According to C 1, the formation of anorthite should now follow:

$$15.6\ Cal + 10.4\ Q = 26.0\ An,$$

and, according to C 3, alkali feldspar is formed from Kp and Ne:

$$4.7\ Kp + 3.1\ Q = 7.8\ Or$$
$$21.0\ Ne + 14.0\ Q = 35.0\ Ab.$$

Then silica is added to the orthosilicates Cs, Fa and Fo:

$$1.7\ Cs + 0.6\ Q = 2.3\ Wo$$
$$1.7\ Fa' + 0.6\ Q = 2.3\ Hy$$
$$7.4\ Fo + 2.5\ Q = 9.9\ En.$$

The Q-balance is reckoned in the following way:

Used for	An	$10.4\ Q$
,,	,, Or	$3.1\ Q$
,,	,, Ab	$14.0\ Q$
,,	,, Wo	$0.6\ Q$
,,	,, Hy	$0.6\ Q$
,,	,, En	$2.5\ Q$
		$31.2\ Q$
Gained from Mt-formation		$1.5\ Q$
Total used		$29.7\ Q.$

Thus, an excess of 42.3 - 29.7 = 12.6 Q remains.

The standard katanorm for the quartz-diorite of Spanish Peak which was thus calculated according to the given rules, yields:

Q	Or	Ab	An	Wo	En	Hy	Mt	Ru	Cp	Total
12.6	7.8	35.0	26.0	2.3	9.9	2.3	3.0	0.5	0.6	100,0

The more detailed subdivision of the pyroxenes is a point of importance. The following pyroxene components appear: 2.3 Wo, 9.9 En and 2.3 Hy. Prior to the formation of diopside, titanite should be formed according to the relationship:

$$\underset{1\ Ru}{TiO_2} + \underset{2\ Wo}{SiO_2 \cdot CaO} = \underset{3\ Tit.}{SiO_2 \cdot TiO_2 \cdot CaO}$$

Only the residual Wo' is then used for diopside:

$$0.5\ Ru + 1.0\ Wo = 1.5\ Tit.$$

Thus, 2.3 − 1.0 = 1.3 Wo' is still available for the formation of diopside. This requires an equal amount of $(En + Hy)$. In accordance with the regulations for the calculation of the CPIW-norm, it is customary to set the ratio of $En:Hy$ in the diopside as equal to that of the total available amounts. Since in diopside $Wo = En + Hy$, the amount of diopside can be calculated in the following way:

$$(En + Hy):En = Wo:En',$$

i. e., in the case under consideration 12.2 : 9.9 = 1.3 : x. It hence follows that $x = En' = 1.1$. Accordingly, the resulting composition of diopside is $Wo\,1.3,\,En\,1.1,\,Hy\,0.2$. Undoubtedly, however, this schematic procedure by no means represents the true natural relationships. As was recently emphasized by P. Eskola (1954, 10), it is absolutely true that the evaluation of the $Mg : Fe$ ratio in femic normative components in dependence on the gross chemical composition of the rock, presents a problem as yet unsolved. This holds true both for the weight percentage CPIW-norm as well as for the equivalent norm. This problem will remain unsolved until the fundamental knowledge of the actual relationships is increased significantly.

After the formation of the diopside, $9.9-1.1 = 8.8\,En$ and $2.3-0.2 = 2.1\,Hy$ will still remain for the formation of ortho-augite. Thus, after considering the subdivision of pyroxenes the standard katanorm will yield:

Q	Or	Ab	An	Wo	En	Hy	En	Hy	Mt	Cp	Tit	Total
12.6	7.8	35.0	26.0	1.3	1.1	0.2	8.8	2.1	3.0	0.6	1.5	100.0

Plag 61.0   Di 2.6   Ortho-aug. 10.9   Acces. 5.1

Feldspar 68.8   Pyroxene 13.5

Light components 81.4   Dark components 18.6

The mineral composition thus calculated, in similarity to the CPIW-norm, corresponds to only one of the various possible heteromorphic facies. It corresponds to a **pyroxene-quartz diorite**, according to the quantitative mineralogical classification of the igneous rocks of P. Niggli (1931), as $Q > 1/8$ of the light ingredients and $Or < 1/8$ of the total feldspars, and also $Ab > An$.

The CPIW-norm provides similar results, namely:

Q	or	ab	an	di	hy	mt	ilm	ap	
13.80	7.78	33.01	25.02	4.64	8.42	4.18	1.22	0.67	(after H. S. Washinton, 1917, 381 No. 82).

In contrast to the weight-percentage results, however, the equivalent-based values of the standard katanorm may be **transformed immediately into other heteromorphic facies**. Thus, a gradual approach to the modal ratios becomes possible. This will be elaborated upon in the following chapters.

However, priority will be given to the discussion of two further examples of standard katanorms of different chemism.

The following basis was previously calculated for the **cordierite-andesite** (from Hoyazo, near Nijar, Spain):

Q	Kp	Ne	Cal	Sp	Fs	Fa	Fo	Ru	Cp	Total
49.4	12.3	9.9	9.3	9.3	3.5	3.9	1.5	0.6	0.3	100.0

This combination may serve as an example of case III. According to the stated regulations, the following should be found:

$$3.5\,Fs + 1.7\,Fa = 3.5\,Mt + 1.7\,Q$$
$$9.3\,Cal + 6.2\,Q = 15.5\,An$$
$$12.3\,Kp + 8.2\,Q = 20.5\,Or$$
$$9.9\,Ne + 6.6\,Q = 16.5\,Ab$$
$$9.3\,Sp + 7.7\,Q = 17.0\,Cord$$
$$2.2\,Fa + 0.7\,Q = 2.9\,Hy$$
$$1.5\,Fo + 0.5\,Q = 2.0\,En.$$

An amount of Q equivalent to $6.2 + 8.2 + 6.6 + 7.7 + 0.7 + 0.5 = 29.9$ was used. Available are $49.4 + 1.7 = 51.1$. Thus, $51.1 - 29.9 = 21.2\,Q$ will remain, and the required standard katanorm will be:

Q	Or	Ab	An	Cord	En	Hy	Mt	Ru	Cp	Total
21.2	20.5	16.5	15.5	17.0	2.0	2.9	3.5	0.6	0.3	100.0

This mineral composition clearly shows that the rock is endomorphic. The mineral combination of the standard katanorm corresponds to a hypersthene-bearing feldspar-cordierite hornfels, as described by V. M. Goldschmidt, among others, from the Oslo region (1911). The dependability of such formations was already mentioned (C. Burri and I. Parga-Pondal, 1936).

Additional examples of norm calculations of rocks with high alumina excess will be provided in the treatment of metamorphic rocks.

For the **lujaurite** from Lille Elv, Kangerdluarsuk, Greenland, the basis yields:

Q	Kp	Ne	Ns	Cs	Fs	Fa	Fo	Ru	Z	Hl	Total
22.8	9.1	40.4	10.9	1.0	9.4	5.2	0.2	0.2	0.5	0.3	100.0

132 A rock with an **alkali excess** is considered, corresponding to case V. Firstly, Fs should be eliminated along with the formation of Fns:

$$9.4\,Fs + 9.4\,Ns = 18.8\,Fns.$$

Since $Fs < Ns$, a quantity of $Ns' = 10.9 - 9.4 = 1.5\,Ns$ will remain. The formation of aegirine is now carried out according to C2:

$$18.8\,Fns + 6.3\,Q = 25.1\,Ac,$$

and the formation of alkali feldspar according to C3:

$$9.1\,Kp + 6.0\,Q = 15.1\,Or.$$

Since the following amounts of Q were used: 6.3 for Ac and 6.0 for Or $22.8 - (6.3 + 6.0) = 10.5$ still remains. This does not suffice for the combination of all the available $40.4\,Ne$ with silica. Therefore, the $10.5\,Q$ is used for the following formation: $15.7\,Ne + 10.5\,Q = 26.2\,Ab$

and subsequently, according to:

$$3(2\,SiO_2 \cdot Al_2O_3 \cdot Na_2O) + 2\,NaCl = (6\,SiO_2 \cdot 3\,Al_2O_3 \cdot 3\,Na_2O \cdot 2\,NaCl)$$
$$18\,Ne \qquad\qquad 2\,Hl \qquad\qquad 20\,Sod$$

$2.7\,Ne + 0.3\,Hl = 3.0\,Sod$, and the standard katanorm will yield:

Or	Ab	Ne	Sod	Ac	Ns	Fo	Fa	Cs	Ru	Z	Total
15.1	26.2	22.0	3.0	25.1	1.5	0.2	5.2	1.0	0.2	0.5	100.0
AF		Foids		AP		Olivine		Accessories			

The sparse amount of 1.0 Cs was calculated as olivine, according to the stated regulations, since a rock poor in silica is dealt with. If, however, the following had been formed:

$$0.3\ Cs + 0.2\ Ru + 0.1\ Q = 0.6\ Tit$$
$$\left.\begin{array}{l}0.7\ Cs + 0.2\ Q = 0.9\ Wo\\0.7\ Fa + 0.2\ Q = 0.9\ Hy\end{array}\right\} 1.8\ Hed,$$

an amount of albite equivalent to the consumed 0.5 Q should have been desilicated according to:

$$1.2\ Ab = 0.7\ Ne + 0.5\ Q,$$

thus giving the corresponding variant of the standard katanorm:

Or	Ab	Ne	Sod	Ac	Ns	Hd	Fo	Fa	Tit	Z	Total
15.1	25.0	22.7	3.0	25.1	1.5	1.8	0.2	4.5	0.6	0.5	100.0

b) The formation of variants of the standard katanorm

It was already mentioned that one of the great advantages of the Niggli katanorm over the CPIW-norm lies in the possibility of passing from a once-calculated mineral combination, without recourse to elaborate calculations, to a different mineral combination of the same gross chemism. This may be achieved without recalculating the sum to 100, since the number of equivalents does not change. This principle will be exemplified in greater detail by the quartz-diorite from Spanish Peak, California. Additional examples may be found in the chapter on the application of the equivalent norm to the study of heteromorphic relationships, as well as in the treatment of metamorphic rocks.

The standard katanorm of the quartz-diorite from Spanish Peak, which was calculated previously, gives:

Q	Or	Ab	An	Wo	En	Hy	En	Hy	Mt	Cp	Tit	Total
12.6	7.8	35.0	26.0	1.3	1.1	0.2	8.8	2.1	3.0	0.6	1.5	100.0

A biotite-variant should now be calculated, i.e. the analysis should be calculated to a mineral composition representing a facies heteromorphic to the pyroxene-quartz diorite. In the variant, the K should be incorporated mainly in biotite, instead of in potassium feldspar.

The following equations were previously provided for the construction of biotite from Kp:

$$6\ Kp + 9\ Fo + 1\ Q\ (+ 2\ W) = 16\ Mg\text{-}Bi$$
$$6\ Kp + 9\ Fa + 1\ Q\ (+ 2\ W) = 16\ Fe\text{-}Bi$$
$$\text{or} \quad 6\ Kp + 9\ (Fo + Fa) + 1\ Q\ (+ 2\ W) = 16\ Bi. \tag{1}$$

The calculations can be based directly on the components Or and $En + Hy$ of the katanorm, instead of on the basis components Kp and $Fo + Fa$. For this purpose the following relations should be considered:

$$3\ Kp + 2\ Q = 5\ Or \tag{2}$$
$$3\ Fo + 1\ Q = 4\ En$$
$$3\ Fa + 1\ Q = 4\ Hy$$
$$\text{or} \quad 3\ (Fo + Fa) + 1\ Q = 4\ (En + Hy). \tag{3}$$

These reaction equations should be treated like algebraic equations. They may be added to or subtracted from one another. Terms may be transferred from one side of the equality to the other, making the appropriate change of sign. A welcome possibility of control is provided by the law which requires that the sum of the reaction coefficients on both sides of the equations should be equal.

In order to attain the desired relationship between Or and $(En + Hy)$ on one side, and Bi on the other, the following procedure should be applied: equation (2) multiplied by a factor of 2 should be subtracted from (1) in order to eliminate Kp. From the result (4) equation (3), multiplied by a factor of 3, should be subtracted in order to eliminate $(Fo + Fa)$. Thus, the desired relationship (5) is obtained:

$$6 Kp + 9 (Fo + Fa) + 1 Q (+ 2 W) = 16 Bi \quad (1)$$
$$6 Kp + 4 Q = 10 Or \quad (2)$$

After substraction: $9 (Fo + Fa) - 3 Q (+ 2 W) = 16 Bi - 10 Or$.

After rearrangement so that only positive signs occur:

$$9 (Fo + Fa) + 10 Or (+ 2 W) = 16 Bi + 3 Q, \quad (4)$$
also valid is: $9 (Fo + Fa) + 3 Q = 12 (En + Hy). \quad (3)$

(3) should be subtracted from (4), and after rearrangment the following is obtained:

$$10 Or + 12 (En + Hy) (+ 2 W) = 16 Bi + 6 Q. \quad (5)$$

Starting with 7.8 Or of the standard katanorm, one obtains:

$$7{,}8\ Or + 9{.}4\ (En + Hy) = 12{.}5\ Bi + 4{.}7\ Q$$

Thus, the sought after **biotite variant** will be:

Q	Ab	An	Bi	Di	En + Hy	Mt	Cp	Tit	Total
17.3	35.0	26.0	12.5	2.6	1.5	3.0	0.6	1.5	100.0

This corresponds to a **biotite-pyroxene quartz-diorite**.

If the above-stated idealized composition of common hornblende is taken as a starting point:

$$1\ Ho = {}^1/_{15}\ (7\ SiO_2 \cdot Al_2O_3 \cdot 4(Mg, Fe)O \cdot 2\ CaO \cdot H_2O)$$

corresponding to: $\begin{bmatrix} Si_7AlO_{22} & (Mg, Fe)_4 \\ (OH)_2 & Al \end{bmatrix} Ca_2$,

then the calculation of a **hornblende-biotite variant** is also possible.

From the relationship previously introduced for the formation of hornblende from basis components:

$$6\ Cal + 3\ Cs + 12\ (Fo + Fa) + 9\ Q\ (+ 2\ W) = 30\ Ho,$$

the following equations may be obtained by combining Cal, Cs, Fo and Fa with silica:

$$5\ An + 2\ Wo + 8\ (En + Hy) (+ 1\ W) = 15\ Ho$$
or $\quad 5\ An + 4\ Di + 6\ (En + Hy) (+ 1\ W) = 15\ Ho.$

Starting with the 2.6 Di of the previously calculated biotite variant, one obtains:

$$3{,}3\ An + 2{,}6\ Di + 3{,}9\ (En + Hy) = 9{,}8\ Ho.$$

Since hornblende quartz-diorites generally contain biotite, this mineral should be formed, starting from $10.9 - 3.9 = 7.0\ (En + Hy)$, according to:

$$5.8\ Or + 7.0\ (En + Hy) = 9.3\ Bi + 3.5\ Q.$$

Thus is obtained the following **hornblende-biotite variant**, typified by an *An*-poor plagioclase:

Q	Or	Ab	An	Ho	Bi	Mt	Cp	Tit	Total
16.1	2.0	35.0	22.7	9.8	9.3	3.0	0.6	1.5	100.0

Additional examples of the formation of variants follow in chapter BII8.

6. The Standard Katanorm and Cation Percentages

a) Relationships between cation percentages and the standard katanorm.

The standard katanorm, as well as any other normative mineral combination or variant, may also be arrived at from the metal-atom (cation) percentages instead of from the basis components. This is achieved in a manner similar to that shown for the basis in section 4. This construction may be formulated in the following way, with regard to some important normative components:

$$1\ Or = \tfrac{1}{5} K + \tfrac{1}{5} Al + \tfrac{3}{5} Si$$
$$1\ Ab = \tfrac{1}{5} Na + \tfrac{1}{5} Al + \tfrac{3}{5} Si$$
$$1\ An = \tfrac{1}{5} Ca + \tfrac{2}{5} Al + \tfrac{2}{5} Si$$
$$1\ Lc = \tfrac{1}{4} K + \tfrac{1}{4} Al + \tfrac{1}{2} Si$$
$$1\ Ne = \tfrac{1}{3} Na + \tfrac{1}{3} Al + \tfrac{1}{3} Si$$
$$1\ Fo = \tfrac{2}{3} Mg + \tfrac{1}{3} Si$$
$$1\ Fa = \tfrac{2}{3} Fe^{2+} + \tfrac{1}{3} Si$$
$$1\ Cs = \tfrac{2}{3} Ca + \tfrac{1}{3} Si$$
$$1\ Wo = \tfrac{1}{2} Ca + \tfrac{1}{2} Si$$
$$1\ En = \tfrac{1}{2} Mg + \tfrac{1}{2} Si$$
$$1\ Hy = \tfrac{1}{2} Fe^{2+} + \tfrac{1}{2} Si$$
$$1\ Ac = \tfrac{1}{4} Na + \tfrac{1}{4} Fe^{3+} + \tfrac{1}{2} Si$$
$$1\ Ns = \tfrac{2}{3} Na + \tfrac{1}{3} Si$$
$$1\ Mg\text{-}Bi = \tfrac{1}{8} K + \tfrac{3}{8} Mg + \tfrac{1}{8} Al + \tfrac{3}{8} Si$$
$$1\ Fe\text{-}Bi = \tfrac{1}{8} K + \tfrac{3}{8} Fe^{2+} + \tfrac{1}{8} Al + \tfrac{3}{8} Si$$
$$1\ Ho = \tfrac{2}{15} Ca + \tfrac{4}{15} Mg + \tfrac{2}{15} Al + \tfrac{7}{15} Si$$
$$1\ Cord = \tfrac{2}{11} Mg + \tfrac{4}{11} Al + \tfrac{5}{11} Si$$
$$1\ Fe\text{-}Cord = \tfrac{2}{11} Fe^{2+} + \tfrac{4}{11} Al + \tfrac{5}{11} Si$$
$$1\ Tit = \tfrac{1}{3} Ca + \tfrac{1}{3} Ti + \tfrac{1}{3} Si$$
$$1\ Pf = \tfrac{1}{2} Ca + \tfrac{1}{2} Ti$$
$$1\ Cp = \tfrac{3}{5} Ca + \tfrac{2}{5} P$$
$$1\ Mt = \tfrac{1}{3} Fe^{2+} + \tfrac{2}{3} Fe^{3+}$$
$$1\ Hm = 1\ Fe^{3+}$$
$$1\ Ilm = \tfrac{1}{2} Fe^{2+} + \tfrac{1}{2} Ti$$
$$1\ Ru = 1\ Ti$$
$$1\ Z = \tfrac{1}{2} Zr + \tfrac{1}{2} Si$$
$$1\ C = 1\ Al$$
$$1\ Q = 1\ Si.$$

b) Calculation of the standard katanorm from given cation percentages.

In principle, this is carried out in a manner analogous to that used for calculating the basis from given cation percentages.

Examples:

The following cation percentages were previously calculated from the quartz-diorite from Spanish Peak, California, from the weight-percentage analyses:

Si^{4+}	Al^{3+}	Fe^{3+}	Fe^{2+}	Mg^{2+}	Ca^{2+}	Na^+	K^+	Ti^{4+}	P^{5+}	Total
56.0	18.9	2.0	2.1	5.0	6.7	7.0	1.6	0.5	0.2	100.0

The following standard katanorm may be calculated therefrom:

starting from

0.5 Ti	0.5 Ti + 0.5 Ca + 0.5 Si	= 1.5 *Tit*
0.2 P	0.2 P + 0.3 Ca	= 0.5 *Cp*
1.6 K	1.6 K + 1.6 Al + 4.8 Si	= 8.0 *Or*
7.0 Na	7.0 Na + 7.0 Al + 21.0 Si	= 35.0 *Ab*
18.9 − (1.6 + 7.0) = 10.3 Al . . .	10.3 Al + 5.1 Ca + 10.3 Si	= 25.7 *An*
6.7 − (0.5 + 0.3 + 5.1) = 0.8 Ca .	0.8 Ca + 0.8 Si	= 1.6 *Wo*
2.0 Fe^{3+}	2.0 Fe^{3+} + 1.0 Fe^{2+}	= 3.0 *Mt*
5.0 Mg	5.0 Mg + 5.0 Si	= 10.0 *En*
2.1 − 1.0 = 1.1 Fe^{2+}	1.1 Fe^{2+} + 1.1 Si	= 2.2 *Hy*.

The remaining Si is given by: $56.0 - (0.5 + 4.8 + 21.0 + 10.3 + 0.8 + 5.0 + 1.1) = 12.5\,Q$.

The desired katanorm thus yields:

Q	Or	Ab	An	Wo	En	Hy	Mt	Cp	Tit	Total
12.5	8.0	35.0	25.7	1.6	10.0	2.2	3.0	0.5	1.5	100.0
(12.6)	(7.8)	(35.0)	(26.0)	(1.3)	(9.9)	(2.3)	(3.0)	(0.6)	(1.5)	(100.0)

The standard katanorm of the cordierite-andesite from Hoyazo, near Nijar, Spain, may be obtained in an analogous way:

		Ru	Cp	Or	Ab	An	Cord	Mt	Hy	En	Q
Si^{4+}	59.7			12.3	9.9	6.2	7.7		1.4	1.0	21.2
Al^{3+}	19.8			4.1	3.3	6.2	6.2				
Fe^{3+}	2.4							2.4			
Fe^{2+}	2.6							1.2	1.4		
Mg^{2+}	4.1						3.1			1.0	
Ca^{2+}	3.3		0.2			3.1					
Na^+	3.3				3.3						
K^+	4.1			4.1							
Ti^{4+}	0.6	0.6									
P^{5+}	0.1		0.1								Total
	100.0	0.6	0.3	20.5	16.5	15.5	17.0	3.6	2.8	2.0	21.2	100.0
		(0.6)	(0.3)	(20.5)	(16.5)	(15.5)	(17.0)	(3.5)	(2.9)	(2.0)	(21.2)	(100.0)

For the lujaurite from Lille Elv, Kangerdluarsuk, Greenland, one obtains:

	Z	Ru	Or	Ab	Ac	Ns	Fo	Fa	Cs	Q
Si^{4+} 48.4	0.3		9.0	40.5	12.6	0.6	(0.05)	1.7	0.3	−16.6
Al^{3+} 16.5			3.0	13.5						
Fe^{3+} 6.3					6.3					
Fe^{2+} 3.5								3.5		
Mg^{2+} 0.1							0.1			
Ca^{2+} 0.7									0.7	
Na^+ 21.0				13.5	6.3	1.2				
K^+ 3.0			3.0							
Ti^{4+} 0.2		0.2								
Zr^{4+} 0.3	0.3									
100.0	0.6	0.2	15.0	67.5	25.2	1.8	0.1	5.2	1.0	−16.6

Throughout the calculation no consideration was given to question whether the amount of Si necessary for the construction of the highly silicated alkali aluminosilicates is indeed available. Thus, 65.0 Si were used up and only 48.4 were available. In order to arrive at the sum of 100, the 16.6 Si-deficit should be indicated as negative in the calculation. Alternatively, this amount must be gained by converting a corresponding amount of albite according to:

$$41.5\ Ab = 24.9\ Ne + 16.6\ Q.$$

Thus, the desired standard katanorm will yield (Cl-content not considered):

Or	Ab	Ne	Ac	Ns	Fo	Fa	Cs	Ru	Z	Total
15.0	26.0	24.9	25.2	1.8	0.1	5.2	1.0	0.2	0.6	100.0

If the 0.3 Cl^- is considered, it will yield, together with an equal amount of Na^+, 0.3 Hl. Thus only 1.2−0.3 = 0.9 Na^+ will remain available for Ns, resulting in $Ns = 1.3$ and $Q = -16.4$. In addition, the elimination of the Q-deficit by desilication of Ab will be given by: $41.0\ Ab = 24.6\ Ne + 16.4\ Q$. If sodalite is formed according to: $0.3\ Hl + 2.7\ Ne = 3.0\ Sod$, then $24.6 - 2.7 = 21.9\ Ne$ will still remain, and the desired standard katanorm will give:

Or	Ab	Ne	Sod	Ac	Ns	Fo	Fa	Cs	Ru	Z	Total
15.0	26.5	21.9	3.0	25.2	1.3	0.1	5.2	1.0	0.2	0.6	100.0
(15.1)	(26.2)	(22.0)	(3.0)	(25.1)	(1.5)	(0.2)	(5.2)	(1.0)	(0.2)	(0.5)	(100.0)

Comparison with the values (provided in brackets) calculated directly from the weight percentage via the basis shows a satisfactory agreement in all three examples.

c) Calculation of the cation percentages from the standard katanorm

This calculation is carried out in a manner principally analogous to that used for the derivation of the cation percentages from the basis, dealt with in section II4. The calculation of the required cation percentages is made possible by the following expressions, which are obtained by the compilation of the same terms which were mentioned under a).

$$Si^{4+} = Q + {}^3/_5(Or + Ab) + {}^1/_2(Wo + En + Hy + Ac + Lc + Z) + {}^5/_{11}(Cord$$
$$+ Fe\text{-}Cord) + {}^2/_5 An + {}^3/_8(Mg\text{-}Bi + Fe\text{-}Bi) + {}^1/_3(Ne + Kp + Fo + Fa$$
$$+ Cs + Ns + Tit)$$
$$Al^{3+} = C + {}^2/_5 An + {}^4/_{11}(Cord + Fe\text{-}Cord) + {}^1/_3(Ne + Kp) + {}^1/_4 Lc + {}^1/_5(Or + Ab)$$
$$+ {}^1/_8(Mg\text{-}Bi + Fe\text{-}Bi)$$
$$Fe^{3+} = {}^2/_3 Mt + {}^1/_4 Ac$$
$$Fe^{2+} = {}^2/_3 Fa + {}^1/_2(Hy + Ilm) + {}^3/_8 Fe\text{-}Bi + {}^1/_3(Mt + Pr) + {}^2/_{11} Fe\text{-}Cord$$
$$Mg^{2+} = {}^2/_3 Fo + {}^1/_2 En + {}^3/_8 Mg\text{-}Bi + {}^2/_{11} Cord$$
$$Ca^{2+} = {}^3/_5 Cp + {}^2/_3 Cs + {}^1/_2(Wo + Pf) + {}^1/_3 Tit + {}^1/_5 An$$
$$Na^+ = Hl + {}^2/_3 Ns + {}^1/_3 Ne + {}^1/_4 Ac + {}^1/_5 Ab$$
$$K^+ = {}^2/_3 Ks + {}^1/_3 Kp + {}^1/_4 Lc + {}^1/_5 Or + {}^1/_8(Mg\text{-}Bi + Fe\text{-}Bi)$$
$$Ti^{4+} = 1 Ru + {}^1/_2(Ilm + Pf) + {}^1/_3 Tit$$
$$P^{5+} = {}^2/_5 Cp$$
$$Zr^{4+} = {}^1/_2 Z$$

All the normative components which may be present in igneous rock were introduced into these equations, irrespective of whether they may occur together or not. In practice the expressions are considerably simplified, since a number of the given compounds preclude one another.

139 Examples:

Quartz-diorite from Spanish Peak, California.

The following standard katanorm was calculated from the analysis in section II6:

Q	Or	Ab	An	Wo	En	Hy	Mt	Cp	Tit	Total
12.6	7.8	35.0	26.0	1.3	9.9	2.3	3.0	0.6	1.5	100.0,

whence the following cation percentages are obtained:

$$Si^{4+} = Q + {}^3/_5(Or + Ab) + {}^1/_2(Wo + En + Hy) + {}^2/_5 An + {}^1/_3 Tit$$
$$= 12.6 + {}^3/_5(7.8 + 35.0) + {}^1/_2(1.3 + 9.9 + 2.3) + {}^2/_5(26.0)$$
$$+ {}^1/_3(1.5) = \ldots \ldots \ldots \ldots \ldots \ldots \ldots \quad 56.0 \quad (56.0)$$
$$Al^{3+} = {}^2/_5 An + {}^1/_5(Or + Ab) = {}^2/_5(26.0) + {}^1/_5(7.8 + 35.0) = \quad 19.0 \quad (18.9)$$
$$Fe^{3+} = {}^2/_3 Mt = {}^2/_3(3.0) = \ldots \ldots \ldots \ldots \ldots \quad 2.0 \quad (2.0)$$
$$Fe^{2+} = {}^1/_2 Hy + {}^1/_3 Mt = {}^1/_2(2.3) + {}^1/_3(3.0) = \ldots \ldots \quad 2.1 \quad (2.1)$$
$$Mg^{2+} = {}^1/_2 En = {}^1/_2(9.9) = \ldots \ldots \ldots \ldots \ldots \quad 5.0 \quad (5.0)$$
$$Ca^{2+} = {}^3/_5 Cp + {}^1/_2 Wo + {}^1/_3 Tit + {}^1/_5 An = {}^3/_5(0.6) + {}^1/_2(1.3)$$
$$+ {}^1/_3(1.5) + {}^1/_5(26.0) = \ldots \ldots \ldots \ldots \quad 6.7 \quad (6.7)$$
$$Na^+ = {}^1/_5 Ab = {}^1/_5(35.0) = \ldots \ldots \ldots \ldots \ldots \quad 7.0 \quad (7.0)$$
$$K^+ = {}^1/_5 Or = {}^1/_5(7.8) = \ldots \ldots \ldots \ldots \ldots \quad 1.6 \quad (1.6)$$
$$Ti^{4+} = {}^1/_3 Tit = {}^1/_3(1.5) = \ldots \ldots \ldots \ldots \ldots \quad 0.5 \quad (0.5)$$
$$P^{5+} = {}^2/_5 Cp = {}^2/_5(0.6) = \ldots \ldots \ldots \ldots \ldots \quad 0.2 \quad (0.2)$$
$$\overline{\quad 100.1 \quad (100.0)}$$

For the cordierite-andesite from Hoyazo, near Nijar, Spain, the standard katanorm yields:

Q	Or	Ab	An	Cord	En	Hy	Mt	Ru	Cp	Total
21.2	20.5	16.5	15.5	17.0	2.0	2.9	3.5	0.6	0.3	100.0

The cation percentages may be calculated in the following way:

$$\begin{aligned}
Si^{4+} &= Q + {}^3/_5(Or + Ab) + {}^1/_2(En + Hy) + {}^5/_{11}\,Cord + {}^2/_5\,An \\
&= 21.2 + {}^3/_5(20.5 + 16.5) + {}^1/_2(2.0 + 2.9) + {}^5/_{11}(17.0) \\
&\quad + {}^2/_5(15.5) = \ldots\ldots\ldots\ldots\ldots\ldots\ldots\ldots\ldots\ldots\ldots\ 59.8\ \ (59.7)
\end{aligned}$$

$Al^{3+} = {}^2/_5\,An + {}^4/_{11}\,Cord + {}^1/_5(Or + Ab) = {}^2/_5(15.5) + {}^4/_{11}(17.0)$
$\quad + {}^1/_5(20.5 + 16.5) = \ldots\ldots\ldots\ldots\ldots\ldots\ldots\ldots\ \ 19.8\ \ (19.8)$
$Fe^{3+} = {}^2/_3\,Mt = {}^2/_3(3.5) = \ldots\ldots\ldots\ldots\ldots\ldots\ldots\ldots\ \ \ 2.3\ \ \ (2.4)$
$Fe^{2+} = {}^1/_2\,Hy + {}^1/_3\,Mt = {}^1/_2(2.9) + {}^1/_3(3.5) = \ldots\ldots\ \ \ 2.6\ \ \ (2.6)$
$Mg^{2+} = {}^1/_2\,En + {}^2/_{11}\,Cord = {}^1/_2(2.0) + {}^2/_{11}(17.0) = \ldots\ \ \ 4.1\ \ \ (4.1)$
$Ca^{2+} = {}^3/_5\,Cp + {}^1/_5\,An = {}^3/_5(0.3) + {}^1/_5(15.5) = \ldots\ldots\ \ \ 3.3\ \ \ (3.3)$
$Na^{+} = {}^1/_5\,Ab = {}^1/_5(16.5) = \ldots\ldots\ldots\ldots\ldots\ldots\ldots\ \ \ 3.3\ \ \ (3.3)$
$K^{+} = {}^1/_5\,Or = {}^1/_5(20.5) = \ldots\ldots\ldots\ldots\ldots\ldots\ldots\ \ \ 4.1\ \ \ (4.1)$
$Ti^{4+} = 1\,Ru = \ldots\ldots\ldots\ldots\ldots\ldots\ldots\ldots\ldots\ldots\ \ \ 0.6\ \ \ (0.6)$
$P^{5+} = {}^2/_5\,Cp = {}^2/_5(0.3) = \ldots\ldots\ldots\ldots\ldots\ldots\ldots\ \ \ 0.1\ \ \ (0.1)$
$\quad\quad\quad\quad\quad\quad\quad\quad\quad\quad\quad\quad\quad\quad\quad\quad\quad\quad\overline{100.0\ \ (100.0)}$

After the conversion of 3.0 *Sod* to 2.7 *Ne* and 0.3 *Hl*, the standard katanorm for the lujaurites from Lille Elv, Kangerdluarsuk, Greenland, will yield:

Or	Ab	Ne	Hl	Ac	Ns	Fo	Fa	Cs	Ru	Z	Total
15.1	26.2	24.7	0.3	25.1	1.5	0.2	5.2	1.0	0.2	0.5	100.0

The cation percentages are obtained as follows:

$Si^{4+} = {}^3/_5(Or + Ab) + {}^1/_2(Ac + Z) + {}^1/_3(Ne + Fo + Fa + Cs$
$\quad + Ns) = {}^3/_5(15.1 + 26.2) + {}^1/_2(25.1 + 0.5) + {}^1/_3(24.7$
$\quad + 0.2 + 5.2 + 1.0 + 1.5) = \ldots\ldots\ldots\ldots\ldots\ldots\ \ 48.5\ \ (48.4)$
$Al^{3+} = {}^1/_3\,Ne + {}^1/_5(Or + Ab) = {}^1/_3(24.7) + {}^1/_5(15.1 + 26.2) = \ \ 16.5\ \ (16.5)$
$Fe^{3+} = {}^1/_4\,Ac = {}^1/_4(25.1) = \ldots\ldots\ldots\ldots\ldots\ldots\ \ \ 6.3\ \ \ (6.3)$
$Fe^{2+} = {}^2/_3\,Fa = {}^2/_3(5.2) = \ldots\ldots\ldots\ldots\ldots\ldots\ldots\ \ \ 3.5\ \ \ (3.5)$
$Mg^{2+} = {}^2/_3\,Fo = {}^2/_3(0.2) = \ldots\ldots\ldots\ldots\ldots\ldots\ldots\ \ \ 0.1\ \ \ (0.1)$
$Ca^{2+} = {}^2/_3\,Cs = {}^2/_3(1.0) = \ldots\ldots\ldots\ldots\ldots\ldots\ldots\ \ \ 0.7\ \ \ (0.7)$
$Na^{+} = 1\,Hl + {}^2/_3\,Ns + {}^1/_3\,Ne + {}^1/_4\,Ac + {}^1/_5\,Ab = 0.3 + {}^2/_3(1.5)$
$\quad + {}^1/_3(24.7) + {}^1/_4(25.1) + {}^1/_5(26.2) = \ldots\ldots\ldots\ \ 21.0\ \ (21.0)$
$K^{+} = {}^1/_5\,Or = {}^1/_5(15.1) = \ldots\ldots\ldots\ldots\ldots\ldots\ldots\ \ \ 3.0\ \ \ (3.0)$
$Ti^{4+} = 1\,Ru = \ldots\ldots\ldots\ldots\ldots\ldots\ldots\ldots\ldots\ldots\ \ \ 0.2\ \ \ (0.2)$
$Zr^{4+} = {}^1/_2\,Z = {}^1/_2(0.5) = \ldots\ldots\ldots\ldots\ldots\ldots\ldots\ \ \ 0.2\ \ \ (0.3)$
$\quad\quad\quad\quad\quad\quad\quad\quad\quad\quad\quad\quad\quad\quad\quad\quad\quad\quad\overline{100.0\ \ (100.0)}$

Here, again the agreement with the directly calculated values (in brackets) is very good for all three examples.

7. Standard Katanorm and Niggli-Values

The calculation of a simplified mineral composition from the Niggli-values, disregarding *ti*, *p* and the oxidation-degree of the iron, i. e. forgoing the formation of *Mt*, was provided in chapter BI 2c, where the essential considerations in the derivation of a simplified standard katanorm from the Niggli-values were dealt with. Below it will be shown that even a complete standard katanorm may be derived from the Niggli-values; the reverse also holds true, namely, the derivation of Niggli-values from the standard katanorm.

a) Calculation of the standard katanorm from given Niggli-values

In principle, the procedure is the same as for the simplified standard katanorm given under section BI 2 c. However, a difference exists in that after the construction of An, not all the Al which is not combined to Ca will be available for Wo. Instead, the amount bound to Cp should be subtracted. In forming Cp each unit of 3 Ca requires 2 P. The amount of Ca required for the formation of Cp is given by 3 P, because the p-value was calculated on the basis of P_2O_5. Only half the amount of Fe^{3+}, expressed in Fe^{2+}, is required for the formation of Mt, and thus Mt is given by $^3/_2\,fm\,(1-mg)\,w$ The amount of Fe^{2+} which is required for Mt should be subtracted prior to the formation of Hy. It should also be mentioned that a considerable amount of Si is saved by forming Cp and Mt, as compared to the simplified calculation procedure. The specific procedure is illustrated by means of the following examples.

Examples:

Quartz-diorite from Spanish Peak, California.

si	al	fm	c	alk	k	mg	ti	p	w
190	32.1	30.9	22.5	14.5	0.18	0.54	1.5	0.4	0.49

Recalculated to the sum of 100

Ru: ti	1.5	0.4	(0.5)
Cp: $5p$	2.0	0.6	(0.6)
Or: $10\,k\cdot alk$	26.1	7.7	(7.8)
Ab: $10\,(1-k)\,alk$	119.0	35.1	(35.0)
An: $5\,(al-alk)$	88.0	26.0	(26.0)
Wo: $2\,[c-(al-alk)-3p]$	7.4	2.2	(2.3)
En: $2\,mg\cdot fm$	33.4	9.8	(9.9)
Mt: $^3/_2\,[fm\,(1-mg)\,w]$	10.4	3.1	(3.0)
Hy: $2\,[fm\,(1-mg)\,(1-w)-^1/_2\,fm\,(1-mg)\,w]$ . . .	7.6	2.2	(2.3)
Q: $si-[100+4\,alk-^3/_2\,fm\,(1-mg)\,w-3p]$ . . .	43.6	12.9	(12.6)
$si+100+al+alk+ti+2p$	339.0 (338.9)	100.0	(100.0)

Cordierite-andesite from Hoyazo, near Nijar, Spain.

si	al	fm	c	alk	k	mg	ti	p	w
229	38.2	34.9	12.7	14.2	0.55	0.45	2.3	0.2	0.48

		Recalculated to the sum of 100	
$Ru: ti$	2.3	0.6	(0.6)
$Cp: 5p$	1.0	0.3	(0.3)
$Or: 10 k \cdot alk$	78.1	20.3	(20.5)
$Ab: 10(1-k) alk$	63.9	16.6	(16.5)
$An: 5(c-3p)$	60.5	15.8	(15.5)
$Cord: {}^{11}/_2 [al-(alk+c-3p)]$	65.5	17.0	(17.0)
$Mt: {}^3/_2 fm(1-mg) w$	13.8	3.6	(3.5)
$En: 2[fm \cdot mg-(al-(alk+c-3p))]$	7.6	2.0	(2.0)
$Hy: 2[fm(1-mg)(1-w)-{}^1/_2 fm(1-mg) w]$	10.8	2.8	(2.9)
$Q: si-si' = 229-148.3 =$	80.7	21.0	(21.2)
$si+100+al+alk+ti+2p$	384.2	100.0	(100.0)
	(384.1)		

Due to the alumina excess $al-(alk+c)$, in this case the An-content will not be given by $5(al-alk)$, but by $5(c-3p)$, i. e. by the c-value from which an appropriate amount was subtracted for Cp. The alumina excess calls for the formation of $Cord$. The remaining amount of alumina, not needed for the feldspars, serves for this purpose. The amount of Mg introduced into $Cord$ should be subtracted before forming enstatite. The consumption of si may be calculated as follows:

$$si' = \{6 alk + 2(c-3p) + {}^5/_2[al-(alk+c-3p)] + fm - {}^3/_2 fm(1-mg) w \\ -[al-(alk+c-3p)]\} \\ = 100 + {}^1/_2[al+7 alk-c-3p-3 fm(1-mg) w] = 148.3.$$

Arfvedsonite-lujaurite from Lille Elv, Kangerdluarsuk, Greenland:

si	al	fm	c	alk	k	mg	ti	zr	cl_2	w
157	26.8	32.0	2.1	39.1	0.13	0.01	0.7	0.9	0.5	0.65

The sum of the alkali feldspars or their substitutes is not given by $10\ alk$ or $6\ alk$, but by $10\ al$ or $6\ al$, and Ac should also be formed, as in this example $alk > al$. As in this case $(K+Na) > Al+Fe^{3+}$, or $2\ alk > 2\ al + fm \cdot (1-mg) w$, Ns will also occur. An additional demand is implied by the alkali excess $al < alk$, i. e. it is not $k = K^+/(K^+ + Na^+) = 0.13$ which is taken as a measure of the alkali feldspar relationships, but the ratio of K to the amount of the alkalis bound to Al: $k' = K/Al = k \cdot alk/al = 0.19$.

Since, moreover, the rock is very low in silica $(qz = -62.5)$, the calculation may be performed in either of the two following ways: i) calculation of the highly silicated components, disregarding the actual amount of si present. The si-deficit is then eliminated by the stepwise desilication of a suitable amount of the highly silicated components; ii) components poor in silica are formed at first and are later silicated, to the extent permitted by the available amount of free Q. The second alternative will be followed here:

		Recalculated to the sum of 100	
$Ru: ti$		0.7	0.2
$Z: 2 zr$		1.8	0.6
$Hl: 2 cl_2$		1.0	0.3
$Kp: 6 k' al$		30.6	9.4
$Ne: 6(1-k') al$		130.2	40.2
$Ac: 4 fm(1-mg) w$		82.4	25.4

$Ns: {}^3/_2 [2(alk - al) - fm(1 - mg)w - 2cl_2]$ 4.2   1.3
$Cs: {}^3/_2 c$ 3.2   1.0
$Fo: {}^3/_2 fm \cdot mg$ 0.5   0.1
$Fa: {}^3/_2 fm(1 - mg)(1 - w)$ 16.6   5.1
$Q: \ 157 - 104.0$ 53.0 16.4
$\quad si + 100 + al + alk + ti + zr =$ 324.2 (100.0)
$\qquad\qquad\qquad\qquad\qquad\qquad\qquad\qquad\qquad$ (324.5)

143 The consumed si may be calculated in the following way:

$$si' = 2\,al + 2\,fm(1-mg)w + {}^1/_2[2(alk-al) - fm(1-mg)w - 2cl_2]$$
$$\quad + {}^1/_2 c + {}^1/_2 fm \cdot mg + {}^1/_2 fm(1-mg)(1-w) + zr$$
$$= al + alk - {}^1/_2 c - cl_2 + zr + {}^1/_2 fm[2w(1-mg) + 1]$$
$$= 100 - {}^1/_2\{c + 2(cl_2 - zr) - fm[2w(1-mg) - 1]\} = 104.0.$$

The resulting Q, which occurs along with the poorly silicated components, should be removed as follows:

$$9.4\,Kp + 6.3\,Q = 15.7\,Or$$
$$15.2\,Ne + 10.1\,Q = 25.3\,Ab.$$

In addition. $0.3\,Hl + 2.7\,Ne = 3.0\,Sod$ should be formed, so that the standard katanorm will yield:

Or	Ab	Ne	Sod	Ac	Ns	Cs	Fo	Fa	Ru	Z	Total
15.7	25.3	22.3	3.0	25.4	1.3	1.0	0.1	5.1	0.2	0.6	100.0
(15.1)	(26.2)	(22.0)	(3.0)	(25.1)	(1.5)	(1.0)	(0.2)	(5.2)	(0.2)	(0.5)	(100.0)

The three explained and calculated examples show that it is entirely possible to calculate the standard katanorm directly from the Niggli-values by means of the given method. A comparison of the results and the values calculated via the basis (given in brackets) shows a satisfactory agreement. This agreement could be improved by consideration of additional decimal places. Some control over the calculation is provided by the fact that the sum of the equivalents, before the recalculation to the sum of 100, should always equal $si + 100 + al + alk + ti + 2p + zr$. The small deviations which were observed here result from the consideration of an insufficient number of decimal places.

The examples also show, however, that the calculations are considerably complex and tedious, especially in cases typified by the last example. This is especially true for the primary investigations intended for the determination of the calculation procedure to be adopted. Thus, in complicated cases, it is recommended that the Niggli-values be first r e s o l v e d   i n t o   t h e   c o r r e s p o n d i n g   c a t i o n   e q u i v a l e n t s , and that the standard katanorm be based on these.

The relationships between Niggli-values and cation percentages were dealt with in greater detail in section B I 1 b, p. 56. The derivation of the standard katanorm from the cation percentages may be found in section B II 6 b, p. 136. Below, the already-discussed example of the a r f v e d s o - n i t e - l u j a u r i t e   f r o m   L i l l e   E l v will be recalculated according to this procedure. The simplicity of this procedure in contrast to the direct calculation from the Niggli-values, is clearly evident. Especially distinct is the difference in the calculation of Q. This value can only be obtained from the Niggli-values as $si - si'$, where the expression si' is highly complex.

144   Arfvedsonite-lujaurite from Lille Elv, Kangerdluarsuk, Greenland

	Niggli-Values		Cation equivalents	
si	= 157		Si^{4+} = 157	
al	= 26.8		$2\,al = Al^{3+}$ = 53.6	
fm	= 32.0	$(1-mg)\,w\cdot fm = Fe^{3+}$	= 20.6	
		$(1-mg)(1-w)\,fm = Fe^{2+}$	= 11.1	
		$mg\cdot fm = Mg^{2+}$	= 0.3	
c	= 2.1		Ca^{2+} = 2.1	
alk	= 39.1	$2\,k\cdot alk = K^+$	= 10.2	
		$2(1-k)\,alk = Na^+$	= 68.0	
ti	= 0.7		Ti^{4+} = 0.7	
p	= Sp.		P^{5+} = —	
zr	= 0.9		Zr^{4+} = 0.9	
$(cl_2 =\,)$	(0.5)		(Cl^-) = (1.0)	

$si + 100 + al + alk + ti + 2\,p + zr$ 324.5

		Ru	Zr	Hl	Kp	Ne	Ac	Ns	Cs	Fo	Fa	Q
Si^{4+}	157		0.9		10.2	43.4	41.2	1.5	1.05	0.15	5.5	53.1
Al^{3+}	53.6				10.2	43.4						
Fe^{3+}	20.6						20.6					
Fe^{2+}	11.1										11.1	
Mg^{2+}	0.3									0.3		
Ca^{2+}	2.1								2.1			
K^+	10.2				10.2							
Na^+	68.0			1.0		43.4	20.6	3.0				
Ti^{4+}	0.7	0.7										
Zr^{4+}	0.9		0.9									
(Cl^-)	(1.0)			(1.0)								
	324.5	0.7	1.8	1.0	30.6	130.2	82.4	4.5	3.15	0.45	16.6	53.1

Recalculated to 100:

Q	Kp	Ne	Ac	Ns	Cs	Fo	Fa	Hl	Ru	Z	Total
16.4	9.4	40.1	25.4	1.4	1.0	0.1	5.1	0.3	0.2	0.6	100.0

$$9.4\,Kp + 6.3\,Q = 15.7\,Or$$
$$15.0\,Ne + 10.1\,Q = 25.1\,Ab$$
$$0.3\,Hl + 2.7\,Ne = 3.0\,Sod.$$

Standard katanorm:

Or	Ab	Ne	Sod	Ac	Ns	Cs	Fo	Fa	Ru	Z	Total
15.7	25.1	22.4	3.0	25.4	1.4	1.0	0.1	5.1	0.2	0.6	100.0
(15.1)	(26.2)	(22.0)	(3.0)	(25.1)	(1.5)	(1.0)	(0.2)	(5.2)	(0.2)	(0.5)	(100.0)

b) Calculation of the Niggli-values from the standard katanorm

From the relationships existing between the Niggli-values, the cation percentages and the standard components, it follows that:

145 si' = Si^{4+} = $Q + \frac{3}{5}(Or + Ab) + \frac{1}{2}(Wo + En + Hy + Ac + Lc + Z)$
$+ \frac{5}{11}(Cord + Fe\text{-}Cord) + \frac{2}{5}An + \frac{3}{8}(Mg\text{-}Bi + Fe\text{-}Bi) + \frac{1}{3}(Ne + Kp$
$+ Fo + Fa + Cs + Ns + Tit)$,
al' = $\frac{1}{2}Al^{3+}$ = $\frac{1}{2}C + \frac{1}{5}An + \frac{2}{11}(Cord + Fe\text{-}Cord) + \frac{1}{6}(Kp + Ne) + \frac{1}{8}Lc$
$+ \frac{1}{10}(Or + Ab) + \frac{1}{16}(Mg\text{-}Bi + Fe\text{-}Bi)$,
fm' = $Fe^{3+} + Fe^{2+} + Mg^{2+}$
Fe^{3+} = $\frac{2}{3}Mt + \frac{1}{4}Ac$,
Fe^{2+} = $\frac{2}{3}Fa + \frac{1}{2}(Hy + Ilm) + \frac{3}{8}Fe\text{-}Bi + \frac{1}{3}(Mt + Pr) + \frac{2}{11}Fe\text{-}Cord$,
Mg^{2+} = $\frac{2}{3}Fo + \frac{1}{2}En + \frac{3}{8}Mg\text{-}Bi + \frac{2}{11}Cord$,
c' = Ca^{2+} = $\frac{3}{5}Cp + \frac{2}{3}Cs + \frac{1}{2}(Wo + Pf) + \frac{1}{3}Tit + \frac{1}{5}An$,
alk' = $\frac{1}{2}(K^+ + Na^+)$,
$\frac{1}{2}K^+$ = $\frac{1}{3}Ks + \frac{1}{6}Kp + \frac{1}{8}Lc + \frac{1}{10}Or + \frac{1}{16}(Mg\text{-}Bi + Fe\text{-}Bi)$,
$\frac{1}{2}Na^+$ = $\frac{1}{2}Hl + \frac{1}{3}Ns + \frac{1}{6}Ne + \frac{1}{8}Ac + \frac{1}{10}Ab$,
ti' = Ti^{4+} = $1\,Ru + \frac{1}{2}(Ilm + Pf) + \frac{1}{3}Tit$,
p' = $\frac{1}{2}P^{5+}$ = $\frac{1}{5}Cp$,
z' = Zr^{4+} = $\frac{1}{2}Z$.

The values al, fm, c, and alk are obtained as usual by recalculation to the sum of 100. si, ti, etc, should be derived on the same basis.

In the review provided previously, several equivalent-normative components generally considered in igneous rocks, were listed, irrespective of whether they may occur together or not. The following examples will illustrate that in practice these expressions become simplified.

Examples:

Quartz-diorite from Spanish Peak, California.

Standard katanorm:

Q	Or	Ab	An	Wo	En	Hy	Mt	Cp	Tit	Total
12.6	7.8	35.0	26.0	1.3	9.9	2.3	3.0	0.6	1.5	100.0

si' = $Q + \frac{3}{5}(Or + Ab) + \frac{1}{2}(Wo + En + Hy) + \frac{2}{5}An + \frac{1}{3}Tit$ = $12.6 + 25.7$
$+ 6.75 + 10.4 + 0.5 = 55.95$,
al' = $\frac{1}{5}An + \frac{1}{10}(Or + Ab) = 5.2 + 4.3 = 9.5$,
Fe^{3+} = $\frac{2}{3}Mt = 2.0$,
Fe^{2+} = $\frac{1}{2}Hy + \frac{1}{3}Mt = 1.15 + 1.0 = 2.15$,
Mg^{2+} = $\frac{1}{2}En = 4.95$,
fm' = $Fe^{3+} + Fe^{2+} + Mg^{2+} = 9.1$,
c' = $\frac{3}{5}Cp + \frac{1}{2}Wo + \frac{1}{3}Tit + \frac{1}{5}An = 0.35 + 0.65 + 0.5 + 5.2 = 6.7$,
$\frac{1}{2}K^+$ = $\frac{1}{10}Or = 0.78$,
$\frac{1}{2}Na^+$ = $\frac{1}{10}Ab = 3.5$,
alk' = 4.28,
ti' = $\frac{1}{3}Tit = 0.5$,
p' = $\frac{1}{5}Cp = 0.12$.

$$
\begin{array}{rlrl}
 & & \text{Recalculated} \\
 & & \text{to 100} \\
al' & = 9.5 & al & = 32.1 \quad (32.1) \\
fm' & = 9.1 & fm & = 30.8 \quad (30.9) \\
c' & = 6.7 & c & = 22.6 \quad (22.5) \\
alk' & = 4.28 & alk & = 14.5 \quad (14.5) \\
\hline
 & 29.58 & & 100.0 \quad (100.0)
\end{array}
$$

$$si = \frac{55.95 \cdot 100}{29.58} = 189 \ (190), \qquad ti = \frac{0.5 \cdot 100}{29.58} = 1.7 \ (1.5),$$

$$p = \frac{0.12 \cdot 100}{29.58} = 0.4 \ (0.4),$$

$$k = \frac{0.78}{0.78 + 3.5} = 0.18 \ (0.18), \qquad mg = \frac{4.95}{9.1} = 0.54 \ (0.54),$$

$$w = \frac{2.0}{2.0 + 2.15} = 0.48 \ (0.49).$$

Cordierite-andesite from Hoyazo, near Nijar, Spain.

Standard katanorm.

Q	Or	Ab	An	Cord	En	Hy	Mt	Ru	Cp	Total
21.2	20.5	16.5	15.5	17.0	2.0	2.9	3.5	0.6	0.3	**100.0**

$si' = Q + {}^3/_5(Or + Ab) + {}^2/_5 An + {}^1/_2 (En + Hy) + {}^5/_{11} Cord = 21.2 + 22.2$
$\qquad + 6.2 + 2.45 + 7.75 = 59.8,$
$al' = {}^1/_5 An + {}^2/_{11} Cord + {}^1/_{10}(Or + Ab) = 3.1 + 3.1 + 3.7 = 9.9.$
$\qquad Fe^{3+} = {}^2/_3 Mt = 2.3,$
$\qquad Fe^{2+} = {}^1/_2 Hy + {}^1/_3 Mt = 1.45 + 1.15 = 2.6,$
$\qquad Mg^{2+} = {}^1/_2 En + {}^2/_{11} Cord = 1.0 + 3.1 = 4.1,$
$fm' = 2.3 + 2.6 + 4.1 = 9.0,$
$c' = {}^3/_5 Cp + {}^1/_5 An = 0.2 + 3.1 = 3.3,$
$\qquad {}^1/_2 K^+ = {}^1/_{10} Or = 2.05, \qquad ti' = 1 Ru = 0.6,$
$\qquad {}^1/_2 Na^+ = {}^1/_{10} Ab = 1.65, \qquad p' = {}^1/_5 Cp = 0.06,$
$alk' = 2.05 + 1.65 = 3.7$

$$
\begin{array}{rlrl}
 & & \text{Recalculated} \\
 & & \text{to 100} \\
al' & = 9.9 & al & = 38.2 \quad (38.2) \\
fm' & = 9.0 & fm & = 34.8 \quad (34.9) \\
c' & = 3.3 & c & = 12.7 \quad (12.7) \\
alk' & = 3.7 & alk & = 14.3 \quad (14.2) \\
\hline
 & 25.9 & & 100.0 \quad (100.0)
\end{array}
$$

$si = 231 \ (229), \qquad ti = 2.3 \ (2.3), \qquad p = 0.2 \ (0.2),$
$k = 0.55 \ (0.55), \qquad mg = 0.46 \ (0.45), \qquad w = 0.47 \ (0.48).$

Arfvedsonite-lujaurite, Lille Elv, Kangerdluarsuk, Greenland.

Standard katanorm:

Or	Ab	Ne	Hl	Ac	Ns	Fo	Fa	Cs	Ru	Z	Total
15.1	26.2	24.7	0.3	25.1	1.5	0.2	5.2	1.0	0.2	0.5	100.0

$si' = {}^3/_5(Or + Ab) + {}^1/_2(Ac + Z) + {}^1/_3(Ne + Fo + Fa + Cs + Ns)$
$= 24.8 + 12.8 + 10.9 = 48.5$,
$al' = {}^1/_6 Ne + {}^1/_{10}(Or + Ab) = 4.1 + 4.1 = 8.2$,
$\quad Fe^{3+} = {}^1/_4 Ac = 6.3$,
$\quad Fe^{2+} = {}^2/_3 Fa = 3.45$,
$\quad Mg^{2+} = {}^2/_3 Fo = 0.15$,
$fm' = 6.3 + 3.45 + 0.15 = 9.9$,
$c' = {}^2/_3 Cs = 0.65$,
$\quad {}^1/_2 K^+ = {}^1/_{10} Or = 1.5$,
$\quad {}^1/_2 Na^+ = {}^1/_2 Hl + {}^1/_3 Ns + {}^1/_6 Ne + {}^1/_8 Ac + {}^1/_{10} Ab$
$\quad = 0.15 + 0.5 + 4.1 + 3.15 + 2.6 = 10.5$,
$alk' = 1.5 + 10.5 = 12.0$,
$ti' = 1\, Ru = 0.2$,
$z' = {}^1/_2 Z = 0.25$,

Recalculated to 100

$al' = 8.2$	$al = 26.7$	(26.8)
$fm' = 9.9$	$fm = 32.2$	(32.0)
$c' = 0.65$	$c = 2.1$	(2.1)
$alk' = 12.0$	$alk = 39.0$	(39.1)
30.75	100.0	(100.0)

$si = 158\ (157)$, $\quad k = 0.125\ (0.13)$,
$ti = 0.7\ (0.7)$, $\quad mg = 0.01\ (0.01)$,
$z = 0.8\ (0.9)$, $\quad w = 0.65\ (0.65)$.

8. The Application of the Equivalent Norm to the Study of Heteromorphic Relations

In the discussions found in the literature, concerning the relationships between the mineral composition and the gross chemism of igneous rocks, the fact that a certain gross-chemical composition may find expression in different mineral combinations, is often somewhat neglected. For example, much significance is often attributed to the presence or absence of nepheline disregarding the fact that both cases may exist for one and the same chemical composition. In addition, quartz-bearing igneous rocks are commonly designated as "oversaturated". Such rocks are contrasted with quartz-free types, and no attention is paid to the fact that both of these mineral compositions may possess the same chemism. The main reason that these phenomena, the importance of which was especially recognized by A. Lacroix (1920) and designated by him as heteromorphism, were not given due attention, was that until now no suitable means for their study were available. In principle, it was possible to recalculate a given chemism of a rock to different mineral compositions. In practice, however, it was very difficult, since it was impossible to pass from one mineral composition to another. Niggli's equivalent norm proved to be a suitable tool, as will be shown hereafter by a few selected examples. At the same time opportunity will be taken of clarifying certain points in the calculation of the equivalent norm, which remained unconsidered as yet.

a) The vaugnerite from Vaugneray (Dép. du Rhône, France)

This rock was already mentioned and designated as such in 1861 by Fournet. In several aspects it resembles a durbachite, and comprises vein-like deposits, 400 m in thickness, in the gneisses of the French Massif Central. Outcrops of this rock are traced over a distance of about 4 km (A. Michel-Levy and A. Lacroix, 1887; A. Lacroix, 1917; Q. A. Palm, 1954; compare also with A. Johannsen, II, 1932, 405). This rock's fame mainly arises from the various difficulties encountered in its classification. The modus, which was measured by W. E. Tröger (1935, 57-58), is considerably different from that determined by A. Johannsen. However, both fall, with regard to the ratio orthoclase/plagioclase, within the variation range determined by Q. A. Palm (1954) for 8 samples:

	W. E. Tröger	A. Johannsen	Q. A. Palm
Biotite	34	21	14-43
Hornblende	26	42	6-56
Plagioclase	22 ⎫ 30	30 ⎫ 31	20-55 ⎫ 26-66
Orthoclase	8 ⎭	1 ⎭	6-11 ⎭
Quartz	8	5	1-19
Apatite, ore, titanite, zircon	2	1	3-6

Of the two published analyses (Analysts: Pisani and Raoult), only the more recent one of Raoult was here considered (in Lacroix, 1917). The following Niggli-values were calculated from this analysis:

si	114	k	0.67	
al	21.9	mg	0.74	
fm	51.3	ti	3.6	Magma: lamprosommaitic
c	16.4	p	0.1	
alk	10.4			

According to P. Niggli's (1931) quantiative mineralogical classification of igneous rocks, it is either considered as a biotite-hornblende-melagranodiorite or a granogabbro, depending on whether the average composition attributed to the zoned plagioclase, determined as ranging over An_{60-25}, is higher or lower than $50\% An$. According to W. E. Tröger (1935, 57-58), this is a mesotype granodiorite.

Calculation of the basis from the analysis:

	wt.%	Atomic equivalent number	Fr	Ru	Cp	Kp	Ne	Cal	Cs	Fs	Fo	Fa	Q
SiO_2	49.50	824				100	50		9	7	136.5	42	479.5
Al_2O_3	16.10	316				100	50	166					
Fe_2O_3	1.14	14								14			
FeO	6.04	84											84
MgO	11.01	273									273		
CaO	6.70	119	15		3			83	18				
Na_2O	1.55	50					50						
K_2O	4.71	100				100							
TiO_2	2.10	26		26									
P_2O_5	0.11	2			2								
H_2O+	0.89												
H_2O-	0.05												
F	0.57	(30)	(30)										
	100.47	1808	15	26	5	300	150	249	27	21	409.5	126	479.5

Basis:
Q	Kp	Ne	Cal	Cs	Fs	Fo	Fa	Cp	Ru	Fr	Total
26.5	16.6	8.3	13.8	1.5	1.2	22.6	7.0	0.3	1.4	0.8	100.0

Calculation of the standard katanorm:

$$1.2\,Fs + 0.6\,Fa = 1.2\,Mt + 0.6\,Q$$
$$13.8\,Cal + 9.2\,Q = 23.0\,An$$
$$16.6\,Kp + 11.1\,Q = 27.7\,Or$$
$$8.3\,Ne + 5.5\,Q = 13.8\,Ab$$
$$1.5\,Cs + 0.5\,Q = 2.0\,Wo$$
$$6.4\,Fa + 2.1\,Q = 8.5\,Hy$$
$$22.6\,Fo + 7.5\,Q = 30.1\,En$$

Q-consumption: $9.2 + 11.1 + 5.5 + 0.5 + 2.1 + 7.5 = 35.9\,Q$

From Mt-formation	$0.6\,Q$
Total consumption	$35.3\,Q$
Available	$26.5\,Q$
Deficit	$8.8\,Q$

150 Therefore, a part of the ($En + Hy$) should be desilicated for ($Fo + Fa$). Before doing this, the $2.0\,(En + Hy)$, corresponding to the $2.0\,Wo$ intended for the formation of diopside, should be set aside. If the ratio Mg/Fe in the diopside is taken as being equal to that in the total available amount, as is commonly done, the resulting composition of the diopside is given by $Wo\;2.0\;En\;1.6\;Hy\;0.4$. An amount of $28.5\,En + 8.1\,Hy = 36.6$ ortho-augite will remain.

In order to free $8.8\,Q$, the equation $3\,(Fo + Fa) + 1\,Q = 4\,(En + Hy)$ is utilized. For this purpose a 4-fold amount, i. e. $35.2\,(En + Hy)$, is required. This corresponds, in accordance with the original Mg/Fe ratio, to $27.4\,En$ and $7.8\,Hy$. Through desilication the following is obtained:

$$27.4\,En = 20.6\,Fo + 6.8\,Q$$
$$7.8\,Hy = 5.8\,Fa + 2.0\,Q.$$

The standard katanorm will yield:

Or	Ab	An	Wo	En	Hy	En	Hy	Fo	Fa	Mt	Cp	Ru	Fr		
27.7	13.8	23.0	2.0	1.6	0.4	1.1	0.3	20.6	5.8	1.2	0.3	1.4	0.8	Total	100.0

36.8 *Plag*   4.0 *Di*   1.4 *Orthaug* 26.4 *Ol*   3.7 *Accessories*

64.5 Feldspar.

In contrast to the modus, this standard katanorm not only does not show any quartz, but indicates the presence of 26.4% olivine, i.e. a quarter of its total composition. The mineral composition corresponds to an **olivine monzonite**, because the ratio of orthoclase: plagioclase lies within the range of 5:3 to 3:5. This olivine-monzonite thus represents a **heteromorphic facies of the vaugnerite**.

The high *Fr*-content of 0.8% is noteworthy. The formation of apatite from the 0.3 *Cp* will only require 0.02 *Fr*, according to the equation $15\,Cp + 1\,Fr = 16\,Ap$. Since, however, the presence of **fluorite** was not stated, it may be assumed that the rest of the fluorine is incorporated in biotite in place of OH⁻. It thus seems reasonable to eliminate all the fluorite, and to calculate a new standard katanorm, disregarding the fluorine. Since $1\,Fr + 1\,Q = 2\,Wo$ or $0.8\,Fr + 0.8\,Q = 1.6\,Wo$, the *Wo*-content will accordingly be raised to 3.6. Thus, a higher *Di*-content of 7.2 will be obtained, which will call for the desilication of some *Ab* to *Ne*, in order to balance the *Q*-deficit. Finally, the new *Fr*-free standard katanorm will give:

Or	Ab	An	Ne	Wo	En	Hy	Fo	Fa	Mt	Ru	Cp		
27.7	11.5	23.0	1.4	3.6	2.8	0.8	20.5	5.8	1.2	1.4	0.3	Total	100.0

34.5 *Plag*   7.2 *Di*   26.3 *Ol*   2.9 *Accessories*

62.2 Feldspar.

This facies should have been designated as a **nepheline-bearing olivine-monzonite**. It is extremely interesting to note that the vaugnerite, containing 8% of modal quartz, is heteromorphic with a nepheline-bearing (even if in only a small amount) rock.

The nepheline may be removed in part, if, according to the following equations, the modally-present titanite is formed instead of ***Ru***:

$$1\,Ru + 2\,Wo = 3\,Tit$$
$$1.4\,Ru + 2.8\,Wo = 4.2\,Tit$$

The *Di* is accordingly reduced to 1.6. According to the original Mg/Fe ratio its composition is given by 0.8 *Wo*; 0.6 *En*; 0.2 *Hy*. By desilication of the remaining (*En* + *Hy*) according to:

$$2.8 - 0.6 = 2.2\,En = 1.7\,Fo + 0.5\,Q,$$
$$0.8 - 0.2 = 0.6\,Hy = 0.45\,Fa + 0.15\,Q,$$

0.65 *Q* are set free for the silication of a part of the nepheline, according to:

$$1.0\,Ne + 0.65\,Q = 1.65\,Ab.$$

Thus, the **titanite variant** will be given by:

Or	Ab	An	Ne	Wo	En	Hy	Fo	Fa	Mt	Tit	Cp	
27.7	13.15	23.0	0.4	0.8	0.6	0.2	22.2	6.25	1.2	4.2	0.3	Total 100.0

36.15 Plag   1.6 Di   28.45 Ol   5.7 Accessories

63.85 Feldspar

This *Tit*-variant, as well as other variants which were discussed previously, correspond to a very specific distribution of SiO_2, conventionally determined, between the light and the dark basis components. However, *Kp* could have been considered to be silicated only to the *Lc*-stage, and the amount of SiO_2 thus liberated could have been used for the silication of a part of the olivine to the ortho-augite stage. This would bring into consideration the well-known heteromorphism:

$$\text{Orthoclase + Olivine} \rightleftarrows \text{Leucite + Ortho-augite}$$

The reaction equations required for the construction of such a **leucite variant** are obtained as follows:

$$6\,SiO_2 \cdot Al_2O_3 \cdot K_2O = 4\,SiO_2 \cdot Al_2O_3 \cdot K_2O + 2\,SiO_2 \quad (1)$$
$$10\,Or \qquad\qquad 8\,Lc \qquad\qquad 2\,Q$$
$$\text{or} \quad 5\,Or \qquad\qquad 4\,Lc \qquad\qquad 1\,Q$$

$$SiO_2 \cdot 2(Mg, Fe)O + SiO_2 = 2(SiO_2 \cdot (Mg, Fe)O), \quad (2)$$
$$3(Fo + Fa) \qquad 1\,Q \qquad 4(En + Hy).$$

By addition of (1) and (2) it follows that

$$5\,Or + 3(Fo + Fa) = 4\,Lc + 4(En + Hy),$$
$$27.7\,Or + 16.7(Fo + Fa) = 22.2\,Lc + 22.2(En + Hy).$$

The nepheline may be removed as follows:

$$3\,Ne + 2\,Q = 5\,Ab \quad (3)$$
$$4(En + Hy) = 3(Fo + Fa) + 1\,Q. \quad (4)$$

Doubling (4) and adding it to (3) yields:

$$3\,Ne + 8(En + Hy) = 5\,Ab + 6(Fo + Fa),$$
$$0.4\,Ne + 1.1(En + Hy) = 0.7\,Ab + 0.8(Fo + Fa).$$

The desired **leucite variant** is given as:

Ab	An	Lc	Wo	En	Hy	(En+Hy)	(Fo+Fa)	Mt	Tit	Cp	Total
13.85	23.0	22.2	0.8	0.6	0.2	21.1	12.55	1.2	4.2	0.3	100.0

It should, however, be stated that such ortho-augite-rich **leucite-basanites** are not known, and that in such rock types it is the Al-bearing augites which are common. For this purpose an amount of the *An* may be converted to *Ts*, according to:

$$2\,SiO_2 \cdot Al_2O_3 \cdot CaO = SiO_2 \cdot Al_2O_3 \cdot CaO + SiO_2$$
$$An \qquad\qquad 4\,Ts \qquad + 1\,Q,$$

and the SiO_2 which is set free may be used for the silication of *Lc*.

The calculation of a **biotite variant**:

Since no olivine and only 8% orthoclase is included in the modus of the vaugnerite, but 34% biotite is present, the calculation of a biotite variant is desirable. From the equation:

$$6\,Kp + 9(Fo + Fa) + 1\,Q = 16\,Bi$$

by addition of $3\,Q$ to both sides of the equation, one obtains:

$$10\,Or + 9\,Fo = 16\,Mg\text{-}Bi + 3\,Q,$$
$$10\,Or + 9\,Fa = 16\,Fe\text{-}Bi + 3\,Q.$$

Accordingly, starting from the available 27.7 Or, the following may be formed:

$$6.95\,Or + 6.25\,Fa = 11.1\,Fe\text{-}Bi + 2.1\,Q,$$
$$20.7\,Or + 18.6\,Fo = 33.1\,Mg\text{-}Bi + 6.2\,Q.$$

The Q which was set free is used for the silication of the Ne and the rest of the Fo:

$$0.4\,Ne + 0.25\,Q = 0.65\,Ab,$$

and starting with $22.2 - 18.6 = 3.6 = 3.6\,Fo$,

$$3.6\,Fo + 1.2\,Q = 4.8\,En,$$

so that the b i o t i t e - v a r i a n t is represented by:

Q	Ab	An	Bi	Wo	En	Hy	En	Mt	Tit	Cp	Total
6.9	13.8	23.0	44.2	0.8	0.6	0.2	4.8	1.2	4.2	0.3	100.0

36.8 Plag   1.6 Di   5.7 Accessories

Without the formation of titanite, because $3\,Tit = 2\,Wo + 1\,Ru$ or $4.2\,Tit = 2.8\,Wo + 1.4\,Ru$, the following variant will be obtained:

Q	Ab	An	Bi	Wo	En	Hy	En	Mt	Ru	Cp	Total
6.9	13.8	23.0	44.2	3.6	3.4	0.2	2.0	1.2	1.4	0.3	100.0

36.8 Plag   7.2 Di   2.9 Accessories

A rock with such a composition should be designated as b i o t i t e - q u a r t z g a b b r o, because $An > Ab$ and Q amounts to more than 1/8 of the light constituents. It represents an additional heteromorphic facies of the vaugnerite.

Calculation of a hornblende variant

Since 26% hornblende is included in the modus of the vaugnerite, a h o r n b l e n d e v a r i a n t should also be calculated. Because, according to experience, the hornblendes of gabbroidal rocks are sesquioxide-bearing, no usage should be made of an idealized composition. The calculation should instead be based directly on the weight-percentage composition of an average hornblende of a gabbroidal to noritic rock.

A good knowledge of the chemism of such c o m p l e x h o r n b l e n d e s and the resulting possibilities of heteromorphism may be obtained from the schematic calculation of a standard katanorm, in a manner similar to that employed for the igneous rocks. In this way, the mineral combination which may substitute for the hornblende, may be seen at once.

A v e r a g e h o r n b l e n d e f r o m g a b b r o t o n o r i t e (Tröger, 1931, 344, No. 19)

Calculation of the basis:

	Wt. %	Atomic equivalent numbers	Ru	Kp	Ne	Cal	Cs	Fs	Fo	Fa	Q
SiO_2	44.88	747		8	47		69	30	160	71.5	361.5
Al_2O_3	10.83	212		8	47	157					
Fe_2O_3	4.85	60						60			
FeO	10.28	143								143	
MgO	12.87	320							320		
CaO	12.18	217					78.5	138.5			
Na_2O	1.44	47				47					
K_2O	0.39	8		8							
TiO_2	0.80	10	10								
H_2O	1.49										
	100.01	1764	10	24	141	235.5	207.5	90	480	214.5	361.5

Basis:

Q	Kp	Ne	Cal	Cs	Fs	Fo	Fa	Ru	Total
20.4	1.4	8.0	13.3	11.8	5.1	27.2	12.2	0.6	100.0

The following katanorm is calculated therefrom:

Or	Ab	An	Wo	En	Hy	Fo	Fa	Mt	Ru	Total
2.3	13.3	22.2	15.7	11.6	4.1	18.5	6.6	5.1	0.6	100.0

35.5 Plag   31.4 Di   25.1 Ol   5.7 Accessories

A rock possessing such a composition should be defined as an olivine gabbro, or even as a mafitite-olivine gabbro, because the fractional content of the dark components ranges between 5/8 and 6/8.

The calculation of the hornblende variant of the vaugnerite may be started from the basis. The hornblende is the only Cs-bearing constituent, as no modal pyroxene is present. Thus, the Cs-content of the basis may be taken as a measure of the amount of the hornblende, i. e. hornblende is formed until all Cs is consumed. Since the Cs-content of the basis of the vaugnerite is 2.7, whereas that of the hornblende is 11.8, the amount of the hornblende can be calculated from the ratio $11.8 : 100 = 2.7\ x$, i. e., $x = 22.9\%$. This amount should be subtracted from the basis of the rock, and the rest should be calculated in accordance with the usual regulations. The calculation should be started from the Fr-free basis, which may be derived from the previously given one as follows, taking into consideration that $0.8\ Fr + 0.4\ Q = 1.2\ Cs$:

Q	Kp	Ne	Cal	Cs	Fs	Fo	Fa	Ru	Cp	Total
26.1	16.6	8.3	13.8	2.7	1.2	22.6	7.0	1.4	0.3	100.0

Calculation of a hornblende variant of the vaugnerite from the basis:

	Basis of vaugnerite	Basis of hornblende	22.9 % hornblende	Remainder
Q	26.1	20.4	4.7	21.4
Kp	16.6	1.4	0.3	16.3
Ne	8.3	8.0	1.8	6.5
Cal	13.8	13.3	3.0	10.8
Cs	2.7	11.8	2.7	—
Fs	1.2	5.1	1.2	—
Fo	22.6	27.2	6.2	16.4
Fa	7.0	12.2	2.8	4.2
Ru	1.4	0.6	0.1	1.3
Cp	0.3	—	—	0.3
	100.0	100.0	22.8	77.2

155 Calculation of the remainder:

$$16.3\ Kp + 10.9\ Q = 27.2\ Or$$
$$6.5\ Ne + 4.3\ Q = 10.8\ Ab$$
$$10.8\ Cal + 7.2\ Q = 18.0\ An$$

Consumed 22.4 Q
Available 21.4 Q

Q-deficit 1.0 Q, balanced by conversion of *Ab* to *Ne*:
$$2.5\ Ab = 1.5\ Ne + 1.0\ Q.$$

The hornblende variant thus gives:

Or	Ab	An	Ne	Hbl	Fo	Fa	Ru	Cp	Total
27.2	8.3	18.0	1.5	22.8	16.4	4.2	1.3	0.3	100.0

26.3 Plag — 20.6 Ol — 1.6 Accessories

53.5 Feldspar

This new heteromorphic facies should be designated as a nepheline-bearing hornblende-olivine monzonite.

A great advantage of Niggli's method is that it presents the possibility of introducing into the calculation analyzed examples of rock-forming minerals along with idealized components. The same example will also be used to show how the calculation is made with the help of the cation percentages. At first an expression must be found for the amount of Ca' which is not bound to Al, i.e. available for hornblende formation. This is calculated as follows:

$$Ca' = Ca - \tfrac{1}{2}[Al - (Na + K)] - \tfrac{3}{2}P = Ca - \tfrac{1}{2}[(Al + 3P) - (Na + K)].$$

By this procedure, $Ca'_V = 1.85$ is obtained for the vaugnerite, and $Ca'_H = 7.9$ for the hornblende. Should x be the percentage-content of hornblende, assuming that the total of Ca' of the vaugnerite is introduced therein, then $100:7.9 = x:1.85$, i.e., $x = 23.4\%$ hornblende.

The following is thus obtained by calculation:

	Vaugnerite cations %	Hornblende cations %	23.4 % hornblende	Remaining cations %
Si^{4+}	45.6	42.3	9.9	35.7
Al^{3+}	17.5	12.0	2.8	14.7
Fe^{3+}	0.8	3.4	0.8	—
Fe^{2+}	4.6	8.1	1.9	2.7
Mg^{2+}	15.1	18.1	4.2	10.9
Ca^{2+}	6.6	12.3	2.9	3.7
Na^{+}	2.8	2.7	0.6	2.2
K^{+}	5.5	0.5	0.1	5.4
Ti^{4+}	1.4	0.6	0.1	1.3
P^{5+}	0.1	—	—	0.1
	100.0	100.0	23.3	76.7

The remainder may be calculated in the following manner:

	Remaining cations %	Cp	Ru	Or	Ab	An	C	Fo	Fa	Q
Si^{4+}	35.7			16.2	6.6	7.0		5.4	1.3	−0.8
Al^{3+}	14.7			5.4	2.2	7.0	0.1			
Fe^{2+}	2.7								2.7	
Mg^{2+}	10.9							10.9		
Ca^{2+}	3.7	0.2				3.5				
Na^{+}	2.2				2.2					
K^{+}	5.4			5.4						
Ti^{4+}	1.3		1.3							
P^{5+}	0.1	0.1								
	76.7	0.3	1.3	27.0	11.0	17.5	0.1	16.3	4.0	−0.8

If the small Q-deficit is eliminated, according to $2.0\,Ab = 1.2\,Ne + 0.8\,Q$, then the hornblende variant calculated from the cations percentages (a small Al-excess is calculated as $0.1\,C$) will yield:

Or	Ab	An	Ne	Ho	Fo	Fa	Ru	Cp	C	Total
27.0	9.0	17.5	1.2	23.3	16.3	4.0	1.3	0.3	0.1	100.0
(27.2)	(8.2)	(18.0)	(1.6)	(22.8)	(16.4)	(4.2)	(1.3)	(0.3)	(—)	(100.0)

Since only one decimal was considered, the agreement with the results which were obtained through the basis is satisfactory.

Calculation of a hornblende-biotite variant

If $Or + Ol$ is substituted by $Bi + Q$ according to:

$$10\,Or + 9\,(Fo + Fa) = 16\,Bi + 3\,Q,$$

and the available $20.6\,Ol$ are taken as a basis for the calculation, then we obtain:

$$22.9\,Or + 20.6\,(Fo + Fa) = 36.6\,Bi + 6.9\,Q.$$

The Q which has been set free should be used for the silication of the nepheline:

$$1.6\,Ne + 1.0\,Q = 2.6\,Ab,$$

and the sought variant will be given by:

Q	Or	Ab	An	Bi	Ho	Ru	Cp	Total
5.9	4.3	10.8	18.0	36.6	22.8	1.3	0.3	100.0

28.8 Plag (under Ab, An)

33.1 Feldspar (under Or, Ab, An)

157 This variant should be designated as a biotite-hornblende grasno-gabbro, because $Q > 1/8$ of the light constituents. This variant agrees well with Tröger's modus of the vaugnerite, as is shown by the following comparison:

	Bi	Ho	Plag	Or	Q	Accessories
Calculated biotite-hornblende variant:	36.6	22.8	28.8	4.3	5.9	1.6
Modus:	34	26	22	8	8	2

b) Selected examples of the heteromorphic possibilities of alkali-gabbroidal magmas

a) Mafraite, luscladite, and berondrite. Among the holocrystalline rocks which are connected with the post-Turonian volcanic eruptions in the area north of Lisbon, the mafraites, described first by P. Choffat (1916) and later by A. Lacroix (1920), are of special interest.

The analysis of the mafraite from Tifao de Mafra (analyzed by Raoult) yields:

SiO_2	Al_2O_3	Fe_2O_3	FeO	MgO	CaO	Na_2O	K_2O	TiO_2	P_2O_5	H_2O+	H_2O-	Total
45.72	15.62	3.15	9.51	5.31	10.92	4.13	1.75	3.01	0.46	0.55	0.11	100.24

From this analysis the following Niggli-values are calculated:

si	al	fm	c	alk	k	mg	ti	p	Magma
103	20.8	41.0	26.5	11.7	0.22	0.44	5.2	0.4	theralite-gabbroidal

Modus, according to Tröger: 44 basaltic hornblende
　　　　　　　　　　　　　　32 plagioclase $An_{53}Ab_{43}Or_4$.
　　　　　　　　　　　　　　10 pyroxene
　　　　　　　　　　　　　　 8 Na-sanidine (rims around plagioclase)
　　　　　　　　　　　　　　 5 Ore
　　　　　　　　　　　　　　 1 apatite

This rock may be defined as a foid-free amphibole essexite or theralite. The following basis was calculated:

Q	Kp	Ne	Cal	Cs	Fs	Fa	Fo	Cp	Ru	Total
21.4	6.1	22.6	11.5	10.0	3.2	11.1	11.2	0.8	2.1	100.0

The standard katanorm is obtained in the following manner:

$$6.1\ Kp + 4.0\ Q = 10.1\ Or$$
$$22.6\ Ne + 15.1\ Q = 37.7\ Ab$$
$$11.5\ Cal + 7.7\ Q = 19.2\ An$$
$$10.0\ Cs + 3.3\ Q = 13.3\ Wo$$
$$3.2\ Fs + 1.6\ Fa = 3.2\ Mt + 1.6\ Q.$$

13.3 $(En + Hy)$ are required for the formation of diopside. These are obtained as follows:

$$\left.\begin{array}{l}5.4\ Fo + 1.8\ Q = 7.2\ En \\ 4.6\ Fa + 1.5\ Q = 6.1\ Hy\end{array}\right\} 13.3\ (En + Hy).$$

Calculation of the Q-balance:

Consumed for	Or	4.0
,, ,,	Ab	15.1
,, ,,	An	7.7
,, ,,	Wo	3.3
,, ,,	En	1.8
,, ,,	Hy	1.5
		33.4
Gained from Mt		1.6
Total consumption		31.8
Available		21.4
Deficit		10.4 Q

The Q-deficit is eleminated by the desilication of Ab:

$$26.0\ Ab = 15.6\ Ne + 10.4\ Q.$$

Thus, the standard katanorm yields:

Or	Ab	An	Ne	Wo	En	Hy	Fo	Fa	Mt	Ru	Cp	Total
10.1	11.7	19.2	15.6	13.3	7.2	6.1	5.8	4.9	3.2	2.1	0.8	100.0

30,9 Plag   26,6 Di   10,7 Ol   6,1 Accessories

This mineral combination does not match the modus of the **mafraite**. Instead, it corresponds to the modus of an **olivine theralite gabbro** of the **luscladite** type (named after the Ravin de Lusclade, Mt. Dore), as known from the same area, north of Lisbon. According to Tröger, a typical **luscladite** from Ravin de Lusclade has the following composition:

> 39 Ti-augite
> 35 plagioclase An_{70} with marginal orthoclase
> 8 olivine
> 7 ore
> 5 biotite
> 5 nepheline
> 1 apatite.

However, in contrast to the previously calculated standard katanorm of the mafraite, stands the fact that feldspar < augite. This may be due to the fact that the augite is not diopside, but is an Al- and alkali-bearing Ti-augite.

In the following an attempt will be made to obtain a better agreement with the modus of the luscladite, by calculation of a Ti-augite— biotite variant.

The following composition may be taken to represent the average chemism of the Ti-augite (W. E. Tröger, 1935, 343, No. 12):

SiO_2	Al_2O_3	Fe_2O_3	FeO	MgO	CaO	Na_2O	K_2O	TiO_2	H_2O	Total
49.28	6.35	8.11	5.06	10.39	17.39	2.80	0.23	1.17	0.02	100.80

From this the following Niggli-values are calculated:

si	al	fm	c	alk	k	mg
97	7.3	50.7	36.5	5.5	0.04	0.60

As might be expected, these values do not correspond to any known magma type, but they do approach, however, calc-rich pyroxenitic to polzenitic magma types. The calculation of the standard katanorm provides an instructive insight into the mineral combination which might have substituted an augite of such a composition. This is obtained as follows:

Basis:	Q	Kp	Ne	Cal	Cs	Fs	Fa	Fo	Ru	Total
	20.2	0.7	15.0	2.5	24.7	8.5	5.9	21.7	0.8	100.0

From this the standard katanorm is obtained:

Or	Ab	An	Ne	Wo	En	Hy	Mt	Hm	Ru	Total
1.1	14.0	4.1	6.6	32.9	28.9	4.0	5.8	1.8	0.8	100.0

Since care was taken throughout the calculation to maintain the relation $Wo = En + Hy$ in diopside, insufficient Fa remained available for converting all of the Fs to Mt. Thus, a part of the Fs must be resolved into $Hm + Q$.

In contrast to the pure diopside, which up to now has been used in the calculations, the Ti-augite under consideration contains 19.2% of potential feldspar and 6.6% of potential nepheline.

Taking the basis of the mafraite as a reference point, a new mineral composition should now be calculated, using this Ti-augite. Primarily it should be assumed that the total amount of Cs was introduced into the Ti-augite. Since the mafraite basis contains 10.0 Cs, and that of the Ti-augite contains 24.7 Cs, the augite amount may be obtained from the ratio 24.7 : 10.0 = 100 : x where $x = 40.5$ Ti-augite.

	Mafraite (basis)	Ti-augite (basis)	40.5% Ti-augite (basis)	Remainder (basis)
Q	21.4	20.2	8.2	13.2
Kp	6.1	0.7	0.3	5.8
Ne	22.6	15.0	6.1	16.5
Cal	11.5	2.5	1.0	10.5
Cs	10.0	24.7	10.0	—
Fs	3.2	8.5	3.4	−0.2
Fa	11.1	5.9	2.4	8.7
Fo	11.2	21.7	8.8	2.4
Cp	0.8	—	—	0.8
Ru	2.1	0.8	0.3	1.8
	100.0	100.0	40.5	59.5

During the subtraction of the amount of augite, as dictated by the Cs-content, from the mafraite basis, a very small deficit of $-0.2\,Fs$ was obtained. This can easily be eliminated by a correspondingly small change in the oxidation degree of the Fe. However, taking into account the extremely small size of the deficit, it may be neglected. If the available $13.2\,Q$ are used for the silication of Cal, Kp and, if sufficient, of Ne according to:

$$10.5\,Cal + 7.0\,Q = 17.5\,An$$
$$5.8\,Kp + 3.9\,Q = 9.7\,Or$$
$$3.5\,Ne + 2.3\,Q = 5.8\,Ab,$$

then the following norm will be obtained as an additional heteromorphic facies of the mafraite:

Or	Ab	An	Ne	Ti-Aug	Fo	Fa	Cp	Ru	Total
9.7	5.8	17.5	13.0	40.5	2.4	8.7	0.8	1.8	100.2
								Fs	−0.2
									100.0

A rock with such a composition should be designated as an **olivine-nepheline gabbro** or as an **olivine theralite**.

In order to approximate the modus of the **luscladite**, which contains biotite in addition to orthoclase, the orthoclase should be eliminated as follows:

$$9\,(Fo + Fa) + 10\,Or = 16\,Bi + 3\,Q$$

If $5.0\,Bi$ are formed, in agreement with the modus of the luscladite:

$$2.9\,(Fo + Fa) + 3.1\,Or = 5.0\,Bi + 1.0\,Q,$$

and the gained Q amount is utilized for the silication of an equivalent amount of Ne:

$$1.5\,Ne + 1.0\,Q = 2.5\,Ab,$$

then the following **Ti-augite — biotite variant of the mafraite** is obtained, which is close to the modus of the luscladite:

Or	Ab	An	Ne	Ti-Aug	Ol	Bi	Ru	Cp	Total
6.6	8.3	17.5	11.5	40.5	8.2	5.0	1.8	0.8	100.2
								Fs	−0.2
									100.0

The fact that this norm does not contain any ore, in contrast to the modus of the luscladite, can be explained by the fact that the oxidation degree of the Fe was taken from the mafraite analysis. Such deviations may be corrected by changing the oxidation degree, bearing in mind in general that $2\,FeO \cdot SiO_2 = 3\,Fa$ are equivalent to $Fe_2O_3 \cdot SiO_2 = 3\,Fs$. For conversion to magnetite only half the amount of Fa is required [as compared to the available trivalent molecule], liberating an equal amount of Q. The formation of Mt from Fa by the partial oxidation of the Fe may thus be represented by the equation:

$$3\,(2\,FeO \cdot SiO_2) = 2\,(Fe_2O_3 \cdot FeO) + 3\,SiO_2$$
$$9\,Fa \qquad\qquad 6\,Mt \qquad\qquad 3\,Q$$
$$\text{or}\quad 3\,Fa \qquad\qquad 2\,Mt \qquad\qquad 1\,Q$$

In the special case considered herewith, $0.2\,Fs$ must be subtracted prior to the conversion of Fa to Mt. If $4.0\,Fa$ are taken, arbitrarily, as the basis for the calculation, they only correspond to $4.0 - 0.2 = 3.8\,Fs$, and the following will hold true:

$$3.8\,Fs + 1.9\,Fa = 3.8\,Mt + 1.9\,Q$$
$$\text{as well as } 2.8\,Ne + 1.9\,Q = 4.7\,Ab,$$

and the following **ore-bearing Ti-augite—biotite variant** results:

	Or	Ab	An	Ne	Ti-Aug	Bi	Ol	Mt	Ru	Cp	Total
	6.6	13.0	17.5	8.7	40.5	5.0	2.3	3.8	1.8	0.8	100.0
		37.1 Feldspar						5.6			
Modus of the luscladite	35 Plag. with 5			39		5	8	7 ore		1	
from Ravin de Lusclade	Or-rims										

This is quite close to the modus of the **luscladite** from Ravin de Lusclade (Mont Dore) according to Tröger.

Attempt at calculating the modus of the mafraite

From the calculations which were carried out till now, and the comparison between mafraite and luscladite, it may be clearly seen that the mafraites' hornblende must be rich in Al and alkali, thus being the more adaptable for the potential construction of Ne. Therefore, an attempt should be made at calculating the analysis of the mafraite, using such a hornblende. Such a hornblende, according to W. E. Tröger (loc. cit. 1935, 344, No. 23), may be typified by the following analysis of essexitic rocks:

SiO_2	Al_2O_3	Fe_2O_3	FeO	MgO	CaO	Na_2O	K_2O	TiO_2	H_2O	Total
38.93	13.17	3.41	10.73	11.61	12.25	3.10	1.24	4.79	0.35	99.58

Niggli-values:

si	al	fm	c	alk	k	mg	ti
73	14.5	54	24.5	7	0.21	0.60	6.7

Similar chemical relationships are quite common in basic-sodic rocks. Due to the compositional similarity between such magmas and these hornblendes, these magmas are designated, as is well known, as **hornblenditic**.

The basis may be calculated as follows:

Q	Kp	Ne	Cal	Cs	Fs	Fa	Fo	Ru	Total
11.5	4.4	16.7	11.0	12.8	3.6	12.5	24.2	3.3	100.0

From this the following standard katanorm may be derived:

Or	Kp	Ne	An	Wo	En	Hy	Fo	Fa	Mt	Pf	Total
6.0	0.8	16.7	18.3	10.5	7.3	3.2	18.7	8.3	3.6	6.6	100.0
	17.5				Di 21.0		Ol 27.0		Access. 10.2		

Since a considerable amount of the Kp could not be silicated during the usual calculation, a **perovskite variant** was constructed. Thus, an

additional amount of SiO₂ must be gained, according to the equation
$2\,Wo + 1\,Ru = 2\,Pf + 1\,Q$, reducing the non-silicated amount of Kp to 0.8.
The calculated mineral composition corresponds to a foid-olivine
gabbro. It is thus verified that Al- and alkali-rich hornblendes, having
the exemplified composition, do indeed contain potentially poorly silicated
components, such as olivine or nepheline, in a significantly larger amount
than do Ti-augites. In the discussed example, the normative Di-content
only amounts to 21.0%, in contrast to 65.8% in the case of the Ti-augite.
It thus becomes evident that the heteromorphism which was established —
mafraite (nepheline- and olivine-free)-luscladite (nepheline- and olivine-
bearing), — may be explained by the special chemism of the hornblende
of the mafraite.

Therefore, an attempt should now be made at calculating a variant corresponding to the modus of the mafraite, using the mentioned hornblende. If an amount of 45% hornblende is taken, in conformity with the
modus, then the following is obtained:

	Mafraite (basis)	Hornblende (basis)	45% hornblende (basis)	Remainder (basis)
Q	21.4	11.5	5.2	16.2
Kp	6.1	4.4	2.0	4.1
Ne	22.6	16.7	7.5	15.1
Cal	11.5	11.0	5.0	6.5
Cs	10.0	12.8	5.8	4.2
Fs	3.2	3.6	1.6	1.6
Fa	11.1	12.5	5.5	5.6
Fo	11.2	24.2	10.9	0.3
Cp	0.8	—	—	0.8
Ru	2.1	3.3	1.5	0.6
	100.0	100.0	45.0	55.0

The remainder left after the formation of hornblende may be calculated in
the following manner:

$$4.1\,Kp + 2.7\,Q = 6.8\,Or$$
$$15.1\,Ne + 10.1\,Q = 25.2\,Ab$$
$$6.5\,Cal + 4.3\,Q = 10.8\,An$$
$$1.6\,Fs + 0.8\,Fa = 1.6\,Mt + 0.8\,Q$$
$$4.2\,Cs + 1.4\,Q = 5.6\,Wo$$
$$3.9\,Fa + 1.3\,Q = 5.2\,Hy$$
$$0.3\,Fo + 0.1\,Q = 0.4\,En.$$

According to this calculation mode, the amount of consumed Q is given by:

$$2.7 + 10.1 + 4.3 + 1.4 + 1.3 + 0.1 = 19.9\,Q$$
Gained from Mt-formation $\quad 0.8\,Q$
Total consumption $\quad 19.1\,Q$

Since only 16.2 Q are available, the deficit of 19.1 - 16.2 = 2.9 Q must be
balanced by the desilication of Ab:

$$7.2\,Ab = 4.3\,Ne + 2.9\,Q.$$

The sought hornblende variant is thus:

Or	Ab	An	Ne	Ho	Wo	En	Hy	Fa	Mt	Cp	Ru	Total
6.8	18.0	10.8	4.3	45.0	5.6	0.4	5.2	0.9	1.6	0.8	0.6	100.0

35.6 Feldspar   11.2 Di

Although this norm is close to the **mafraite**, it is not completely identical to it. The norm is, however, also similar to an additional heteromorphic facies, as yet undiscussed, of an alkali gabbroidal magma, i.e., the facies described by A. Lacroix (1920) as **berondrite** (amphibole-theralite gabbro). A typical **berondrite** from the Berondra River in the Bezavona massif, North Madagascar, has the following composition, according to W. E. Tröger (loc. cit. 1935, 229-230, No. 545):

Basaltic hornblende	44 %
Titan augite	18 %
Plagioclase An_{70} with orthoclase rims	25 %
Nepheline	7 %
Ore	6 %

Compared with the calculated hornblende variant, this composition is somewhat richer in augite and nepheline, and contains less feldspar. In addition, the plagioclase is richer in *An*, and the pyroxene is Ti-augite instead of the diopside which was considered in our calculation for the sake of simplicity.

In order to obtain a variant which will completely match the *Ne*-free mafraite composition, a hornblende, even more enriched in alumina, should be considered in the calculation. Also, Ti-augite should be taken into account rather than the ideally composed diopside. Thus, the very small amount of nepheline still available, can be eliminated. Because of the complexity of the two aforementioned components, and the absolute lack of knowledge of their actual compositions, this problem can only be solved experimentally. This, however, will not be dealt with here.

β) **The fasinite.** It was again A. Lacroix (1916 and 1922, 644-646) who described this rock, from Ambahila, Ampasindava, Northern Madagascar, and designated it as a **fasinite**. According to W. C. Brögger (1920), the rock is characterized by the following mineral composition (wt %):

Ti-augite	67 %
Nepheline	15 %
Biotite	5 %
Olivine	4 %
Na-microline	2 %
Ore, apatite	7 %

According to Tröger (loc. cit. 1935, 251, No. 611), this is a *Ne*-rich jacupirangite.

The following chemical analysis is given for this rock (Analyst: Raoult, in A. Lacroix, 1922):

SiO_2	Al_2O_3	Fe_2O_3	FeO	MgO	CaO	Na_2O	K_2O	TiO_2	P_2O_5	H_2O	Total
40.10	15.50	6.35	7.29	8.41	12.40	3.87	1.67	2.98	1.28	0.87	100.22

165 The following Niggli-values were calculated for this rock:

si	al	fm	c	alk	k	mg	ti	p	Magma
80	18.2	46.6	26.5	8.7	0.25	0.54	4.6	1.1	gabbrotheralitic

The basis is:

Q	Kp	Ne	Cal	Cs	Fs	Fa	Fo	Ru	Cp	Total
15.1	6.1	18.4	13.3	9.7	6.6	8.5	17.7	2.1	2.5	100.0,

from which we obtain the following standard katanorm:

Or	Ab	An	Ne	Cs	Fa	Fo	Ru	Mt	Cp	Total
10.2	13.5	22.2	10.3	9.7	5.2	17.7	2.1	6.6	2.5	100.0

45,9 Feldspar

The almost feldspar-free fasinite is thus heteromorphic with a type which has 45.9% feldspar, 10.3% Ne, 32.6% orthosilicate components and 11.2% accessories. The high Cs-content may be reduced somewhat by the construction of a **titanite variant**:

$$3\,Cs + 2\,Ru + 1\,Q = 6\;Tit$$
$$3.2\,Cs + 2.1\,Ru + 1.0\,Q = 6.3\;Tit.$$

The required amount of Q is provided by desilication of a corresponding quantity of albite:

$$2.5\,Ab = 1.5\,Ne + 1.0\,Q.$$

The remainder of the Cs may be combined with an equal amount of $(Fo + Fa)$ to obtain monticellite:

$$6.5\,Cs + 6.5\,(Fo + Fa) = 13.0\;Moni.$$

Thus, the **titanite-monticellite variant** of the standard katanorm of the fasinite is:

Or	Ab	An	Ne	Mont	Ol	Tit	Mt	Cp	Total
10.2	11.0	22.2	11.8	13.0	16.4	6.3	6.6	2.5	100.0

Because of the high c-content and the low degree of silication, the formation of a **melilite variant** can eventually be considered. From the relations:

$$\underset{10\,An}{2\,(2\,SiO_2 \cdot Al_2O_3 \cdot CaO)} + \underset{3\,Cs}{SiO_2 \cdot 2\,CaO} = \underset{10\,Ge}{2\,(SiO_2 \cdot Al_2O_3 \cdot 2\,CaO)} + \underset{3\,Q}{3\,SiO_2}$$

$$\underset{3\,Fo}{SiO_2 \cdot 2\,MgO} + \underset{6\,Cs}{2\,(SiO_2 \cdot 2\,CaO)} + \underset{1\,Q}{SiO_2} = \underset{10\,Ak}{2\,(2\,SiO_2 \cdot MgO \cdot 2\,CaO)},$$

166 by addition, the following is obtained for average melilites:

$$10\,An + 9\,Cs + 3\,Fo = 10\,Ge + 10\,Ak + 2\,Q.$$

If the Q which was set free is used for the silication of an equivalent amount of Ne, an extraordinary variant is obtained, in which melilite occurs in combination with acid plagioclase + Or or Bi. This hardly ever occurs in nature as such. Thus, the degree of silication should not be so low as to merit the occurrence of melilite.

Whereas the calculated katanorm is mesocratic, the fasinite itself is melanocratic.

In order to use the quantitative mineral composition as a basis for their classification, the light components must be considered in relation to the calculated facies, whereas the dark ingredients should be referred to when considering the fasinite. This could serve as one of the most impressive examples of how heteromorphic rocks may differ not only in their mineral composition, but also in the ratio of the salic/femic components. A. Lacroix has previously indicated that fasinite and berondrite (amphibole theralite gabbro) are heteromorphs. Thus, an attempt should now be made at proving this, using the equivalent norm. For a change, the cation percentages should be taken as a basis for the calculation.

Fasinite from Ambahila, calculation of the metal atom (cation) percentages:

	Wt. %	Atomic equivalent numbers × 1000		Cation %
SiO_2	40.10	668	Si^{4+}	37.4
Al_2O_3	15.50	304	Al^{3+}	17.0
Fe_2O_3	6.35	78	Fe^{3+}	4.4
FeO	7.29	101	Fe^{2+}	5.7
MgO	8.41	210	Mg^{2+}	11.8
CaO	12.40	221	Ca^{2+}	12.4
Na_2O	3.37	110	Na^+	6.2
K_2O	1.67	36	K^+	2.0
TiO_2	2.98	38	Ti^{4+}	2.1
P_2O_5	1.28	18	P^{5+}	1.0
H_2O	0.87			
	100.22	1784		100.0

The cation percentages of the Al-rich hornblende and the Ti-augite, which were considered in the treatment of mafraite, may be calculated in a way analogous to the following:

	Si^{4+}	Al^{3+}	Fe^{3+}	Fe^{2+}	Mg^{2+}	Ca^{2+}	Na^+	K^+	Ti^{4+}	Total
Hornblende	36.2	14.4	2.4	8.3	16.1	12.2	5.6	1.4	3.4	100.0
Ti-augite	45.7	6.9	5.7	3.9	14.4	17.3	5.0	0.3	0.8	100.0

167 If 45% of this hornblende and 17% of the augite are subtracted:

	Fasinite (cation %)	45% hornblende (cation %)	17% Ti-augite (cation %)	Remainder (cation %)
Si^{4+}	37.4	16.3	7.8	13.3
Al^{3+}	17.0	6.5	1.2	9.3
Fe^{3+}	4.4	1.1	1.0	2.3
Fe^{2+}	5.7	3.7	0.7	1.3
Mg^{2+}	11.8	7.3	2.5	2.0
Ca^{2+}	12.4	5.5	2.9	4.0
Na^+	6.2	2.5	0.8	2.9
K^+	2.0	0.6	—	1.4
Ti^{4+}	2.1	1.5	0.1	0.5
P^{5+}	1.0	—	—	1.0
	100.0	45.0	17.0	38.0

a remainder exists, which may be recalculated as follows:

	Cp	Ru	Or	Ab	An	Mt	Fa	Fo	Q
Si^{4+}	13.3		4.2	8.7	5.0		0.05	1.0	−5.65
Al^{3+}	9.3		1.4	2.9	5.0				
Fe^{3+}	2.3					2.3			
Fe^{2+}	1.3					1.2	0.1		
Mg^{2+}	2.0							2.0	
Ca^{2+}	4.0	1.5			2.5				
Na^{+}	2.9			2.9					
K^{+}	1.4		1.4						
Ti^{4+}	0.5	0.5							
P^{5+}	1.0	1.0							
	38.0	2.5	0.5	7.0	14.5	12.5	3.5	0.15	3.0 −5.65

The Q-deficit of −5.65 can be eliminated by desilication of the corresponding amount of albite:

$$14.1\,Ab = 8.45\,Ne + 5.65\,Q,$$

so that the following is obtained:

Or	Ab	An	Ne	Fa	Fo	Mt	Ru	Cp	Total
7.0	0.4	12.5	8.45	0.15	3.0	3.5	0.5	2.5	38.0

and the **hornblende-augite variant** of the fasinite is given by:

Or	Ab	An	Ne	Ho	Ti-Aug	Fo	Fa	Mt	Cp	Ru	Total
7.0	0.4	12.5	8.5	45.0	17.0	3.0	0.1	3.5	2.5	0.5	100.0

19.9 Feldspar   3.1 *Ol*   6.5 *Accessories*

According to Tröger, the following composition corresponds to the **berondrite** from the Berondra River, Northern Madagascar:

Zoned plag. with *Or* rims	Ne	Ho	Ti-*Aug*	Ore and apatite
25	7	44	18	6

Thus, the calculated fasinite variant is mainly distinguished by a very small *Ol*-content of 3.1% and can be designated as an **olivine-bearing berondrite**.

Using the following equations, the olivine may be eliminated by the formation of biotite:

$$3.1\,(Fo + Fa) + 3.4\,Or = 5.5\,Bi + 1.0\,Q,$$
$$1.5\,Ne + 1.0\,Q = 2.5\,Ab,$$

and an additional variant, a **biotite-bearing berondrite**, is obtained:

Or	Ab	An	Ne	Ho	Ti-Aug	Bi	Mt	Cp	Ru	Total
3.6	2.9	12.5	7.0	45.0	17.0	5.5	3.5	2.5	0.5	100.0

Such rocks are also known.

The discussed heteromorphic possibilities of the fasinite should be attributed to the unusually high Al-content of its Ti-augite. Thus, it is of interest to consider the composition of this augite in greater detail; no chemical analyses are available. It is impossible to obtain such analyses as no separation can be accomplished due to mutual growth of biotite, caused by magmatic reaction.

The composition was, however, calculated by W. C. Brögger from the rock's analysis, combined with thin-section measurement (W. C. Brögger, 1920). This provided the following weight-percentage values:

SiO_2	Al_2O_3	Fe_2O_3	FeO	MgO	CaO	Na_2O	K_2O	TiO_2	Total
42.86	13.86	6.54	7.36	9.74	15.81	1.04	0.40	2.39	100.0

The basis is calculated from this analysis:

Q	Kp	Ne	Cal	Cs	Fs	Fa	Fo	Ru	Total
21.2	1.4	5.8	19.5	14.2	7.0	8.7	20.6	1.7	100.1

The following standard katanorm is calculated:

		Ru	Or	Ab	An	Wo	Mt	Hy	En	Q
Q	21.2		0.9	3.9	13.0	4.7	(−3.5)	1.7	5.2	−4.7
Kp	1.4		1.4							
Ne	5.8			5.8						
Cal	19.5				19.5					
Cs	14.2					14.2				
Fs	7.0						7.0			
Fa	8.7						(3.5)		5.2	
Fo	20.6								20.6	
Ru	1.7	1.7								
	100.1	1.7	2.3	9.7	32.5	18.9	7.0	6.9	25.8	−4.7

The Q-deficit cannot be balanced merely by the desilication of the ortho-augite, because the high Wo-content of 18.9 requires an equal amount of (En + Hy) for the formation of Di. Thus, some Ne must be formed from Ab. If, for the sake of simplicity, the olivine is taken as pure forsterite, the following norm is obtained:

Or	Ab	An	Ne	Wo	En	Hy	Fo	Mt	Ru	Total
2.3	6.5	32.5	1.9	18.9	12.0	6.9	10.4	7.0	1.7	100.1

41.3 Feldspar — 37.8 Di — 8.7 Accessories

The augite of the fasinite does indeed potentially contain 43.2 light ingredients, of which 41.3 are feldspars. This is explained by the established heteromorphic possibilities of the fasinite.

The following Niggli-values are obtained for the augite:

si	al	fm	c	alk	k	mg	ti
82	15.7	49.3	32.6	2.4	0.19	0.57	3.4

They are closely related to pyroxenitic, and especially to issitic magmas.

Since the composition of the augite was calculated from the rock analysis of the fasinite, as mentioned by W. C. Brögger, it is evident that the mineral composition of the rock may be calculated satisfactorily, using the reverse procedure, from the rock-analysis, in the form of a Ti-augite variant. This may be achieved if the mentioned augite composition is taken as a basis for the calculation. The actual calculation is, however, omitted here.

9. Equivalent Norm, Weight and Volume Percentages

The equivalent norm, in its different variants, provides normative mineral compositions. These are expressed as percentages of the chosen standard minerals, which have a so-called comparable formula size, i.e. the formula unit is based on a sum of the cations = 1. This method has the following principal advantage: if it is required to change from one variant to another, the number of equivalents does not change. Thus, once calculated, the sum is maintained. Actually, however, the mineral composition present (modus) is generally expressed as volume percentages, if it was obtained by measurement of a thin-section, using either integrating or point-counting methods. Thus, it remains desirable to check to what extent equivalent-percentage and volume-percentage results are directly comparable with one another, if at all, or whether the introduction of corrections is eventually necessary. Under certain circumstances, comparisons with weight-percentage results of the CPIW-norm are of interest. In addition, this also pertains to mineral components which were separated by mechanical, magnetic or other physical methods, followed by the weighing of each separated mineral species. In this case too, the results are given as weight percentages. Thus, it is necessary to clarify the mutual relations between equivalent and weight percentages.

The knowledge of the exact mutual relationships between equivalent-, weight- and volume-percentage representation of a mineral composition becomes especially important, if one is required to draw conclusions on the chemism of individual rock-forming minerals from the comparison between an equivalent norm, or its variant, calculated from the chemical analysis, and a volume-percentage modus, determined by thin-section measurement. For this purpose it is insufficient to establish an approximate agreement. Instead, the scale of eventually required corrections must be evaluated. This again requires an exact knowledge of the relationships existing between the three mentioned modes of representing results.

a) Equivalent percentages and weight percentages

The following designations should be introduced:

$A_1, A_2, A_3 \ldots \ldots A_n$ for the equivalent weights of the individual components
$a_1, a_2, a_3 \ldots \ldots a_n$ for the corresponding equivalent percentages
$g_1, g_2, g_3 \ldots \ldots g_n$ for the corresponding weight percentages.

The relationships are provided in the following:

α) Given: weight percentages.

Required: equivalent percentages.

The relative number of equivalents of the individual components can be obtained by dividing the weight-percentage values by the corresponding equivalent weights. If these are recalculated to 100, the required equivalent percentages are obtained.

$$a_1 = \frac{100 g_1}{A_1 S'}, \quad a_2 = \frac{100 g_2}{A_2 S'} \ldots\ldots\ldots\ldots\ldots\ldots a_n = \frac{100 g_n}{A_n S'},$$

where $\quad S' = \frac{g_1}{A_1} + \frac{g_2}{A_2} + \frac{g_3}{A_3} + \ldots\ldots\ldots\ldots\ldots + \frac{g_n}{A_n}.$

β) Given: **equivalent percentages**.
Required: **weight percentages**.

The equivalent percentages should be multiplied by the corresponding equivalent weights, in order to obtain the relative weight amounts of the individual components. Recalculated to 100, the required weight percentages are obtained from these in the following manner:

$$g_1 = \frac{100 a_1 A_1}{S''}, \quad g_2 = \frac{100 a_2 A_2}{S''} \ldots\ldots\ldots\ldots g_n = \frac{100 a_n A_n}{S''},$$

where $\quad S'' = a_1 A_1 + A_2 a_2 + a_3 A_3 + \ldots\ldots\ldots\ldots + a_n A_n.$

It is evident that equivalent and weight percentages approach one another as the equivalent weights become increasingly similar. If the equivalent weights of all components were completely identical, the same would also hold true for the equivalent and weight percentage values a_i and g_i.

The equivalent weights of the important rock-forming minerals, given for their ideal compositions, may be divided into 3 groups, as follows:

I. Equivalent weights of approximately 50-60 and higher:

a) Common rock-forming components:

Q	60.1	Gro	56.3	Mt	77.1
Cp	62.0	Alm	62.2	Hm	79.8
Ac	57.7	Andr	63.5	Ilm	75.9
Tit	65.3	Spe	61.9	Cc	100.1
Ms	56.9	Staur	56.5		
		Zo	56.8		
		Wo	58.1		

b) Rare components, only in trace amounts:

A	68.1	Spur	62.6	Cm	74.6
Fr	78.1	Till	69.8	Pf	68.0
Hz	57.9	Ves	56.9	Z	91.6
Tephr	67.3	Cs	57.4		
Me	57.1–59.1				
Pph	63.0				

c) Fe-bearing end-members of mixed-crystal series, the high equivalent weights of which are reduced in most of the cases by admixture of the Mg-end-members:

Hy	65.9	Fe-Ant	74.3	Ot	63.0
Hed	62.0	Fe-At	68.3		
Pi	60.4	Fe-Cord	58.9		
Fe-Bi	64.0				

II. Equivalent weights in the proximity of 53–56:

Alkf	52.3–55.6	Di	54.1	Zwd	55.6
Or	55.6	Ts	54.5	Nc	53.0
Lc	54.5	Cord	53.2		
An	55.6	Ant	55.4		
Anc	55.0	At	55.7		
Ge	54.8	Gram	54.1		
Ak	54.5	Sil	54.0		

III. Equivalent weights below 53:

Kp	52.7	Mg-Bi	52.1	Anth	52.0
Ab	52.3	Fo	46.9	Lep	49.6
Ne	47.3	Ave. Ol ~	52	Na-Ge	51.6
Ma	52.0–52.8	En	50.2	Sp	47.4
Sod	48.4	Jd	50.5	C	51.0
Nos	47.3	Pyp	50.4	Th	47.3
Hau	51.2	Omph	52.3	Ps	40.3
Canc	52.6	Spod	46.5	Pr	40.0
		Mont	52.1		

From this presentation it may be concluded, as was already shown for the rock-forming oxides and the basis components, that the equivalent weights of the norm-minerals indeed do not show very large differences. Large-scale deviations exist only in the ore-minerals, Mt, Hm, Ilm, as well as in Z and Ru, and especially in Cc. In these cases the equivalent weights are higher than the average. In contrast to these, the equivalent weights of the feldspar substitutes Ne, Hau, Nos and $Canc$, but not of Lc, Anc, Ge and Ak, have values considerably lower than the average of approximately 54. One may thus anticipate that equivalent-percentage norms are directly comparable with mineral compositions, expressed as weight percentages, except in rocks rich in ore, Cc or foids. Ore minerals and calcite will occur in too small an amount, whereas the mentioned foids will occur in too large an amount, in the equivalent norm. Obviously, this fact also has an influence on the rest of the components.

The following examples of simulated mineral compositions should clarify this point. It is assumed that various weight-percentage contents of Mt, Cc and Ne are associated with components of an average equivalent weight 54. The corresponding equivalent percentages are calculated for each case.

1. Magnetite (equiv. wt 77.1) + components having an equivalent weight of 54.

Wt. % Mt	10	20	30	40	50	60	70	80	90
Eq. % Mt	7.2	14.9	23.0	31.8	41.2	51.2	62.0	73.6	86.4
Eq. % R^*	92.8	85.1	77.0	68.2	58.8	48.8	38.0	26.4	13.6
$\dfrac{\text{Eq. \% } Mt}{\text{Wt. \% } Mt}$	0.72	0.75	0.77	0.80	0.82	0.85	0.89	0.92	0.96

2. Calcite (equiv. wt 100.1) + components having an equivalent weight of 54.

Wt. % Cc	10	20	30	40	50	60	70	80	90
Eq. % Cc	5.6	11.8	18.6	26.3	34.8	44.4	55.4	68.2	82.8
Eq. % R	94.4	88.2	81.4	73.7	65.2	55.6	44.6	31.8	17.2

* [Remainder].

Eq.% Cc / Wt. % Cc	0.56	0.59	0.62	0.66	0.70	0.74	0.79	0.85	0.92

3. Nepheline (equivalent weight 47.3) + components having an equivalent weight of 54.

Wt. % Ne	10	20	30	40	50	60	70	80	90
Eq.% Ne	11.2	22.2	32.8	43.3	53.3	63.1	72.7	82.0	91.1
Eq.% R	88.8	77.8	67.2	56.7	46.7	36.9	27.3	18.0	8.9
Eq.% Ne / Wt. % Ne	1.12	1.11	1.09	1.08	1.07	1.05	1.04	1.03	1.01

The following selected examples show the relationships under consideration, regarding close equivalent weights:

$$3 Kp + 2 Q = 5 Or$$
Weights, for example in g $2.84 + 2.16 = 5.00$

$$3 Ne + 1 Q = 4 Jd$$
Weights, for example in g $2.81 + 1.19 = 4.00$

$$3 Fa + 1 Q = 4.0 Hy$$
Weights, for example in g $3.09 + 0.91 = 4.00$

$$1 Wo + 1 En = 2 Di$$
Weights, for example in g $1.05 + 0.95 = 2.00$

$$5 An + 4 Wo = 8 Gro + 1 Q$$
Weights, for example in g $4.92 + 4.15 = 8.00 + 1.07$.

174   The similar relatively low values of the equivalent weights of *Fo* and *En* will exert no practical influence, as these components generally occur with *Fa* and *Hy* in mixed crystals, in which the higher equivalent weight of the Fe-bearing component acts as a compensator. For the same reason no interference is caused by the higher equivalent weights of *Fa* and *Hy*. The close agreement between equivalent and weight-percentage values is further confirmed (with the exception of the ore minerals) by the good agreement found between the equivalent and CPIW-norms. This is illustrated by the following comparison of the two norms of the previously calculated quartz-diorite from Spanish Peak, California, and of the lujaurite from Lille Elv, Greenland. The CPIW-norms were obtained from Washington's tables. The equivalent norms are slightly modified in comparison to the previously calculated standard katanorms, and are arranged parallel to the CPIW-norms for the sake of easy comparison. For the quartz-diorite from Spanish Peak, *Tit* was eliminated by the formation of *Ilm* according to $1.5 Tit + 1.0 Hy = 1.0 Ilm + 1.0 Wo + 0.5 Q$. In the case of the lujaurite from Lille Elv, *Ru* was eliminated by the formation of *Ilm* according to $0.2 Ru + 0.3 Fa = 0.4 Ilm + 0.1 Q$. In addition, *Cs* was silicated to *Wo* according to $1.0 Cs + 0.3 Q = 1.3 Wo$. The $1.3(En + Hy)$ required for *Di*-formation, were gained by silication of $0.9 Fa$ and $0.1 Fo$. The required Q-amount was obtained by desilicating $1.2 Ab$.

A comparison shows clearly the relationships which were discussed previously, i.e., the values of the equivalent norm are low for the ore minerals, and too high for *Ne* and *Sod*, relative to the CPIW-norm.

	Quartz-diorite Spanish Peak, California				Lujaurite Lille Elv, Greenland		
	Equivalent norm		CPIW-norm		Equivalent norm		CPIW-norm
Q	13.1	Q	13.80	Or	15.1	or	15.57
Or	7.8	or	7.78	Ab	25.0	ab	23.06
Ab	35.0	ab	33.01	Ne	22.7 } 25.7 ⟶	ne	22.15
An	26.0	an	25.02	Sod	3.0		
Di	4.6	di	4.64	Ac	25.1	ac	26.33
En } 8.9		hy	8.42	Ns	1.5	ns	1.46
Hy				Fo	0.1 } 4.1	ol	4.83
Mt	3.0 ⟷	mt	4.18	Fa	4.0		
Ilm	1.0 ⟷	il	1.22	Di	2.6	di	2.97
Cp	0.6	ap	0.67	Ilm	0.4	ilm	0.61
				Z	0.5 ⟷	Z	1.00*

175 γ) Relations between equivalent norm and CPIW-norm. The method of recalculating the equivalent norm to the CPIW-norm and vice versa becomes evident from the former section. In order to obtain the equivalent norm from the CPIW-norm, the weight-percentage results of the latter should be divided by the equivalent weights calculated for a cation sum of unity. The equivalent numbers thus obtained are then recalculated to the sum of 100.

The calculation of the equivalent norm of the quartz-diorite from Spanish Peak, from the CPIW-norm, will serve as an example. In order to obtain comparable relations, the norm calculated by H. S. Washington (1915, 381, No. 82) will not be taken as a basis for the calculation, as this norm was derived through the usage of rounded formula weights. The CPIW-norm was recalculated, using the exact formula weights, according to the usual regulations. The results were, however, only calculated to the first decimal place.

	a)	b)	c)		d)	e)
Q	14.1	60.1	234	Q	13.2	(13.1)
or	7.8	55.6	140	Or	7.9	(7.8)
ab	32.4	52.3	619	Ab	35.0	(35.0)
an	25.6	55.6	460	An	26.0	(26.0)
wo	2.2	58.1	38	Wo	2.1	(2.3)
en	8.8	50.2	175	En	9.9	(9.9)
fs	1.6	65.9	24	Hy	1.4	(1.3)
mt	4.2	77.2	54	Mt	3.0	(3.0)
ap	0.6	62.0	10	Cp	0.6	(0.6)
il	1.2	75.9	16	Ilm	0.9	(1.0)
	98.5		1770		100.0	(100.0)

*H.S. Washington (1917) 555. No. 9, mistakenly gives Z 1.65.

a) CPIW-norm, calculated on the basis of exact formula weights

b) Equivalent weights, corresponding to a cation sum = 1

c) Equivalent numbers multiplied by 100

d) Equivalent norm

e) Equivalent norm (Ilmenite variant), calculated directly from analysis, for the sake of comparison.

A good agreement exists, as is shown by comparison with the equivalent norm directly obtained from the analysis (ilmenite variant).

If, conversely, one is required to recalculate the equivalent norm to the CPIW-norm, the following procedure should be adopted: the equivalent percentages should be multiplied by the corresponding equivalent weights, which are referred to a cation sum = 1. The weight ratios thus obtained should now be recalculated to the sum of 100.

If the ilmenite variant of the standard katanorm of the quartz-diorite from Spanish Peak is taken as a reference, the calculation will proceed thus:

	a)	b)	c)		d)	e)	f)
Q	13.1	60.1	787.3	Q	14.2	(14.1)	14.0
Or	7.8	55.6	433.7	or	7.8	(7.8)	7.7
Ab	35.0	52.3	1830.5	ab	32.9	(32.4)	32.4
An	26.0	55.6	1445.6	an	26.0	(25.6)	25.6
Wo	2.3	58.1	133.6	wo	2.4	(2.2)	2.4
En	9.9	50.2	497.0	en	8.9	(8.8)	8.7
Hy	1.3	65.9	85.7	fs	1.5	(1.6)	1.5
Mt	3.0	77.2	231.6	mt	4.2	(4.2)	4.1
Cp	0.6	62.0	37.2	ap	0.7	(0.6)	0.7
Ilm	1.0	75.9	75.9	il	1.4	(1.2)	1.4
	100.0		5558.1		100.0	(98.5)	98.5

a) Standard katanorm (ilmenite variant), calculated from analysis

b) Equivalent weights, referred to a cation sum = 1

c) Weight ratios

d) CPIW-norm, recalculated to the sum of 100

e) CPIW-norm, calculated from analysis, for sake of comparison

f) CPIW-norm, as under d), however calculated to the sum of 98.5

The CPIW-norm, in contrast to the equivalent norm, conventionally is not calculated to the sum of 100, but to the analysis sum minus H_2O and perhaps other components which are not reckoned in the calculation. Therefore, in order to obtain comparable results, it should be recalculated to the same sum. A comparison of f) with the CPIW-norm directly calculated from the analysis, e), shows a satisfactory agreement.

b) Equivalent percentages and volume percentages

If the following designations are introduced:

$A_1, A_2, A_3 \ldots\ldots A_n$ for the equivalent weights of the individual components

$a_1, a_2, a_3 \ldots\ldots a_n$ for the equivalent percentages

$V_1, V_2, V_3 \ldots\ldots V_n$ for the equivalent volumes

$v_1, v_2, v_3, \ldots\ldots v_n$ for the volume percentages

$d_1, d_2, d_3 \ldots\ldots d_n$ for the densities

then the relationships are as follows:

α) Given: volume percentages

Required: equivalent percentages

The volume percentages should be divided by the corresponding equivalent volumes, thus giving the relative number of equivalents of the individual components. Recalculation to the sum 100 will yield the required equivalent percentages:

$$a_1 = \frac{100\,v_1}{V_1 S'''}, \qquad a_2 = \frac{100\,v_2}{V_2 S'''}, \qquad \ldots\ldots\ldots\ a_n = \frac{100\,v_n}{V_n S'''},$$

where $\quad S''' = \dfrac{v_1}{V_1} + \dfrac{v_2}{V_2} + \dfrac{v_3}{V_3} + \ldots\ldots\ldots\ldots + \dfrac{v_n}{V_n}.$

Since, however, according to definition, $V_i = A_i/d_i$, it also follows that:

$$a_1 = \frac{100\,v_1 d_1}{A_1 S'''}, \qquad a_2 = \frac{100\,v_2 d_2}{A_2 S'''} \ldots\ldots\ a_n = \frac{100\,v_n d_n}{A_n S'''}.$$

β) Given: equivalent percentages

Required: volume percentages

The equivalent percentages should be multiplied by the corresponding equivalent volumes in order to obtain the relative volumes of the individual components. Recalculation to the sum of 100 will yield the required volume percentages:

$$v_1 = \frac{100\,a_1 V_1}{S''''}, \qquad v_2 = \frac{100\,a_2 V_2}{S''''} \ldots\ldots\ v_n = \frac{100\,a_n V_n}{S''''},$$

where $\quad S'''' = a_1 V_1 + a_2 V_2 + a_3 V_3 + \ldots\ldots\ldots\ldots + a_n V_n.$

Since once again, according to definition $V_i = A_i/d_i$, it follows that

$$v_1 = \frac{100\,a_1 A_1}{d_1 S'''''}, \qquad v_2 = \frac{100\,a_2 A_2}{d_2 S'''''} \ldots\ldots\ v_n = \frac{100\,a_n A_n}{d_n S'''''},$$

where $\quad S''''' = a_1 \dfrac{A_1}{d_1} + a_2 \dfrac{A_2}{d_2} + a_3 \dfrac{A_3}{d_3} + \ldots\ldots\ldots\ a_n \dfrac{A_n}{d_n}.$

The less the differences in the equivalent volumes of the components, the better is the agreement between equivalent- and volume-percentage values.

The equivalent volume is defined as:

$$V = \frac{A}{d} = \frac{\text{Formula weight}}{n \cdot d}$$

where n expresses the number of metal atoms (cations) of a single formula unit. Orthoclase, for example, with a formula weight of 556.5, corresponds to the composition $6\,SiO_2 \cdot Al_2O_3 \cdot K_2O$ and $d = 2.55$. On the basis of this data one obtains $V = \frac{556.5}{10 \cdot 2.55} = 21.8$. In this manner the equivalent volumes of the important rock-forming minerals may be calculated. The following equivalent values of the most important standard kata-minerals were given by P. Niggli, who used for this purpose the compilation of W. Biltz (1934). They may be classed in four groups according to their size:

I. Equivalent volumes higher than 24:

Hl 26.3
Fr 24.3

II. Equivalent volumes approximately = 20:

Q 22.6   Ge 18.9
Or 21.8   Cord 21.5
Ab 20.8   Ru 18.9
An 20.1   Nc 20.5
Kp 20.7   Tit 18.5
Lc 22.5   Ap approx. 20
Ne 18.1
Z 19.4

According to E. Niggli (1944, 240), the most important standard epi-minerals, which will be mentioned later, belong likewise to group II:

Ms 20.9   Gram 18.0   Ot 18.3
At 20.0   Akt 21.0   Mg-Ot 17.2
Fe-At 22.8 Zo 17.6   Xon 22.2
Ant 21.7   Kaol 24.7   Bru 24.3

III. Equivalent volumes approximately = 16:

Sil 16.3   En 15.8   Hm 15.2
Ns 15.4   Di 16.5   Mt 14.8
Fo 14.6   Gro 15.7   Ilm 16
Fa 15.8   Alm 14.4
Cs 17.5   Cm 14.5

IV. Extremely low or high equivalent volumes:

Hz 13.5   Cc 36.8
Sp 13.2   Sid 30.1
C 12.9   Mgs 27.4

For many important, mainly leucocratic components, such as quartz, feldspars and feldspathoids (group II), the values are close to 20. Thus, for these components the equivalent percentages at the same time provide the distribution by volume. For a number of components which generally are of only restricted occurrence (group I), the equivalent volumes are approximately 20% higher, whereas a large number of the more important, mainly melanocratic components (group III) have equivalent volumes which are 20% lower than those of group II. Thus, corrections may be introduced for small amounts of minerals of groups I and III in the presence of large amounts of group II minerals. In the presence of larger amounts of minerals of group I, 20% of the equivalent-percentage content should be added. If this be the case with minerals of group III, 20% should, however, be subtracted, in order to obtain volume-percentage values. In addition, it may be shown that in many cases the equivalent-percentage values fall between the weight- and volume-percentage values. Therefore, by merely estimating the fraction by volume of the individual components, instead of exactly measuring the equivalent-percentage mineral composition, a sufficiently good picture is obtained of the volume- as well as the weight-percentage composition of the rock under consideration. If the rock is calcite-rich (group IV), due to the nearly doubled equivalent volume in comparison to that of group II, care should be taken that the content of this mineral in the equivalent norms should be approximately half of that occurring in a volume- or weight-percentage representation. Thus, a suitable correction should be introduced. If the equivalent weight of calcite is set at $1/2\,CaCO_3$, taking into consideration the carbon, the abnormal value of the equivalent volume will be eliminated. This practice is, however, not to be recommended, because thereby a new disadvantage is introduced, namely, by adding or subtracting the highly volatile CO_2, one changes the number of the equivalents or atoms considered in the calculation. Thus, a sum, previously calculated to 100, cannot be maintained.

c) Volume percentages and weight percentages

The relationships between volume and weight percentages should also be mentioned for the sake of completeness.

a) Given: volume percentages

Required: weight percentages

The volume percentages should be divided by the volumes of the weight units, i.e., by the so-called specific volumes, in order to obtain the relative weights of the individual components. Instead, they may be multiplied by the reciprocals of the specific volumes, i.e., by the density d, which generally is better known. By recalculation to the sum of 100, the required weight percentages are obtained.

$$g_1 = \frac{100\,v_1 d_1}{S''''''}, \quad g_2 = \frac{100\,v_2 d_2}{S''''''} \ldots \ldots g_n = \frac{100\,v_n d_n}{S''''''},$$

where $\quad S'''''' = v_1 d_1 + v_2 d_2 + v_3 d_3 + \ldots \ldots \ldots + v_n d_n.$

180 The products $v_i d_i$, the so-called volume products or volproducts of the most important rock-forming minerals, can be obtained from tables (for example H. v. Philipsborn, 1933, 244-293), thus rendering the calculation more easy.

β) Given: **weight percentages**

Required: **volume percentages**

The weight percentages should be divided by the weight of the volume units, i.e. by the density d. Thus, the relative volumes of the individual components are obtained. Recalculation to the sum of 100 provides the required volume percentages.

$$v_1 = \frac{100\, g_1}{d_1\, S''''''''}, \qquad v_2 = \frac{100\, g_2}{d_2\, S''''''''} \ldots \ldots v_n = \frac{100\, g_n}{d_n\, S''''''''},$$

where $\quad S'''''''' = \dfrac{g_1}{d_1} + \dfrac{g_2}{d_2} + \dfrac{g_3}{d_3} + \ldots \ldots \ldots \ldots \ldots + \dfrac{g_n}{d_n}.$

It may be shown that the less the difference between their densities, the closer the weight and volume percentages approach one another, and that these become identical for the same d values.

The densities of the important rock-forming minerals vary only slightly, as is shown in the following compilation:

G r o u p   I ($d = 5,2$–4): magnetite, pyrite, ilmenite, chromite, pyrrhotite, zircon, rutile, and andalusite.

G r o u p   II ($d = 4$–3): corundum, siderite, pyrope, spinel, staurolite, kyanite, topaz, titanite, olivine, epidote, vesuvianite, augite, ortho-augite, diopside, hornblende, andalusite, fluorite, apatite, tourmaline, biotite, actinolite, grammatite, glaucophane.

G r o u p   III ($d = 3$–2): anhydrite, dolomite, muscovite, wollastonite, anorthite, chlorite, calcite, talc, scapolite, quartz, albite, serpentine, cordierite, K-feldspars, nepheline, leucite, sodalite group, kaolin, tridymite, gypsum, analcite, halite.

The minerals of group I, especially the ores, present an exception in that their density is about 2/3 higher than the approximate average, taken as 3, of the other rock-forming minerals. If, for example, a mineral composition expressed as weight percentages is compared with results obtained from a thin-section measurement, a recalculation to volume percentages should be carried out first if ore minerals are present. In contrast, a recalculation to weight percentages is necessary in the analogous case, when the modus obtained by thin-section measurement is intended for comparison with a CPIW-norm or with the results obtained from a separation accomplished by physical methods.

10. Graphic Representation of the Chemism of the Rock Based on the Basis Components

a) The basis-group values Q, L, M and the QLM-triangle

A number of graphic representations based on the basis values were suggested by P. Niggli (1936, 1938, 1941, 1943, 1946, 1950, 1951); C. Burri and P. Niggli (1945). These were especially intended for illustrating the chemical and mineralogical variation of igneous rock series and petrographic provinces, and are exceptionally well adapted for comparative purposes.

The fundamental problem of all petrochemical projection methods, namely, the means of reducing, by suitable combination, the large numbers of variables in order to allow for the clear illustration of the concentration triangle, also exists in this case. A second problem is related hereto, namely, how to clarify the composition of the composite components by the construction of suitable ratios.

Niggli has offered the following suggestions for the combination of the basis components:

1) Feldspar- and feldspathoid-based components: $Kp + Ne + Cal$
2a) Olivine- and augite-based components: $Cs + Fo + Fa$
2b) Corundum- and sillimanite-based components: C
3) Quartz and silication substances: Q
4a) Alkali-augite- and Fe-ore-based components: Ns (possibly Ks) + Fs
4b) Spinel, cordierite and ore-based components: $Sp + Hz + Fs$.

According to the regulations established for the calculation mode, C and Cs, as well as Ns (perhaps Ks) and $Sp + Hz$, mutually exclude each other. Thus, only one of the groups 2a) or 2b) and one of the groups 4a) or 4b) are present. In addition, in many cases in igneous rocks the basis components Ks, Sp, Hz, C are missing altogether, and Fs and Ns may be united with the augite-based component $Cs + Fo + Fa$ (aegirine formation). Thus, the following three composite components are obtained:

$$Q = Q$$
$$Kp + Ne + Cal = L$$
$$Cs + Fo + Fa + Fs + Ns = M *,$$

These may be represented in a concentration triangle in the usual way. In these so-called QLM diagrams the projection points of the components, which were derived from the basis components by silication, may be introduced immediately, if we consider the reaction coefficients previously formulated for silication reactions, on an equivalent basis. From the formula $3 Kp + 2 Q = 5 Or$ it follows, for example, that the projection point of Or, as well as those of Ab and An, is located on the QL side of the triangle, corresponding to 40% Q and 60% L. Leucite, however, is located at 25% Q and 75% L according to the relationship $3 Kp + 1 Q = 4 Lc$, etc (compare Figure 45).

The projection point of a given rock's chemism on the QLM-triangle, immediately provides a preliminary general picture of the possible mineral compositions which correspond to this chemism. In this case, PF represents the line of saturation, i.e., it is the line which separates the quartz-bearing from the quartz-free normative mineral combinations. A number of heteromorphic possibilities may be shown immediately for the area below PF. Projection points whitin the triangle PFR may correspond, for example, to the mineral combinations pyroxene-feldspars-olivine or pyroxene-feldspar-feldspathoid. Projection points in the triangle PMR correspond to the normative parageneses pyroxene-feldspar-olivine, or pyroxene-feldspathoid-olivine, and in the triangle FLR—the parageneses pyroxene-feldspar-feldspathoid or feldspar-feldspathoid-olivine. In the

* Ru and Cp should also be calculated to M, because Ti mainly substitutes for Mg, and in the case of Cp one deals with Ca which is not bound to Al.

triangle MRL olivine and feldspathoids, together with pyroxene or feldspar, are represented.

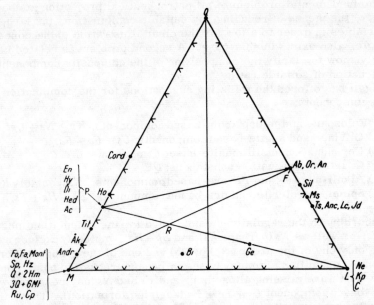

FIGURE 45. Projection points of simple idealized crystal components on a QLM-triangle. After C. Burri and P. Niggli (1945). (The point Ab, Or, An is designated as F in the text)

The **degree of saturation** with respect to SiO_2, i.e., the location of a projection point in relation to the saturation line $P\tilde{F}$, may be expressed mathematically as follows: if all the accessories are neglected, the equality $3Q = 2L$ will be valid for all the points lying on FM. $(3Q - 2L)/M =$ const. will hold for any line radiating from F. Thus, the following expression

$$a = \frac{3Q - 2L}{M}$$

defines the "**saturation coefficient**". This will be equal to $+1$ for points lying on FP and to zero for the points located on FM. Points above FP have $a > +1$ increasing to $+\infty$ when located on line FQ. Points projected below FM will have negative a-values, and the points located on LF will attain values of $-\infty$. If we follow the line $a = -1$ to its contact with the extension of side QM, we arrive at the so-called **oxide point**, which is the projection point of Mt, Hm, Cc, Dol, as can be easily proved by calculation. Taking Mt as an example, according to the relationships $2Fs + Fa = 2Mt + Q$ or, $2Mt = 2Fs + Fa - Q$, it follows that $100\% \ Mt = 100\% \ Fs + 50\% \ Fa - 50\% \ Q$. In a similar calculation it can be shown that from $4Dol = 3Cs + 3Fo - 2Q$, $100\% \ Dol = 75\% \ Cs + 75\% \ Fo - 50\% \ Q$. Therefrom, the coordinates of the oxide point can be derived. They are: $Q = -50$, $M = +150$, and $L = 0$. If these values are substituted in the equation for a, the value -1 is indeed obtained (Figure 46).

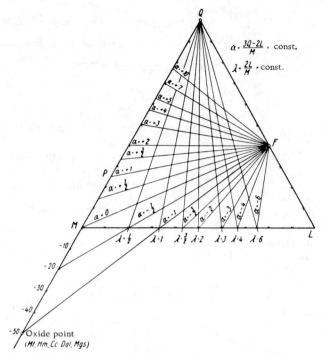

FIGURE 46. QLM-triangle with lines a = const. and λ = const. According to P. Niggli (1943)

For the estimation of the qualitative ratio M/L, i.e., the ratio of the dark basis components to the leucocratic silicates, the value $2L/M = \lambda$ is substituted in the expression of a. The geometric points for which λ = const. lie on straight lines radiating from Q. The side of the triangle QM represents $\lambda = 0$, the height of the triangle has the value $\lambda = 2$; however, for the side QL, $\lambda = +\infty$ (Figure 46). With the exception of points lying on QL, the location of any projection point in the triangle QLM is unequivocally fixed by stating the values (a, λ).

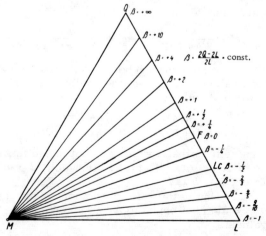

FIGURE 47. QLM-triangle with lines β = const. According to P. Niggli (1943)

If we set $a + \lambda = \frac{3Q-2L}{M} + \frac{2L}{M} = \frac{3Q}{M} = \varkappa$, i.e., $a = \varkappa - \lambda$, then along the line LP $\varkappa = +1$. Along the lines which radiate from the pole M either Q/L or $Q/(Q+L)$ is constant. If we take $\frac{3Q}{2L} - 1 = \beta$, then:

$$\frac{a}{\lambda} = \frac{(3Q-2L)/M}{2L/M} = \frac{3Q-2L}{2L} = \beta.$$

Accordingly, the lines radiating from M are the geometric locii for $\beta =$ const. In particular, $\beta = 0$ for MF, $\beta = -1$ for ML and $\beta = +\infty$ for MQ. For the important connecting line $M-Lc$ $\beta = -1/2$. Since undersaturated rock series of different degrees of silication are projected as points which lie near to the lines radiating from the olivine pole M, β provides a simple indication on the state of silication (Figure 47).

According to their definition, L and M are composite components. A more profound knowledge of their composition may be of great importance. Thus, ratio numbers should be introduced, which may provide this information. These numbers may be selected in different ways, The following has proved to be practical:

$$\pi = \frac{Cal}{Cal + Kp + Ne},$$

$$\gamma = \frac{Cs}{Cs + Fs + Fo + Fa},$$

$$\mu = \frac{Fo}{Cs + Fs + Fo + Fa}.$$

π gives the ratio of the normative Ca bound to Al ("chaux feldspathisable"), to the total sum of Ca + Na + K bound to Al

γ gives the ratio of Ca not bound to Al ("chaux non feldspathisable"), i. e. the Ca available for the femic silicate basis complexes, to the total sum of Ca + Mg + Fe not bound to Al

μ gives the ratio of the normative Mg not bound to Al, to the total sum of Ca + Mg + Fe not bound to Al.

The following are also defined:

$$k' = \frac{Kp}{Kp + Ne}, \qquad mg' = \frac{Fo}{Fo + Fa + Fs},$$

i. e., k' is the ratio of the normative K bound to Al, to the total sum of K+Na bound to Al. These values differ from the earlier values introduced when the Niggli-values were discussed:

$$k = \frac{K_2O}{K_2O + Na_2O} = \frac{Kp + Ks}{Kp + Ks + Ne + Ns},$$

In this analogous value the alkalis which are not bound to Al are also introduced — in acmite or K-acmite. In rocks for which $al \geq alk$, where no normative Ns or Ks occurs, $k' = k$.

Likewise, the mg'-value is the ratio of the normative Mg not bound to Al, to the sum of Mg + Fe also not bound to Al. The analogous term introduced with the Niggli-values:

$$mg = \frac{MgO}{MgO + FeO + 2\,Fe_2O_3 + MnO} = \frac{Fo + Sp}{Fo + Sp + Hz + Fs}$$

corresponds to the total Mg and Fe (including Mn), i.e. the Mg and Fe both bound and not bound to Al. Here, too, *mg'* becomes equal to *mg* in the case when neither *Sp* nor *Hz* occur, i.e., in the case when $(al-alk) \leq c$.

For the calculation of *mg*, as already mentioned previously, the total available Fe should be taken as divalent. Thus, the equivalent number of Fe_2O_3 should be doubled, because $1\,Fe_2O_3$ is equivalent to $2\,FeO$.

b) The KNaCa- and MgFeCa-triangles and the determination of normative feldspar relations.

The values π and k' or γ and *mg'* may be combined to great advantage into two complementary triangles to the QLM-triangle. These provide the required information on the composition of the composite components L and M.

The KNaCa-triangle shows the quantitative relations of the K, Na, and Ca which are bound to Al. For this purpose, however, it is unnecessary to recalculate the amounts of Kp, Ne and Cal to the sum of 100. Instead, the value k' is indicated on the triangle side $K(Kp) - Na(Ne)$. All of the radial lines emanating from the corner $Ca(Cal)$ toward this side, represent geometric points of equal k'. If, in addition, the π-value is indicated on the triangle side $K(Kp) - Ca(Cal)$, then the points of equal π are located along all lines parallel to the triangle side $K(Kp) - Na(Ne)$. The required projection points may thus be introduced directly on the basis of k' and π (Figure 48).

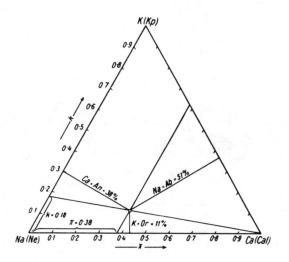

FIGURE 48. Determination of the normative feldspar ratios for the quartz-diorite of Spanish Peak on the basis of $k = 0.18$ and $\pi = 0.38$ (since $al > alk$, $k' = k$). The coordinates of the point, which are fixed by k and π, may be read from the diagram: Or 11.0 %; Ab 51.0 %; An 38.0 %. Thus it follows that $Or/(An+Ab) = 0.12$, and $An/Ab = 0.75$, or $An/(Ab+An) = 0.43$, or An_{43}. These results are in agreement with those which were obtained from Figure 24 or 24b.

If use is made of common triangular coordinate paper, the percentage ratios of the basis components $Kp : Ne : Cal$ can be read off immediately, these being equal to the ratio of K : Na : Ca (all bound to Al). If there is a sufficient amount of SiO_2 present for the construction of feldspars, this ratio represents at the same time the ratio of $Or : Ab : An$. This may be explained by the fact that the corresponding formula units were defined on the basis of isomorphic relationships and thus contain equal amounts of K, Na, and Ca. The "feldspar triangles", which are dealt with in many publications, are thus obtained in a very simple way, taking the basis as a reference.

187   The MgFeCa-triangle illustrates the quantitative relations of the Fe, Mg and Ca which are not bound to Al. If sufficient SiO_2 is present, it corresponds at the same time to the triangle Fe-silicate—enstatite—wollastonite. It should be noted that both divalent and trivalent iron are included under Fe-silicate. The previously introduced value

$$w = \frac{2\,Fe_2O_3}{2\,Fe_2O_3 + FeO} = \frac{Fs}{Fs + Fa} = \frac{Fe^{3+}}{Fe^{3+} + Fe^{2+}}$$

provides information on the degree of oxidation of the iron.

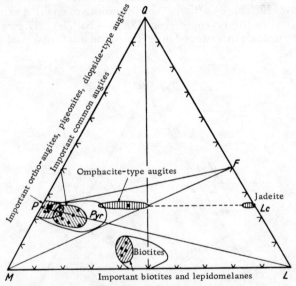

FIGURE 49. Important pyroxenes and biotites of igneous rocks, represented in the QLM-triangle. After C. Burri and P. Niggli (1945)

In this case there is also no need to recalculate to the sum of 100; instead, the introduction of the projection points is carried out in analogy to the KNaCa-triangle, being based on the mg' and γ values.

Figures 49-54 are practical examples of the previous discussion. These figures illustrate the projection fields of the most important rock-forming minerals within the range of their natural variability in the three mentioned triangles, according to C. Burri and P. Niggli (1945). Figures 55-57 show the resulting point fields, in the three triangles, for the lavas of Lassen Peak, California, which were dealt with earlier, and for which the variation diagram of the Niggli-values was provided in Figure 42. The pictures obtained are typical for Pacific provinces with mainly SiO_2-oversaturated rocks. Examples of other province types, and, generally, the distinction between these types by use of illustrations are dealt with, among others, by P. Niggli (1938) and C. Burri and P. Niggli (1945).

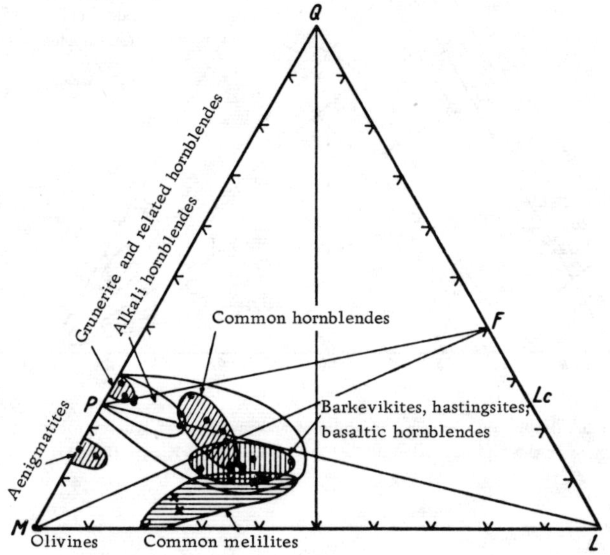

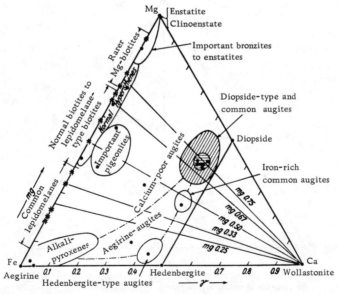

FIGURE 50. Important amphiboles and common melilites of igneous rocks, represented in the **QLM**-triangle. After C. Burri and P. Niggli (1945)

FIGURE 51. Important pyroxenes and biotites represented in the MgFeCa-triangle. After C. Burri and P. Niggli (1945)

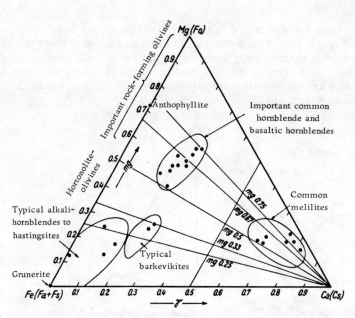

FIGURE 52. Important olivines, amphiboles and melilites of igneous rocks, represented in the MgFeCa-triangle. After C. Burri and P. Niggli (1945)

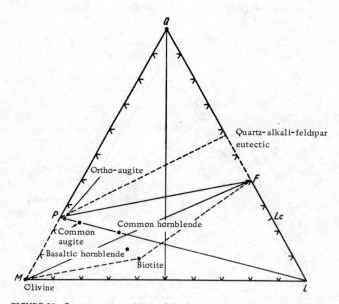

FIGURE 53. Common compositions of the important components of rocks of the calc-alkali series, represented in the MgFeCa-triangle. After C. Burri and P. Niggli (1945)

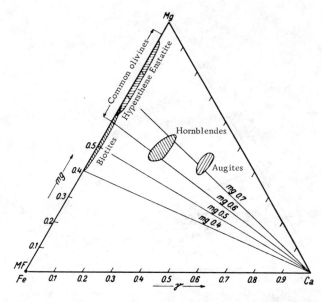

FIGURE 54. Common compositions of important components of rocks of the calc-alkali series, represented in the MgFeCa-triangle. After C. Burri and P. Niggli (1945)

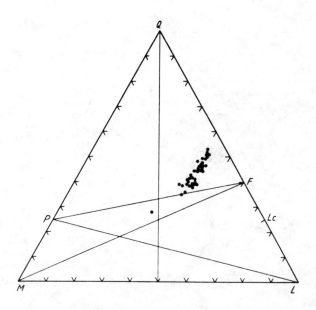

FIGURE 55. Lavas from Lassen Peak, California, projected on the QLM-triangle, as an example of the Pacific province with a predominance of SiO_2-oversaturated rocks (the point field is located above the saturation line PF)

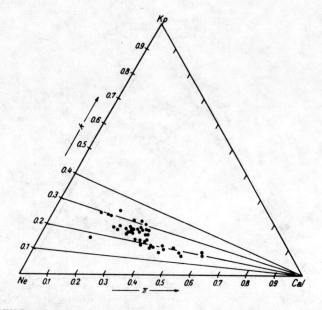

FIGURE 56. Lavas from Lassen Peak, California, projected on the KNaCa-triangle

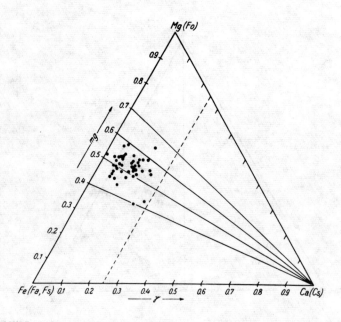

FIGURE 57. Lavas from Lassen Peak, California, projected on the MgFeCa-triangle

The QLM-triangle shown in Figure 58 illustrates the example of an Atlantic province poor in silica. It depicts young lavas of Cape Verde and the Dakar region, which were already mentioned in section BI4b. The silica relationships for these rocks were illustrated in Figure 26 by means of Rittmann's values $Si°$ and $Az°$. In agreement with the very characteristic silication-degree, shown in Figure 16, or rocks which belong to this province, the various rock series illustrated in Figure 58 are clearly distinguished from each other by their Q-level. Series Ia corresponds to the normal association basalt-trachybasalt-trachyandesite-trachyte, or to the corresponding holocrystalline subvolcanic rocks of analogous chemical composition, such as nepheline-monzonite, essexite, etc. Series Ib trends toward the phonolite pole. Both series are well known from numerous Atlantic provinces which developed analogously, and they represent the products of the normal crystallization differentiation of a basaltic-gabbroidal magma of simatic origin. Special conditions govern whether the development will proceed according to Ia — toward trachyte, or according to Ib — toward phonolite. Phonolitic residual melts may result instead of trachytic ones, either due to separation of augite instead of olivine, or due to pneumatolytic alkali enrichment. In the latter case members of the sodalite family occur instead of the nepheline. Series II is decidedly less siliceous than series I, and includes ankaratrite ± melilite, limburgite, tephrite, nephelinite ± melilite, and phonolite. In the last type, a leucite is formed, despite the low K-content, due to the low silication. Ijolites are known as the holocrystalline forms. Lastly, series III represents an association with an extremely low degree of silication, the representatives of which are only rarely found. The rocks meriting mention are: ankaratrite ± melilite, olivine-melilite, K-poor leucite basalt, and the holocrystalline form melteigite.

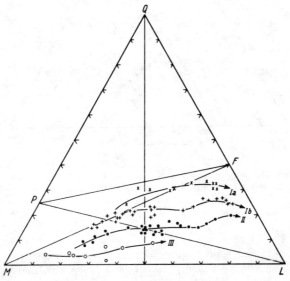

FIGURE 58. The young igneous rocks from Cape Verde and Dakar, plotted on the QLM-triangle. The series of different degrees of silication, which were distinguished in Figure 26, shows up clearly because of their different Q-levels. After C. Burri (1959)

The occurrence of such poorly silicated series as II and III, near the normal series I, may be attributed to desilication, by means of calc assimilation. This assumption is acceptable because, on the one hand, calcareous sediments of great thickness are known on Cape Verde, and on the other hand striking reaction phenomena were observed between limestones and silicate melts. These have resulted, among others, in the rebuilding of cancrinite in the holocrystalline, subvolcanic rocks (C. Burri, 1959). This area has a very complex character. In contrast to the Vesbic volcano, it comprises a multitude of individual volcanoes. It is impossible to establish an exact chronology of the eruptions, especially so as large parts of these are covered by the sea. For these reasons, no conclusions may be reached as to the development of the desilication process with the time. It may also be assumed that principally analogous desilication processes took place in different local magma reservoirs, varying somewhat at times. Thus, it becomes quite possible that at one and the same time magmas of different degrees of silication were supplied by different volcanoes. This is in contrast with the Vesbic volcano with its uniform magma reservoir and its single conduit.

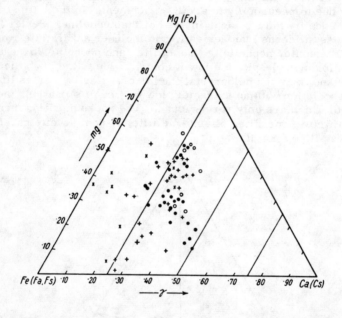

FIGURE 59. The young igneous rocks from Cape Verde and Dakar, represented in the MgFeCa-triangle. The increase of the Ca, which is not bound to Al, with the increase of desilication (series Ia→Ib→II→III corresponding to Figures 26 and 58) is clearly illustrated. After C. Burri (1959)

The MgFeCa-triangle on Figure 59 clearly shows how the portion of Ca not bound to Al, given by Cs or γ, increases from series I through II to III.

c) The method of the remainder triangle

The QLM-triangle is not only a very suitable means of representing graphically the chemism of rock series on petrographic provinces, but it also makes possible the semi-graphic calculation of norm variants which are adapted to the modus. This method is used to advantage when complex minerals are present. Chemically, these minerals deviate considerably from the idealized normative components which were used for the calculation of the standard norm. This, for example, is the case when Al-bearing augites and hornblendes are present. The principle of the method, which is especially well adapted to treatment of metamorphic mineral components and which was firstly used by E. A. Diehl (1938) and H. S. Wang (1939), is given hereafter. At first a list of normative components is calculated, which in compostion essentially correspond to the idealized formulae. Their amount is determined either from the rock's chemism, e. g. Cp from the P_2O_5-content, Ru or Tit from the amount of TiO_2, carbonate from the content of CO_2, etc, or it is determined according to the results obtained from thin-section measurements. Either the basis or cation percentages may be used as references for this calculation. This calculation is carried out until only three components, of mostly complex nature, remain unreckoned. The remainder, R, which is thus obtained, should at first be recalculated to basis components in the usual manner. After recalculation to the sum of 100, these are combined, in the usual way, to give the composite components $Q_R L_R M_R$ (to distinguish them from the equally designated QLM values which refer to the complete analysis). On the basis of these values R may be introduced into the QLM-triangle. The three components A, B and C, which were not calculated in the meanwhile, should now also be introduced into the QLM-triangle. For this purpose one may either use the chemical composition of analyzed samples of the corresponding mineral with equal optical properties, or the chemical analyses of the corresponding components which are isolated from the rock itself. The latter is obviously the ideal solution.

The triangle ABC within the QLM-triangle may be designated as an oblique-angled concentration triangle. In this triangle the percentage compositions of A, B and C may be read for each point (Figure 60).

If the projection point of R falls within the triangle ABC, the ratio $A:B:C$ into which R should be subdivided, immediately reveals how the modus-adapted norm variant should be obtained. A reverse calculation of the composition of R from the read-off ratio $A:B:C$ shows whether or not the previously made assumptions are in agreement with the facts. This control is necessary, as Q, L, M or Q_R, L_R, M_R represent composite components.

If the projection point of R does not fall within the triangle ABC, this may be explained otherwise.

1. The assumptions on the chemism of the complex components A, B, C are incorrect, either because the optical determinations were imperfect, or that they were not sufficiently exactly related to the chemism, or that the possibility of mixed-crystal formation was not sufficiently taken into account.

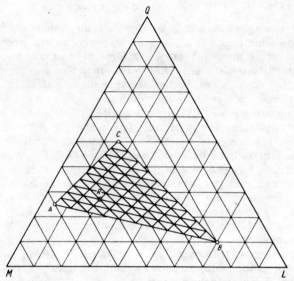

FIGURE 60. Oblique-angled "remainder triangle" within the *QLM*-triangle. The remainder *R QR* 27, *LR* 20, *MR* 53 may be divided, graphically, into *A* 60, *B* 20, *C* 20 where *A*, *B*, and *C*, themselves correspond to the following compositions: *A* (*Q* 25, *L* 5, *M* 70), *B* (*Q* 10, *L* 70, *M* 20) and *C* (*Q* 50, *L* 15, *M* 35).

2. The assumption made on the chemism of *A*, *B*, *C* may indeed be in agreement with the facts. However, errors may have been introduced in the minerals first formed and these errors may have crept into the remainder formation, by subtraction of the eliminated normative minerals. This is especially the case when the amount of these minerals is not unequivocably determined by the chemism, but is fixed by thin-section measurement instead.

If there is reason for assuming that case 1 is at hand, one should attempt to obtain a better result by changing the composition of *A*, *B* or *C*, i.e., by selecting other analyzed examples. In particular, Al-bearing hornblendes vary most markedly in their composition, as is very clearly shown by the work of H. S. Wang (1939, Diagram 17).

If case 2 appears to be at hand, an attempt must be made to move the projection point of the remainder *R* into the triangle *ABC* by changing its composition. E. A. Diehl (1938, especially 357-366) has shown for a number of cases how this task may be accomplished graphically. Some of the mentioned possibilities are as follows (Figure 61):

a) If the remainder is either enriched or empoverished in *Q*, the point of projection of *R* is displaced along a straight line passing through the *Q* apex of the *QLM*-triangle and through *R*.

b) Addition or subtraction of **Hm**, **Mt**, **Cc** or **Dol** will displace the projection point from *R* along a straight line which extends to the "oxide point" (coordinates: *Q* − 50, *L* 0, *M* + 150).

c) On addition or subtraction of feldspar or of **An** only, i. e., on changing the total feldspar contents or the composition of the plagioclase, **R** will be displaced along a straight line running through **F** and **R**.

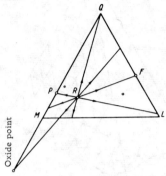

FIGURE 61. Possibilities of displacement of the remainder point by changing the chemism

Obviously, these possibilities may appear in combined form. The method is of importance mainly for metamorphic rocks; for its application reference should be made to the section dealing with the mineral composition of the prasinites. Here we shall deal only with a simpler example, i. e. the ap‑proximate determination of the chemism of the horn‑blende in hornblende rocks of simple composition, e. g. quartz-diorites, diorites, amphibolites, etc. In contrast to the previously outlined procedure for the division of a remainder R into three components A, B and C, we now deal with determination by construction of the third cor‑ner of the remainder triangle, when the position of the two other corners, as well as that of R, is known in the QLM triangle. Let us assume that by mathematical elimination of Cp, Ru or Tit, Bi, etc, the number of components may be reduced to three, namely: quartz, feldspar and horn‑blende. In addition, let us assume that after the recalculation to 100, this remainder R will have the coordinates Q_R 44, L_R 39, M_R 17 (Figure 62). It is also known from the thin-section measurement that the modus which was recalculated to 100 after subtraction of the previously mentioned components yields 20% quartz, 50% feldspar and 30% hornblende. The projection point of the hornblende in the QLM-triangle may now be found in the follow‑ing way. The line QF, which is a side of the component triangle $QFHo$, is divided into 100 divisions, setting the distance $QA = 50$. This corresponds to a feldspar content of 50%. FB is set at 20, corresponding to a quartz content of 20%. The required, third apex Ho of the component triangle $QFHo$ is determined by drawing a line parallel to AR through Q, and a line parallel to BR through F. The coordinates of Ho in the QLM-triangle deter‑mine the composition of the hornblende. The fields of the kata-, meso-, and epi-hornblendes, according to H. S. Wang (1939), are indicated in the figure.

For the discussion of special cases which consider the representation of complex hornblendes in the QLM-projection, reference should also be made to P. Niggli (1946), 34-43.

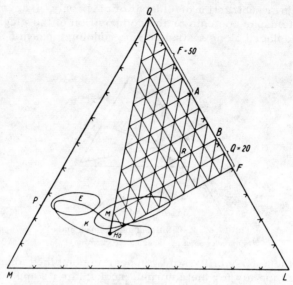

FIGURE 62. Graphic determination of hornblende from dioritic-amphibolitic rocks, the remainder of which is modally composed of 20% Q, 50% feldspar and 30% hornblende. The coordinates of R are $Q_R 44$, $L_R 39$, $M_R 17$. QF is divided into 100 parts with $QA = 50$ and $FB = 20$, in correspondence with the modus. Lines parallel to AR and BR, are drawn through Q and F respectively, thus providing the third apex Ho of the remainder triangle. The coordinates of this apex correspond to the required composition of the hornblende, being $Q\,14$, $L\,29{,}5$, and $M\,56{,}5$. The statistically determined fields of the kata-, meso-, and epi-hornblendes, according to H. Wang (1939), are designated as K, M, and E.

11. The Direct Calculation of the Basis Group Values Q, L, M from the Niggli-Values

In the graphic representation discussed in the previous sections, the used group values Q, L, M, together with the ratios π and γ, are generally obtained from the calculated basis values. One may, however, not be dealing with individual analyses, but with average values expressed as Niggli-values, e.g., those obtained by average-value graphic determination or by interpolation from variation diagrams. In such cases one is confronted with the problem of calculating Q, L, M, π and γ directly from these values. In many cases ti and p may be disregarded. In the following outline, however, the complete procedure will be provided. The simplified expressions may, however, be obtained immediately by setting $ti = p = 0$.

In order to obtain the equations for the conversion, various cases must be distinguished, according to the chemism. These cases may be characterized as follows:

A. $al < alk$, i.e. an excess of alkali over alumina. This corresponds to cases V, Va, VI and VIa of Table 8, p. 122. The possible basis components are: *Kp, Ne, Ns, Ks, Cs, Fs, Fa, Fo, Q, Ru, Cp*.

B. $al \leq (alk + c)$, i.e. neither alkali- nor alumina-excess is present (the so-called normal case). This corresponds to cases IIIa, IV and IVa of Table 8. The possible basis components are: *Kp, Ne, Cal, Cs, Fs, Fa, Fo, Q, Ru, Cp*.

C. Excess of alumina is present. Two cases should be distinguished:

1. $al > alk$, but at the same time $al < (alk + 2c + 2fm)$, corresponding to case Ia, II, IIa, and III of Table 8. The following basis components may occur: *Kp, Ne, Cal, Sp, Hz, Fs, Fa, Fo, Q, Ru, Cp*.

2. $al > alk$, but at the same time $al > (alk + 2c + 2fm)$, corresponding to case I of Table 8. The following basis components may occur: *Kp, Ne, Cal, Sp, Hz, C, Fs, Q, Ru, Cp*.

If the following are taken as a reference:

$$si = Si$$
$$2\,al = Al$$
$$fm = Fe^{3+} + Fe^{2+} + Mg$$
$$c = Ca$$
$$2\,alk = Na + K$$
$$ti = Ti$$
$$2\,p = P$$

$$si + 100 + al + alk + ti + 2p = Si + Al + Fe + Mg + Ca + Na + K + Ti + P = Z,$$

then the following relations may be obtained for the four cases which were distinguished:

Case A: $al < alk$

$$Kp + Ne = 6\,al \qquad\qquad Ru = ti$$
$$Ns + Ks = 3\,(alk - al) \qquad Cp = 5\,p$$
$$Cs = {}^3/_2\,(c - 3\,p)$$
$$Fs + Fa + Fo = {}^3/_2\,fm$$
$$Q = si - [2\,al + (alk - al) + {}^1/_2(c - 3\,p) + {}^1/_2\,fm]$$
$$= si - [100 - {}^1/_2\,(fm + c - 3\,p)].$$

By combination one obtains:

$$\left.\begin{array}{l} L' = (Kp + Ne) = 6\,al \\ M' = (Ns + Ks + Cs + Fs + Fa + Fo + ti + 5p) = 3\,(alk-al) + {}^3/_2(fm+c) + ti + \dfrac{7}{2}p \\ Q' = si - [100 - {}^1/_2(fm + c - 3\,p)]. \end{array}\right\}$$

The values Q, L and M, which by definition are calculated to the sum of 100, can be obtained by the following conversion:

$$L = \frac{L'}{Z}\,100, \qquad M = \frac{M'}{Z}\,100, \qquad Q = \frac{Q'}{Z}\,100,$$

where $Z = si + 100 + al + alk + ti + 2p$.

In addition one obtains:
$$\pi = 0, \quad \gamma = \frac{c-3p}{c+fm-3p}, \quad \mu = \frac{fm}{c+fm-3p} mg.$$

Case B: $al \leq (alk + c)$, the so-called normal case

$Kp + Ne = 6\,alk \qquad\qquad Ru = ti$
$Cal = 3(al - alk) \qquad\quad Cp = 5p$
$Cs = {}^3/_2[c - (al - alk) - 3p]$
$Fs + Fa + Fo = {}^3/_2\,fm$
$\quad Q' = si - \{2\,alk + {}^1/_2[c-(al-alk)-3p] + {}^1/_2 fm\}$
$\quad\;\; = si - [50 + 2\,alk - al - {}^3/_2\,p].$

$\left.\begin{array}{l} L' = 3(al + alk) \\ M' = 3(50 - al) + {}^1/_2 p + ti \\ Q' = si - (50 + 2\,alk - al - {}^3/_2\,p) \end{array}\right\}$ from which Q, L, M are calculated as in A.

In addition one obtains:
$$\pi = \frac{al - alk}{al + alk}, \quad \gamma = \frac{c + alk - al - 3p}{100 - 2\,al - 3p}, \quad \mu = \frac{fm}{100 - 2\,al - 3p}\,mg.$$

Case C 1: $(alk + c) < al < (alk + 2c + 2\,fm)$

$Kp + Ne = 6\,alk \qquad\qquad Fs + Fo + Fa = {}^3/_2(fm - al + c + alk - 3p)$
$Cal = 3(c - 3p) \qquad\qquad\qquad\qquad\;\;\; = {}^3/_2[100 - (2\,al + 3p)]$
$Sp + Hz = 3[(al-alk)-(c-3p)] \qquad Ru = ti$
$\qquad\qquad\qquad\qquad\qquad\qquad\qquad Cp = 5p$
$\quad Q' = si - \{2\,alk + {}^1/_2[100 - (2\,al + 3p)]\}$
$\quad\;\; = si - (50 + 2\,alk - al - {}^3/_2 p)$

$\left.\begin{array}{l} L' = 3[2\,alk + (c - 3p)] \\ M' = 150 - 3(c + alk) + {}^{19}/_2 p + ti \\ Q' = si - (50 + 2\,alk - al - {}^3/_2 p) \end{array}\right\}$ from which Q, L, M are calculated as in A.

and in addition:
$$\pi = \frac{(c-3p)}{(2\,alk + c - 3p)}, \quad \gamma = 0, \quad \mu = mg.$$

Case C 2: $al > (alk + 2c + 2\,fm)$

$Kp + Ne = 6\,alk \qquad C = 2\{al - [alk + (c-3p) + fm]\} = 2[al-(alk+c$
$Cal = 3(c - 3p) \qquad\qquad + fm - 3p)] = 4(al - 50) + 6p$
$Sp + Hz = 3\,fm. \qquad Ru = ti, \qquad Cp = 5p, \qquad Q' = si - 2\,alk.$

$\left.\begin{array}{l} L' = 3(2\,alk + c - 3p) \\ M' = 4\,al + 3\,fm - 200 + 11\,p + ti \\ Q' = si - 2\,alk \end{array}\right\}$ from which Q, L, M are calculated as in A.

and in addition:

$$\pi = \frac{c-3p}{2\,alk+c-3p}, \qquad \gamma = 0, \qquad \mu = 0.$$

The reverse procedure, i.e., the calculation of the Niggli-values from Q, L, M, π and γ, will rarely be called for. For the normal case this problem was solved by C. Burri and P. Niggli (1945), page 97, disregarding ti and p.

An additional means of converting Niggli-values to QLM-values was provided by P. Niggli, 1943, p. 603.

12. The Application of the Equivalent Norm to the Study of Metamorphic Rocks

a) General

The great advantage of the equivalent norm over other similar calculation methods becomes especially apparent when it is applied to **metamorphic rocks**. These rocks are characterized by the fundamental feature that for a given chemism, several different mineral facies may appear, depending on the prevailing PT-conditions. The phenomenon of **heteromorphism**, which admittedly occurs also in igneous rocks (although the full significance of this fact was not recognized for a long time), is the main characteristic of metamorphic rocks. Thus, in principle only one calculation method can be used for the study of these rocks, which allows for a preliminary discussion of the heteromorphic possibilities when applied to igneous rocks. This is entirely true for the method of the **equivalent norm**, as was shown in the respective sections, because this method enables one to pass from one calculated mineral facies to another by means of simply formulated reaction equations. In addition, this method allows for the introduction into the calculation of compositions determined by the chemical analysis, instead of ideal mineral compositions. It should be thus concluded that the equivalent-norm method is highly suitable for the handling of metamorphic rocks, in contrast to the weight percentage CPIW-norm which, however, was never suggested for this purpose by its authors. Particularly at present, when the conventionally accepted boundaries between igneous and metamorphic rocks have partly been eliminated, and the two groups are combined according to P. Niggli's (1948) suggestion under the term "endogenic rocks", a calculation method equally applicable igneous and metamorphic rocks in the conventional sense, is much sought after.

As in the case of the igneous rocks, which until recently were the only ones to be considered, two aims are served by the calculation of the equivalent norm. Firstly, normative mineral components are calculated according to a standardized procedure, thus facilitating the comparison between rocks or rock series. It may be desirable to draw such comparisons between rocks of similar or of different mineral facies. In the latter case it is advantageous, for the sake of comparison, to recalculate the rocks to an equal facies of mineral components. The mentioned method is especially suited for this purpose. Secondly, in metamorphic rocks, as well as in igneous rocks, this method should solve the problem of calculating a normative mineral composition which should correspond as nearly as possible to the modus in question, i.e., this method should allow for the building of variants without resort to lengthy calculation. Since, however, the dependence of mineral composition on the conditions of formation is the main characteristic of metamorphic rocks, it is obvious that different standard norms should be built, corresponding to these different conditions of formation. In contrast to this, only one norm was needed for the igneous rocks. This was the standard katanorm, corresponding to the high temperatures of magmatic crystallization. Initially, F. Becke chose the two extreme cases — of high and of relatively low formation temperature — in order to characterize the conditions of formation of metamorphic rocks. Subsequently, U. Grubenmann introduced a third case, characterizing intermediate formation temperatures, mainly as the result of his observations on the crystalline schists at the southern margin of the Gotthard massif. These are the known kata-, meso- and epi-facies*.

In the meantime it has proved necessary to introduce a standard epinorm, close to the standard katanorm, and to provide the required standardized calculation procedure. In accordance with the previously mentioned considerations, no specific standard norm should be considered for the meso-facies. In certain cases, meso-mineral components should be calculated in accordance with the corresponding microscopic observations. In the future it should prove possible to arrive at standardized calculation methods for the individual facies of the facies classification of P. Eskola and others (sub-facies, in P. Niggli's sense). In the meanwhile it is preferable to wait for the complete clarification of the different views on the definition of exact mutual boundaries between these facies.

---

*   As is known, F. Becke did not speak of f a c i e s, but of "T i e f e n s t u f e n" [depth stages], a term which is still most frequently used in the Austrian or East-Alpine literature. U. Grubenmann spoke of "z o n e s", i.e. kata-, meso-, and epi-zones. In both cases it was stressed that kata-, meso-, and epi-conditions normally prevail in the deepest, middle and uppermost levels of the earth's crust, respectively. Despite the aforementioned clear definitions provided by F. Becke, U. Grubenmann, and other outstanding investigators of rock metamorphism, among which only P. Niggli and P. Eskola will be mentioned here, these designations have repeatedly lead to numerous misunderstandings. For example, the opinion was repeatedly expressed that kata-rocks can only be formed at great depths, or that the deepest parts of the now exposed Fenno-Scandian or North American basement complexes are composed solely of kata-rocks. As is well known, this is in absolute contradiction to the facts. Apparently, the fact was overlooked that the Becke-Grubenmann definition dealt intentionally with "normal" conditions. Such conditions are often not existent, especially during periods of mountain building, during which time metamorphic rocks are formed. In order to prevent such misunderstandings, P. Niggli has introduced the terms kata-, meso-, and epi-"f a c i e s", instead of kata-, meso-, or epi-"z o n e s". These terms designate the kata-(high-temperature), meso-(intermediate-temperature), or epi-(low temperature) conditions under which the rocks were formed. He intended these three facies to be recognized as "c h i e f   f a c i e s" types, including the numerous other facies distinguished in the facies classification of P. Eskola, C. E. Tilley, Th. Vogt, Fr. J. Turner, and others. In P. Niggli's view, the other facies serve as s u b - f a c i e s. It was, however, immediately recognized that the PT-ranges of the various sub-facies grouped under the meso-facies, do not coincide completely. Thus, while not discarding the term m e s o - f a c i e s as such, P. Niggli did not attach the same significance to it as to the kata- and epi-facies. The justification for such a treatment is also supported by actual facts already presented by U. Grubenmann. This author has proved that his "meso-zone", in the sense of a t r a n s i t i o n a l   z o n e between the kata- and the epi-zones, may be characterized by an a s s o c i a t i o n of typomorphic minerals of the kata- and the epi-zone, as well as by its own typomorphic mineral-components.

Nothing new need be added for the norm calculation of the metamorphic rocks, as compared with the procedure adopted and already explained in detail for the igneous rocks. This norm may either be calculated via the basis or directly from the cation percentages; its derivation from the Niggli-values is less recommended. The basis components Sp, Hz and C which only rarely occur in fresh, non-endomorphic igneous rocks, may now appear in larger amounts due to the possible presence of great alumina excesses in rocks of sedimentary origin. These cases correspond to the groups I, Ia, II, IIa, and III of the compilation presented in Table 8 (p. 122).

Whereas almost no new normative components need be introduced for the **katanorm** of metamorphic rocks, provided that para-rocks very rich in Al or Ca are disregarded, this is not the case with the **epinorm**, for which a large number of such components are required. The following compilation provides a list of the normative components to be considered in metamorphic rocks, in addition to the components which were already enlisted in section II 5, Table 7, p. 116 for the igneous rocks. These are arranged in a completely general order, disregarding the facies.

Table IX. Equivalent normative components of metamorphic rocks*

Hornblendes and augites

Grammatite	$1\ Gram = 1/15\ (8\ SiO_2 \cdot 5\ MgO \cdot 2\ CaO \cdot H_2O)$	(54.8)
Fe-actinolite	$1\ Fe\text{-}Akt = 1/15\ (8\ SiO_2 \cdot 5\ FeO \cdot 2\ CaO \cdot H_2O)$	(64.7)
Actinolite (average)	$1\ Akt = 1/30\ (16\ SiO_2 \cdot 8{,}5\ MgO \cdot 1{,}5\ FeO \cdot 4\ CaO \cdot 2\ H_2O)$	(55.7)
Anthophyllite₁	$1\ Anth_1 = 1/15\ (8\ SiO_2 \cdot 7\ MgO \cdot H_2O)$	(52.0)
Anthophyllite₂	$1\ Anth_2 = 1/30\ (16\ SiO_2 \cdot 7\ MgO \cdot 7\ FeO \cdot 2\ H_2O)$	(59.4)
Glaucophane	$1\ Glph = 1/15\ (8\ SiO_2 \cdot Al_2O_3 \cdot 3\ (Mg,Fe)O \cdot Na_2O \cdot H_2O)$	(52.2–58.5)
Omphacite (average)	$1\ Omph = 1/16\ (8\ SiO_2 \cdot Al_2O_3 \cdot 2\ MgO \cdot 2\ CaO \cdot Na_2O)$	(52.1)

Sheet-silicates

Antigorite	$1\ Ant = 1/5\ (2\ SiO_2 \cdot 3\ MgO \cdot 2\ H_2O)$	(55.4)
Fe-antigorite	$1\ Fe\text{-}Ant = 1/5\ (2\ SiO_2 \cdot 3\ FeO \cdot 2\ H_2O)$	(74.3)
Amesite	$1\ At = 1/5\ (SiO_2 \cdot Al_2O_3 \cdot 2\ MgO \cdot 2\ H_2O)$	(55.7)
Fe-amesite	$1\ Fe\text{-}At = 1/5\ (SiO_2 \cdot Al_2O_3 \cdot 2\ FeO \cdot 2\ H_2O)$	(68.3)
Fe-ottrelite	$1\ Ot = 1/4\ (SiO_2 \cdot Al_2O_3 \cdot FeO \cdot H_2O)$	(63.0)
Mg-ottrelite	$1\ Mg\text{-}Ot = 1/4\ (SiO_2 \cdot Al_2O_3 \cdot MgO \cdot H_2O)$	(55.1)
Kaolin	$1\ Kaol = 1/4\ (2\ SiO_2 \cdot Al_2O_3 \cdot 2\ H_2O)$	(64.5)
Talc	$1\ Tc = 1/7\ (4\ SiO_2 \cdot 3\ MgO \cdot 2\ H_2O)$	(56.7)
Pyrophyllite	$1\ Pph = 1/6\ (4\ SiO_2 \cdot Al_2O_3 \cdot 2\ H_2O)$	(63.0)
Muscovite₁	$1\ Ms_1 = 1/14\ (6\ SiO_2 \cdot 3\ Al_2O_3 \cdot K_2O \cdot 2\ H_2O)$	(56.9)
Muscovite₂ (phengitic)	$1\ Ms_2 = 1/14\ (7\ SiO_2 \cdot 2\ Al_2O_3 \cdot (Mg,Fe)O \cdot K_2O \cdot 2\ H_2O)$	(55.4–57.9)
Paragonite	$1\ Pg = 1/14\ (6\ SiO_2 \cdot 3\ Al_2O_3 \cdot Na_2O \cdot 2\ H_2O)$	(54.6)
Margarite	$1\ Marg = 1/7\ (2\ SiO_2 \cdot 2\ Al_2O_3 \cdot CaO \cdot H_2O)$	(57.0)

* This table compliments Table VII, p. 116 in which the minerals occurring both in igneous and metamorphic rocks were already included.

Epidote-zoisite group

Pistacite (end-member)	$1\ Pi = {}^1/_{16}(6\ SiO_2 \cdot 2\ Al_2O_3 \cdot Fe_2O_3 \cdot 4\ CaO \cdot H_2O)$	(63.9)
Zoisite	$1\ Zo = {}^1/_{16}(6\ SiO_2 \cdot 3\ Al_2O_3 \cdot 4\ CaO \cdot H_2O)$	(56.7)

Garnet-group

Almandine	$1\ Alm = {}^1/_8(3\ SiO_2 \cdot Al_2O_3 \cdot 3\ FeO)$	(62.2)
Grossularite	$1\ Gro = {}^1/_8(3\ SiO_2 \cdot Al_2O_3 \cdot 3\ CaO)$	(56.3)
Pyrope	$1\ Pyp = {}^1/_8(3\ SiO_2 \cdot Al_2O_3 \cdot 3\ MgO)$	(50.4)
Spessartite	$1\ Spe = {}^1/_8(3\ SiO_2 \cdot Al_2O_3 \cdot 3\ MnO)$	(61.9)
Andradite	$1\ Andr = {}^1/_8(3\ SiO_2 \cdot Fe_2O_3 \cdot 3\ CaO)$	(63.5)

Alumina- and calc-silicates

Sillimanite, andalusite, kyanite	$1\ Sil = {}^1/_3(SiO_2 \cdot Al_2O_3)$	(54.0)
Staurolite	$1\ Staur = {}^1/_7(2\ SiO_2 \cdot 2\ Al_2O_3 \cdot FeO \cdot H_2O)$	(56.5)
Cordierite	$1\ Cord = {}^1/_{11}(5\ SiO_2 \cdot 2\ Al_2O_3 \cdot 2\ MgO)$	(58.2)
Fe-cordierite	$1\ Fe\text{-}Cord = {}^1/_{11}(5\ SiO_2 \cdot 2\ Al_2O_3 \cdot 2\ FeO)$	(58.9)
Vesuvianite	$1\ Ves = {}^1/_{25}(9\ SiO_2 \cdot 2\ Al_2O_3 \cdot 10\ CaO \cdot 2\ MgO \cdot 2\ H_2O)$	(56.9)
Larnite	$1\ Cs = {}^1/_3(SiO_2 \cdot 2\ CaO)$	(57.4)
Xonotlite	$1\ Xon = {}^1/_{12}(6\ SiO_2 \cdot 6\ CaO \cdot H_2O)$	(58.1)
Spurrite	$1\ Spur = {}^1/_7(2\ SiO_2 \cdot 5\ CaO \cdot CO_2)$	(63.5)
Tilleyite	$1\ Till = {}^1/_7(2\ SiO_2 \cdot 5\ CaO \cdot 2\ CO_2)$	(69.8)

Carbonates and sulfates

Calcite	$1\ Cc = 1(CaO \cdot CO_2)$	(100.1)
Magnesite	$1\ Mgs = 1(MgO \cdot CO_2)$	(84.3)
Siderite	$1\ Sid = 1(FeO \cdot CO_2)$	(115.8)
Na-carbonate	$1\ Nc = {}^1/_2(Na_2O \cdot CO_2)$	(53.0)
Anhydrite	$1\ A = {}^1/_2(CaO \cdot SO_3)$	(68.1)

The relationships provided in section B II 6 a for the calculation of the equivalent norm from the cation percentages may be applied to the newly introduced normative components of metamorphic rocks in the following, analogous way:

Hornblendes and augites

$1\ Gram = {}^8/_{15}\ Si + {}^5/_{15}\ Mg + {}^2/_{15}\ Ca$
$1\ Fe\text{-}Akt = {}^8/_{15}\ Si + {}^5/_{15}\ Fe + {}^2/_{15}\ Ca$
$1\ Anth_1 = {}^8/_{15}\ Si + {}^7/_{15}\ Mg$
$1\ Anth_2 = {}^{16}/_{30}\ Si + {}^7/_{30}\ Mg + {}^7/_{30}\ Fe$
$1\ Glph = {}^8/_{15}\ Si + {}^2/_{15}\ Al + {}^3/_{15}(Mg, Fe) + {}^2/_{15}\ Na$
$1\ Omph = {}^8/_{16}\ Si + {}^2/_{16}\ Al + {}^2/_{16}\ Mg + {}^2/_{16}\ Ca + {}^2/_{16}\ Na$.

Sheet-silicates

$1\ Ant = {}^2/_5\ Si + {}^3/_5\ Mg$
$1\ Fe\text{-}Ant = {}^2/_5\ Si + {}^3/_5\ Fe$
$1\ At = {}^1/_5\ Si + {}^2/_5\ Al + {}^2/_5\ Mg$
$1\ Fe\text{-}At = {}^1/_5\ Si + {}^2/_5\ Al + {}^2/_5\ Fe$

$1\ Ot = {}^1/_4\,Si + {}^2/_4\,Al + {}^1/_4\,Fe$
$1\ Mg\text{-}Ot = {}^1/_4\,Si + {}^2/_4\,Al + {}^1/_4\,Mg$
$1\ Kaol = {}^1/_2\,Si + {}^1/_2\,Al$
$1\ Tc = {}^4/_7\,Si + {}^3/_7\,Mg$
$1\ Pph = {}^2/_3\,Si + {}^1/_3\,Al$
$1\ Pg = {}^3/_7\,Si + {}^3/_7\,Al + {}^1/_7\,Na$
$1\ Marg = {}^2/_7\,Si + {}^4/_7\,Al + {}^1/_7\,Ca$.

Epidote-Zoisite group

$1\ Pi = {}^3/_8\,Si + {}^2/_8\,Al + {}^1/_8\,Fe^{3+} + {}^2/_8\,Ca$
$1\ Zo = {}^3/_8\,Si + {}^3/_8\,Al + {}^2/_8\,Ca$.

Garnet group

$1\ Alm = {}^3/_8\,Si + {}^2/_8\,Al + {}^3/_8\,Fe$
$1\ Gro = {}^3/_8\,Si + {}^2/_8\,Al + {}^3/_8\,Ca$
$1\ Pyp = {}^3/_8\,Si + {}^2/_8\,Al + {}^3/_8\,Mg$
$1\ Spe = {}^3/_8\,Si + {}^2/_8\,Al + {}^3/_8\,Mn$
$1\ Andr = {}^3/_8\,Si + {}^2/_8\,Fe^{3+} + {}^3/_8\,Ca$.

Alumina-silicates

$1\ Sil = 1\ Dist = {}^1/_3\,Si + {}^2/_3\,Al$
$1\ Staur = {}^2/_7\,Si + {}^4/_7\,Al + {}^1/_7\,Fe$
$1\ Cord = {}^5/_{11}\,Si + {}^4/_{11}\,Al + {}^2/_{11}\,Mg$
$1\ Fe\text{-}Cord = {}^5/_{11}\,Si + {}^4/_{11}\,Al + {}^2/_{11}\,Fe$
$1\ Ves = {}^9/_{25}\,Si + {}^4/_{25}\,Al + {}^2/_{25}\,Mg + {}^{10}/_{25}\,Ca$
$1\ Cs = {}^1/_3\,Si + {}^2/_3\,Ca$
$1\ Xon = {}^1/_2\,Si + {}^1/_2\,Ca$
$1\ Spur = {}^2/_7\,Si + {}^5/_7\,Ca$
$1\ Till = {}^2/_7\,Si + {}^5/_7\,Ca$.

Carbonates, Sulfates, Halides

$1\ Cc = 1\ Ca$
$1\ Mgs = 1\ Mg$
$1\ Sid = 1\ Fe$
$1\ Nc = {}^2/_2\,Na = 1\ Na$
$1\ A = {}^1/_2\,S + {}^1/_2\,Ca$.

If one has need of carrying out the reverse procedure, i. e. calculating the cation percentages from the norm of metamorphic rocks, the expressions provided in section B II 6 c must be supplemented by the newly introduced normative components. This should, however, cause no difficulties, in view of the data provided. Thus, it is considered unnecessary to cite these expressions in detail.

b) The standard katanorm of metamorphic rocks

In addition to the reaction equations given for the building of the standard katanorm of the igneous rocks, a number of others are required for the construction of normative kata-components, in order to calculate the standard katanorm and its variants for metamorphic rocks from the basis. The most important of these are given in the following compilation:

Sillimanite (kyanite):
$$Al_2O_3 + SiO_2 = SiO_2 \cdot Al_2O_3.$$
$$2\,C \quad\ 1\,Q \quad\ 3\,Sil\,(Dist)$$

Cordierite:
$$2(Al_2O_3 \cdot MgO) + 5\,SiO_2 = 5\,SiO_2 \cdot 2\,Al_2O_3 \cdot 2\,MgO,$$
$$6\,Sp \qquad\qquad 5\,Q \qquad\qquad 11\,Cord$$

$$2(Al_2O_3 \cdot FeO) + 5\,SiO_2 = 5\,SiO_2 \cdot 2\,Al_2O_3 \cdot 2\,FeO,$$
$$6\,Hz \qquad\qquad 5\,Q \qquad\qquad 11\,Fe\text{-}Cord$$

$$2(SiO_2 \cdot MgO) + 2(SiO_2 \cdot Al_2O_3) + SiO_2 = 5\,SiO_2 \cdot 2\,Al_2O_3 \cdot 2\,MgO,$$
$$4\,En \qquad\qquad 6\,Sil \qquad\qquad 1\,Q \qquad\qquad 11\,Cord$$

$$2(SiO_2 \cdot FeO) + 2(SiO_2 \cdot Al_2O_3) + SiO_2 = 5\,SiO_2 \cdot 2\,Al_2O_3 \cdot 2\,FeO,$$
$$4\,Hy \qquad\qquad 6\,Sil \qquad\qquad 1\,Q \qquad\qquad 11\,Fe\text{-}Cord$$

$$SiO_2 \cdot 2\,MgO + 2\,Al_2O_3 + 4\,SiO_2 = 5\,SiO_2 \cdot 2\,Al_2O_3 \cdot 2\,MgO,$$
$$3\,Fo \qquad\quad 4\,C \qquad\quad 4\,Q \qquad\qquad 11\,Cord$$

$$SiO_2 \cdot 2\,FeO + 2\,Al_2O_3 + 4\,SiO_2 = 5\,SiO_2 \cdot 2\,Al_2O_3 \cdot 2\,FeO.$$
$$3\,Fa \qquad\quad 4\,C \qquad\quad 4\,Q \qquad\qquad 11\,Fe\text{-}Cord$$

Garnet group:
$$Al_2O_3 \cdot CaO + SiO_2 \cdot 2\,CaO + 2\,SiO_2 = 3\,SiO_2 \cdot Al_2O_3 \cdot 3\,CaO,$$
$$3\,Cal \qquad\quad 3\,Cs \qquad\quad 2\,Q \qquad\qquad 8\,Gro$$

$$2\,SiO_2 \cdot Al_2O_3 \cdot CaO + SiO_2 \cdot 2\,CaO = 3\,SiO_2 \cdot Al_2O_3 \cdot 3\,CaO,$$
$$5\,An \qquad\qquad\qquad 3\,Cs \qquad\qquad 8\,Gro$$

$$Al_2O_3 \cdot MgO + SiO_2 \cdot 2\,MgO + 2\,SiO_2 = 3\,SiO_2 \cdot Al_2O_3 \cdot 3\,MgO,$$
$$3\,Sp \qquad\qquad 3\,Fo \qquad\qquad 2\,Q \qquad\qquad 8\,Pyp$$

$$Al_2O_3 \cdot MgO + 2(SiO_2 \cdot MgO) + SiO_2 = 3\,SiO_2 \cdot Al_2O_3 \cdot 3\,MgO,$$
$$3\,Sp \qquad\qquad 4\,En \qquad\qquad 1\,Q \qquad\qquad 8\,Pyp$$

$$Al_2O_3 \cdot FeO + SiO_2 \cdot 2\,FeO + 2\,SiO_2 = 3\,SiO_2 \cdot Al_2O_3 \cdot 3\,FeO,$$
$$3\,Hz \qquad\qquad 3\,Fa \qquad\qquad 2\,Q \qquad\qquad 8\,Alm$$
$$\text{bzw. } 3\,Hz \qquad 4\,Hy \qquad\qquad 1\,Q \qquad\qquad 8\,Alm$$

$$2\,Fe_2O_3 + 3(SiO_2 \cdot 2\,CaO) + 3\,SiO_2 = 2(3\,SiO_2 \cdot Fe_2O_3 \cdot 3\,CaO),$$
$$4\,Hm \qquad\quad 9\,Cs \qquad\qquad 3\,Q \qquad\qquad 16\,Andr$$

$$Fe_2O_3 + 3(SiO_2 \cdot CaO) = 3\,SiO_2 \cdot Fe_2O_3 \cdot 3\,CaO.$$
$$2\,Hm \qquad 6\,Wo \qquad\qquad 8\,Andr$$

207 Miscellaneous

$$SiO_2 \cdot 2\,MgO + SiO_2 \cdot 2\,CaO = 2(SiO_2 \cdot MgO \cdot CaO),$$
$$3\,Fo \qquad\qquad 3\,Cs \qquad\qquad 6\,Mont$$

$$2\,SiO_2 \cdot Al_2O_3 \cdot Na_2O + SiO_2 \cdot 2\,MgO + SiO_2 \cdot 2\,CaO + 4\,SiO_2$$
$$6\,Ne \qquad\qquad\qquad 3\,Fo \qquad\qquad 3\,Cs \qquad\qquad 4\,Q$$
$$= 8\,SiO_2 \cdot Al_2O_3 \cdot 2\,MgO \cdot 2\,CaO \cdot Na_2O,$$
$$16\,Omph$$

$$2\,SiO_2 \cdot Al_2O_3 \cdot Na_2O + 2(SiO_2 \cdot MgO) + 2(SiO_2 \cdot CaO) + 2\,SiO_2$$
$$6\,Ne \qquad\qquad\qquad 4\,En \qquad\qquad 4\,Wo \qquad\qquad 2\,Q$$
$$= 8\,SiO_2 \cdot Al_2O_3 \cdot 2\,MgO \cdot 2\,CaO \cdot Na_2O,$$
$$16\,Omph$$

$$6\,SiO_2 \cdot Al_2O_3 \cdot Na_2O + SiO_2 \cdot 2\,MgO + SiO_2 \cdot 2\,CaO$$
$$10\,Ab \qquad\qquad 3\,Fo \qquad\qquad 3\,Cs$$
$$= 8\,SiO_2 \cdot Al_2O_3 \cdot 2\,MgO \cdot 2\,CaO \cdot Na_2O\,,$$
$$16\,Omph$$

$$2(SiO_2 \cdot 2\,CaO) + CaO \cdot CO_2 = 2\,SiO_2 \cdot 5\,CaO \cdot CO_2,$$
$$6\,Cs \qquad\qquad 1\,Cc \qquad\qquad 7\,Spur$$

$$SiO_2 \cdot 2\,CaO + SiO_2 \cdot CaO + 2(CaO \cdot CO_2) = 2\,SiO_2 \cdot 5\,CaO \cdot 2\,CO_2,$$
$$3\,Cs \qquad\quad 2\,Wo \qquad\quad 2\,Cc \qquad\qquad 7\,Till$$

$$SiO_2 \cdot 2\,MgO + 5(SiO_2 \cdot 2\,CaO) + 2(SiO_2 \cdot Al_2O_3) + SiO_2(+\,2\,W)$$
$$3\,Fo \qquad\quad 15\,Cs \qquad\qquad 6\,Sil \qquad\qquad 1\,Q$$
$$= 9\,SiO_2 \cdot 2\,Al_2O_3 \cdot 2\,MgO \cdot 10\,CaO \cdot 2\,H_2O\,.$$
$$25\,Ves.$$

The formation of epi-minerals will be considered together with the epinorm.
Examples:

Since there is no essential difference between the standard-katanorm calculation of metamorphic and igneous rocks, reference may be made to the examples provided for igneous rocks, and all the more so as, in the case of the cordierite-andesite from Hoyozo, we are dealing with a rock with an alumina-excess. The regulations stated in sections II 4 and II 5 are thus also valid for the metamorphic rocks, since during their formulation a possible Al-excess was taken into account. If this excess becomes larger in derivatives of clayey and clayey-marly sediments, the basis components Sp, Hz and C, as well as the standard components which are formed therefrom, are also present in greater amounts. The examples to be presented here should thus preferentially consider types of extreme composition.

1. The corundum-cordierite-spinel hornfels from Tom a'Mhinn, Glen Lednock, Comrie area, Perthshire, Scotland (C. E. Tilley, 1924, 46), may serve as the first example of an extremely alumina-rich rock (analyzed by C. E. Tilley).

The basis is calculated in the following manner:

	Wt. %		Ru	Cp	Kp	Ne	Cal	Sp	Hz	C	Q
SiO_2	39.40	656			82	68					506
Al_2O_3	35.86	703			82	68	40	176	244	93	
Fe_2O_3	Sp.										
FeO	8.79	122							122		
MnO	Sp.										
MgO	3.57	88						88			
CaO	2.15	38		18			20				
Na_2O	2.10	68				68					
K_2O	3.84	82			82						
TiO_2	2.00	2ɔ	25								
P_2O_5	0.82	12		12							
H_2O+	1.02										
H_2O-	0.29										
S	0.09										
	99.93	1794	25	30	246	204	60	264	366	93	506

Q	Kp	Ne	Cal	Sp	Hz	C	Cp	Ru	Total
28.2	13.7	11.4	3.3	14.7	20.4	5.2	1.7	1.4	100.0

Therefrom the standard katanorm is calculated:

$$3.3\,Cal + 2.2\,Q = 5.5\,An$$
$$13.7\,Kp + 9.1\,Q = 22.8\,Or$$
$$11.4\,Ne + 7.6\,Q = 19.0\,Ab$$
$$11.2\,Hz + 9.3\,Q = 20.5\,\text{Fe-}Cord.$$

Standard katanorm:

Or	Ab	An	Fe-Cord	Hz	Sp	C	Cp	Ru	Total
22.8	19.0	5.5	20.5	9.2	14.7	5.2	1.7	1.4	100.0

With the exception that Fe-cordierite occurs instead of the normal Mg-cordierite, this standard katanorm closely corresponds to the modus of the rock. This results from the silication of *Hz* prior to *Sp*, which was carried out according to the accepted regulations. A variant showing closer correspondence to the modus can be formulated as follows:

Or	Ab	An	Cord+Fe-Cord	Hz+Sp	C	Cp	Ru	Total
22.8	19.0	5.5	20.5	23.9	5.2	1.7	1.4	100.0

According to its description, the rock also contains 0.5% biotite, which may be obtained as follows:

$$6\,SiO_2 \cdot Al_2O_3 \cdot K_2O + 6\,[Al_2O_3 \cdot (Mg, Fe)O]\ (+\ 2\,H_2O)$$

	10 Or	18 (Sp + Hz)	(2 W)
or	5 Or	9 (Sp + Hz)	(1 W)

$$= 6\,SiO_2 \cdot Al_2O_3 \cdot 6\,(Mg, Fe)O \cdot K_2O \cdot 2\,H_2O + 6\,Al_2O_3,$$

	16 Bi	12 C
	8 Bi	6 C

$$0.3\,Or + 0.6\,(Sp + Hz) = 0.5\,Bi + 0.4\,C.$$

Thus, the *Bi*-variant of the standard katanorm, which was adapted to the modus, will yield:

Or	Ab	An	Cord+Fe-Cord	Bi	Hz+Sp	C	Cp	Ru	Total
22.5	19.0	5.5	20.5	0.5	23.3	5.6	1.7	1.4	100.0

Here, again, it is possible to calculate the standard katanorm from the cation percentages instead of via the basis. The cation percentages are as follows:

Si^{4+}	Al^{3+}	Fe^{2+}	Mg^{2+}	Ca^{2+}	Na^+	K^+	Ti^{4+}	P^{5+}	Total
36.5	39.2	6.8	4.9	2.1	3.8	4.6	1.4	0.7	100.0

$$1.4\,Ti = 1.4\,Ru$$
$$0.7\,P + 1.0\,Ca = 1.7\,Cp$$
$$1.1\,Ca + 2.2\,Al + 2.2\,Si = 5.5\,An$$
$$4.6\,K + 4.6\,Al + 13.8\,Si = 23.0\,Or$$
$$3.8\,Na + 3.8\,Al + 11.4\,Si = 19.0\,Ab$$
$$4.9\,Mg + 9.8\,Al = 14.7\,Sp$$
$$6.8\,Fe + 13.6\,Al = 20.4\,Hz.$$

The remaining Al will be: $39.2 - (2.2 + 4.6 + 3.8 + 9.8 + 13.6) = 5.2 = 5.2\,C$, and that of Si: $36.5 - (2.2 + 13.8 + 11.4) = 9.1\,Q$. The latter should be used in part for the silication of Hz, according to: $10.9\,Hz + 9.1\,Q = 20.0$ Fe-Cord. Thus, the standard katanorm calculated from the cation percentages yields the following values, which are in close agreement with those obtained from the basis (values in brackets):

Or	Ab	An	Fe-Cord	Hz	Sp	C	Cp	Ru	Total
23.0	19.0	5.5	20.0	9.5	14.7	5.2	1.7	1.4	100.0
(22.8)	(19.0)	(5.5)	(20.5)	(9.2)	(14.7)	(5.2)	(1.7)	(1.4)	(100.0)

2. Cordierite-spinel hornfels, Creag, near Iolaire, Glen Lednock, Comrie area, Perthshire. Scotland. The basis of this rock, which was also described and analyzed by C. E. Tilley (1924, 52), is calculated as follows:

	Wt. %		Ru	Kp	Ne	Cal	Sp	Hz	Fs	Fa	Q
SiO_2	48.35	805		49	59				10.5	22	664.5
Al_2O_3	28.29	555		49	59	48	176	223			
Fe_2O_3	1.67	21						21			
FeO	10.86	151 } 156						111.5		44.5	
MnO	0.34	5									
MgO	3.54	88					88				
CaO	1.35	24				24					
Na_2O	1.84	59			59						
K_2O	2.28	49		49							
H_2O+	0.76										
H_2O-	0.04										
P_2O_5	0.05										
TiO_2	0.60	8	8								
S	0.07										
	100.04	1765	8	147	177	72	264	334.5	31.5	66.5	664.5

Basis	Q	Kp	Ne	Cal	Sp	Hz	Fs	Fa	Ru	Total
	37.6	8.3	10.0	4.1	15.0	18.9	1.8	3.8	0.5	100.0

The standard katanorm is calculated as follows, according to the previously stated regulations:

$$1.8\,Fs + 0.9\,Fa = 1.8\,Mt + 0.9\,Q$$
$$4.1\,Cal + 2.7\,Q = 6.8\,An$$
$$8.3\,Kp + 5.5\,Q = 13.8\,Or$$
$$10.0\,Ne + 6.7\,Q = 16.7\,Ab$$
$$18.9\,Hz + 15.7\,Q = 34.6\,Fe\text{-}Cord.$$

Up to this point an amount equal to: $2.7 + 5.5 + 6.7 + 15.7 = 30.6\,Q$ was consumed. Since $37.6 + 0.9 = 38.5\,Q$ are available, a remainder of $7.9\,Q$ is still available. This is used for the silication of a part of the Sp:

$$9.5\,Sp + 7.9\,Q = 17.4\,Cord,$$

and the standard katanorm yields:

Or	Ab	An	Fe-Cord	Cord	Sp	Fa	Mt	Ru	Total
13.8	16.7	6.8	34.6	17.4	5.5	2.9	1.8	0.5	100.0

According to the description of the rock, 2% biotite are present. In order to adapt the norm to the modus and in view of the high Fe-content of the rock, pure Fe-biotite may be formed:

$$1.3\,Or + 1.1\,Fa = 2.0\,Fe\text{-}Bi + 0.4\,Q.$$

The remainder of the Fa should be converted into Fe-Ant, in agreement with the modus:

$$1.8\,Fa + 0.2\,Q = 2.0\,Fe\text{-}Ant,$$

and the $0.2\,Q$ excess allows for the silication of $0.3\,Sp$, according to:

$$0.3\,Sp + 0.2\,Q = 0.5\,Cord,$$

so that the variant of the standard katanorm adapted to the modus yields:

Or	Ab	An	Cord+Fe-Cord	Sp+Hz	Fe-Ant	Bi	Mt	Ru	Total
12.5	16.7	6.8	52.5	5.2	2.0	2.0	1.8	0.5	100.0

Here again the standardized norm-calculation procedure provides a very Fe-rich cordierite, because Hz was silicated prior to Sp, whereas the spinel occurs as pure Mg-spinel. By combining **Cord + Fe-Cord** on the one hand, and $Sp + Hz$ on the other, in the variant which was adapted to the modus, it was expressed that the distribution of Mg and Fe among the two mineral species may be changed at choice.

In addition, the calculation of the standard katanorm from the cation percentages will be provided. The following cation percentages were calculated from the analysis:

Si^{4+}	Al^{3+}	Fe^{3+}	Fe^{2+}	Mn^{2+}	Mg^{2+}	Ca^{2+}	Na^+	K^+	Ti^{4+}	Total
45.6	31.4	1.2	8.8		5.0	1.4	3.3	2.8	0.5	100.0

The following are formed first:

$$1.2\,Fe^{3+} + 0.6\,Fe^{2+} = 1.8\,Mt$$
$$0.5\,Ti = 0.5\,Ru$$
$$1.4\,Ca + 2.8\,Al + 2.8\,Si = 7.0\,An$$
$$2.8\,K + 2.8\,Al + 8.4\,Si = 14.0\,Or$$
$$3.3\,Na + 3.3\,Al + 9.9\,Si = 16.5\,Ab.$$

In order to maintain the distribution of Mg and Fe, according to the standard regulations, Sp should be formed first, and Hz should only be formed subsequently, provided that there is sufficient Al available:

$$5.0\,Mg + 10.0\,Al = 15.0\,Sp.$$

An amount of $31.4 - (2.8 + 2.8 + 3.3 + 10.0) = 12.5$ Al is still available. This may be used to form the following:

$$6.2\,Fe + 12.5\,Al = 18.7\,Hz.$$

The remainder of Fe should be converted into Fa:

$$2.0\,Fe + 1.0\,Si = 3.0\,Fa.$$

An amount of $45.6 - (2.8 + 8.4 + 9.9 + 1.0) = 23.5$ Si is still available, and should be used for the silication of Hz and, if sufficient, of Sp:

$$18.7\,Hz + 15.6\,Q = 34.3\,\text{Fe-Cord}$$
$$9.5\,Sp + 7.9\,Q = 17.4\,Cord.$$

The required standard katanorm thus yields (the values obtained from the basis are given in brackets for comparison):

Or	Ab	An	Fe-Cord	Cord	Sp	Fa	Mt	Ru	Total
14.0	16.5	7.0	34.3	17.4	5.5	3.0	1.8	0.5	100.0
(13.8)	(16.7)	(6.8)	(34.6)	(17.4)	(5.5)	(2.9)	(1.8)	(0.5)	(100.0)

3. **Light calc-silicate fels from Castione, near Bellinzona, Tessin, Switzerland**

An additional example of rock of extreme composition is provided by the well-known calc-silicate fels from Castione, near Bellinzona (light variety). The following basis is calculated from A. Mittelholer's (1936, 122) analysis (analyzed by J. Jakob):

	Wt. %		Ru	Kp	Ne	Cal	Cs	Fa	Fo	Q
SiO_2	45.25	754		13	24		186.5	37	19.5	474
Al_2O_3	7.31	143		13	24	106				
Fe_2O_3	0.00									
FeO	5.33	} 74						74		
MnO	0.02									
MgO	1.58	39							39	
CaO	23.91	426				53	373			
Na_2O	0.72	24			24					
K_2O	0.63	13		13						
H_2O+	0.34									
H_2O-	0.00									
TiO_2	0.25	3	3							
P_2O_5	0.02									
CO_2	14.57	(331)								
	99.93	1476	3	39	72	159	559.5	111	58.5	474

	Q	Kp	Ne	Cal	Cs	Fa	Fo	Ru	Total
	32.1	2.6	4.9	10.8	37.9	7.5	4.0	0.2	100.0

213 The katanorm is obtained from the basis as follows:

$$10.8\ Cal + 7.2\ Q = 18.0\ An$$
$$2.6\ Kp + 1.7\ Q = 4.3\ Or$$
$$4.9\ Ne + 3.3\ Q = 8.2\ Ab$$
$$37.9\ Cs + 12.6\ Q = 50.5\ Wo$$
$$7.5\ Fa + 2.5\ Q = 10.0\ Hy$$
$$4.0\ Fo + 1.3\ Q = 5.3\ En.$$

Thus, the standard katanorm yields:

Q	Or	Ab	An	Wo	En	Hy	Wo	Ru	Total
3.5	4.3	8.2	18.0	15.3	5.3	10.0	35.2	0.2	100.0,
		Plag 26.2		Di 30.6					

the plagioclase composition being:

$$\frac{18.0 \cdot 100}{26.2} = 68.7\%\ An.$$

The kata- standard-norm thus corresponds to a **wollastonite-diopside-plagioclase fels**. The analysis, however, shows the presence of 14.57 wt. % CO_2, disregarded in the standard-katanorm calculation. This should be taken into account during the attempt at calculating a modus-adapted variant. The calcite is gained by the reconversion of Wo, thus liberating Q according to:

$$SiO_2 \cdot CaO\ (+\ CO_2) = CaO \cdot CO_2 + SiO_2.$$
$$2\ Wo \quad (CO_2) \quad\quad 1\ Cc \quad\ 1\ Q$$

The equivalent number of 14.57 wt. % CO_2, multiplied by 1000, equals:

$$\frac{14.57}{44.01} \cdot 1000 = 331,$$

and the equivalent - percentage amount, which at the same time equals the equivalent percentage Cc-content, is equal to:

$$\frac{331}{1476} \cdot 100 = 22.4\%$$

(1476 = the sum of the atomic-equivalent numbers, cf. basis calculation). Accordingly, one obtains:

$$44.8\ Wo = 22.4\ Cc + 22.4\ Q,$$

214 and the **calcite variant** of the standard katanorm yields:

Q	Or	Ab	An	Wo	En	Hy	Cc	Ru	Total
25.9	4.3	8.2	18.0	5.7	5.3	10.0	22.4	0.2	100.0.

This is the mineral composition of a **calcite-bearing calc-silicate fels** or **calc-silicate gneiss**.

The following were measured by A. Mittelholzer:

Calcite	Quartz	Calc-silicates (basic plagioclase, scapolite, garnet, epidote, titanite)	Diopside hornblende	Potassium feldspar biotite, muscovite
26 %	27 %	33 %	13 %	1 %

Here, also, the calculation may be started from the **cation percentages**. These are obtained as follows:

Si^{4+}	Al^{3+}	Fe^{2+}	Mg^{2+}	Ca^{2+}	Na^+	K^+	Ti^{4+}	P^{5+}	Total
51.1	9.7	5.0	2.6	28.9	1.6	0.9	0.2	Sp.	100.0

$$0.2\,Ti = 0.2\,Ru$$
$$3.6\,Ca + 7.2\,Al + 7.2\,Si = 18.0\,An$$
$$0.9\,K + 0.9\,Al + 2.7\,Si = 4.5\,Or$$
$$1.6\,Na + 1.6\,Al + 4.8\,Si = 8.0\,Ab$$
$$25.3\,Ca + 25.3\,Si = 50.6\,Wo$$
$$5.0\,Fe + 5.0\,Si = 10.0\,Hy$$
$$2.6\,Mg + 2.6\,Si = 5.2\,En$$

An amount of $51.1 - (7.2 + 2.7 + 4.8 + 25.3 + 5.0 + 2.6) = 3.5$ Si remains, so that the standard katanorm yields (the values which were obtained via the basis are given in brackets for comparison):

Q	Or	Ab	An	Wo	En	Hy	Wo	Ru	Total
3.5	4.5	8.0	18.0	15.2	5.2	10.0	35.4	0.2	100.0
		Plag 26.0		Di 30.4					
(3.5)	(4.3)	(8.2)	(18.0)	(15.3)	(5.3)	(10.0)	(35.2)	(0.2)	(100.0)
		Plag 26.2		Di 30.6					

c) The epinorm of metamorphic rocks

α) General. As in the case of the katanorm, a large number of components, previously not considered in detail, should be used in the calculation of the epinorm. All of these, however, were previously introduced in section B II 12 a.

The majority of **epi-minerals** contain H_2O; some are also CO_2-bearing. According to the previous convention, however, neither H_2O nor CO_2 should be introduced into the calculation. This is advantageous in that when H_2O- or CO_2-bearing minerals are converted into H_2O- or CO_2-free species, or vice-versa, the sum of the equivalents, i.e., the sum of 100, is maintained. If, in certain cases, H_2O or CO_2 are not present in sufficient amounts, the required amounts should be simply added.

In order to calculate the standard epi-minerals directly from the cation percentages, the relationships provided on p. 204 under section B II 12a should be used.

If the basis components are used as a reference, then some of the useful required reactions are given in the following:

Muscovite:

$$2\,SiO_2 \cdot Al_2O_3 \cdot K_2O + 2\,(Al_2O_3 \cdot CaO) + 5\,SiO_2\,(+2\,H_2O)$$
$$6\,Kp 6\,Cal 5\,Q (2\,W)$$
$$= 6\,SiO_2 \cdot 3\,Al_2O_3 \cdot K_2O \cdot 2\,H_2O + SiO_2 \cdot 2\,CaO,$$
$$14\,Ms 3\,Cs$$

$$2\,SiO_2 \cdot Al_2O_3 \cdot K_2O + 2\,(Al_2O_3 \cdot MgO) + 5\,SiO_2\,(+2\,H_2O)$$
$$6\,Kp 6\,Sp 5\,Q (2\,W)$$
$$= 6\,SiO_2 \cdot 3\,Al_2O_3 \cdot K_2O \cdot 2\,H_2O + SiO_2 \cdot 2\,MgO,$$
$$14\,Ms 3\,Fo$$

$$2\,SiO_2 \cdot Al_2O_3 \cdot K_2O + 2\,(Al_2O_3 \cdot FeO) + 5\,SiO_2\,(+\,H_2O)$$
$$6\,Kp 6\,Hz 5\,Q (2\,W)$$
$$= 6\,SiO_2 \cdot 3\,Al_2O_3 \cdot K_2O \cdot 2\,H_2O + SiO_2 \cdot 2\,FeO.$$
$$14\,Ms 3\,Fa$$

Zoisite and epidote:

$$6\,(Al_2O_3 \cdot CaO) + SiO_2 \cdot 2\,CaO + 11\,SiO_2\,(+\,2\,H_2O)$$
$$18\,Cal 3\,Cs 11\,Q (2\,W)$$
$$= 2\,(6\,SiO_2 \cdot 3\,Al_2O_3 \cdot 4\,CaO \cdot H_2O),$$
$$32\,Zo$$

$$6\,(Al_2O_3 \cdot MgO) + 4\,(SiO_2 \cdot 2\,CaO) + 11\,SiO_2\,(+\,2\,H_2O)$$
$$18\,Sp 12\,Cs 11\,Q (2\,W)$$
$$= 2\,(6\,SiO_2 \cdot 3\,Al_2O_3 \cdot 4\,CaO \cdot H_2O) + 3\,(SiO_2 \cdot 2\,MgO),$$
$$32\,Zo 9\,Fo$$

$$4\,(Al_2O_3\ CaO) + SiO_2 \cdot 2\,MgO + 6\,SiO_2\,(+\,3\,H_2O)$$
$$12\,Cal 3\,Fo 6\,Q (3\,W)$$
$$= 6\,SiO_2\ 3\,Al_2O_3 \cdot 4\,CaO \cdot H_2O + SiO_2 \cdot Al_2O_3 \cdot 2\,MgO \cdot 2\,H_2O,$$
$$16\,Zo 5\,At$$

$$2\,(Al_2O_3 \cdot CaO) + SiO_2 \cdot 2\,CaO + 5\,SiO_2 + Fe_2O_3\,(+\,H_2O)$$
$$6\,Cal 3\,Cs 5\,Q 2\,Hm (1\,W)$$
$$= 6\,SiO_2 \cdot Fe_2O_3 \cdot 2\,Al_2O_3 \cdot 4\,CaO \cdot H_2O.$$
$$16\,Pi$$

Serpentine, chlorite:

$3\,(SiO_2 \cdot 2\,MgO) + SiO_2\,(+\,4\,H_2O) = 2\,(2\,SiO_2 \cdot 3\,MgO \cdot 2\,H_2O)$,
    $9\,Fo$ $1\,Q$ $(4\,W)$ $10\,Ant\,(Serp)$

$3\,(SiO_2 \cdot 2\,FeO) + SiO_2\,(+\,4\,H_2O) = 2\,(2\,SiO_2 \cdot 3\,FeO \cdot 2\,H_2O)$,
    $9\,Fa$ $1\,Q$ $(4\,W)$ $10\,Fe$-Ant

$2\,(Al_2O_3 \cdot MgO) + SiO_2 \cdot 2\,MgO + SiO_2\,(+\,4\,H_2O) = 2\,(SiO_2 \cdot Al_2O_3 \cdot 2\,MgO \cdot 2\,H_2O)$,
    $6\,Sp$ $3\,Fo$ $1\,Q$ $(4\,W)$ $10\,At$

$2\,(Al_2O_3 \cdot FeO) + SiO_2 \cdot 2\,FeO + SiO_2\,(4\,H_2O) = 2\,(SiO_2 \cdot Al_2O_3 \cdot 2\,FeO \cdot 2\,H_2O)$.
    $6\,Hz$ $3\,Fa$ $1\,Q$ $(4\,W)$ $10\,Fe$-At

Thus, for an average chlorite (clinochlore) one obtains:

$6\,Sp + 6\,Hz + 3\,Fo + 3\,Fa + 2\,Q\,(+\,8\,W) = 20\,Klchl$.

Talc:

$3\,(SiO_2 \cdot 2\,MgO) + 5\,SiO_2\,(+\,4\,H_2O) = 2\,(4\,SiO_2 \cdot 3\,MgO \cdot 2\,H_2O)$.
    $9\,Fo$ $5\,Q$ $(4\,W)$ $14\,Tc$

Chloritoid:

$(Al_2O_3 \cdot FeO) + SiO_2\,(+\,H_2O) = (SiO_2 \cdot Al_2O_3 \cdot FeO \cdot H_2O)$,
    $3\,Hz$ $1\,Q$ $(1\,W)$ $4\,Ot$

$(Al_2O_3 \cdot MgO) + SiO_2\,(+\,H_2O) = (SiO_2 \cdot Al_2O_3 \cdot MgO \cdot H_2O)$.
    $3\,Sp$ $1\,Q$ $(1\,W)$ $4\,Mg$-Ot

Hornblendes:

$2\,(SiO_2 \cdot 2\,CaO) + 5\,(SiO_2 \cdot 2\,MgO) + 9\,SiO_2\,(+\,2\,H_2O)$
      $6\,Cs$ $15\,Fo$ $9\,Q$ $(2\,W)$

$\quad\quad = 2\,(8\,SiO_2 \cdot 5\,MgO \cdot 2\,CaO \cdot H_2O)$
$\quad\quad\quad\quad\quad 30\,Gram$

or $2\,Cs + 5\,Fo + 3\,Q\,(+\,{}^2/_3\,W) = 10\,Gram$,

$2\,(SiO_2 \cdot 2\,CaO) + 5\,(SiO_2 \cdot 2\,FeO) + 9\,SiO_2\,(+\,2\,H_2O)$
      $6\,Cs$ $15\,Fa$ $9\,Q$ $(2\,W)$

$\quad\quad = 2\,(8\,SiO_2 \cdot 5\,FeO \cdot 2\,CaO \cdot H_2O)$
$\quad\quad\quad\quad\quad 30\,Fe$-$Akt$

or $2\,Cs + 5\,Fa + 3\,Q\,(+\,{}^2/_3\,W) = 10\,Fe$-$Akt$
and $2\,Cs + 5\,(Fo + Fa) + 3\,Q\,(+\,{}^2/_3\,W) = 10\,Akt$,

$2\,(2\,SiO_2 \cdot Al_2O_3 \cdot Na_2O) + 3\,[SiO_2 \cdot 2\,(Mg,\,Fe)\,O] + 9\,SiO_2\,(+\,2\,H_2O)$
   $12\,Ne$ $9\,(Fo + Fa)$ $9\,Q$ $(2\,W)$

$\quad\quad = 2\,[8\,SiO_2 \cdot Al_2O_3 \cdot 3\,(Mg,\,Fe)\,O \cdot Na_2O \cdot H_2O]$,
$\quad\quad\quad\quad\quad 30\,Glph$

or $4\,Ne + 3\,(Fo + Fa) + 3\,Q\,(+\,{}^2/_3\,W) = 10\,Glph$.

Miscellaneous:

$$SiO_2 \cdot Fe_2O_3 = Fe_2O_3 + SiO_2,$$
$$3\,Fs \qquad 2\,Hm \quad 1\,Q$$

$$Al_2O_3 \cdot CaO + Al_2O_3 + 2\,SiO_2\,(+\,H_2O) = 2\,SiO_2 \cdot 2\,Al_2O_3 \cdot CaO \cdot H_2O,$$
$$3\,Cal \qquad 2\,C \qquad 2\,Q \quad (1\,W) \qquad\qquad 7\,Marg$$

$$SiO_2 \cdot 2\,CaO\,(+\,2\,CO_2) = 2\,(CaO \cdot CO_2) + SiO_2,$$
$$3\,Cs \qquad (2\,CO_2) \qquad 2\,Cc \qquad 1\,Q$$

$$SiO_2 \cdot 2\,MgO\,(+\,2\,CO_2) = 2\,(MgO \cdot CO_2) + SiO_2,$$
$$3\,Fo \qquad (2\,CO_2) \qquad 2\,Mgs \qquad 1\,Q$$

$$SiO_2 \cdot 2\,FeO\,(+\,2\,CO_2) = 2\,(FeO \cdot CO_2) + SiO_2,$$
$$3\,Fa \qquad (2\,CO_2) \qquad 2\,Sid \qquad 1\,Q$$

$$Al_2O_3 + 2\,SiO_2\,(+\,2\,H_2O) = 2\,SiO_2 \cdot Al_2O_3 \cdot 2\,H_2O,$$
$$2\,C \qquad 2\,Q \quad (2\,W) \qquad\qquad 4\,Kaol$$
$$\text{or}\ \ 1\,C \qquad 1\,Q \quad (1\,W) \qquad\qquad 2\,Kaol$$

$$3\,(SiO_2 \cdot 2\,CaO) + 3\,SiO_2\,(+\,H_2O) = 6\,SiO_2 \cdot 6\,CaO \cdot H_2O.$$
$$9\,Cs \qquad\qquad 3\,Q \quad (1\,W) \qquad\qquad 12\,Xon$$
$$\text{or}\ \ 3\,Cs \qquad\qquad 1\,Q \quad (^1/_3\,W) \qquad\qquad 4\,Xon.$$

A selection of important relationships between the given components is presented below:

$$8\,SiO_2 \cdot Al_2O_3 \cdot 3\,(Mg,Fe)O \cdot Na_2O \cdot H_2O\,(+\,H_2O)$$
$$\qquad 15\,Glph \qquad\qquad\qquad\qquad (1\,W)$$
$$\text{or} \qquad\ \ 3\,Glph \qquad\qquad\qquad\qquad (^1/_5\,W)$$
$$\qquad = 6\,SiO_2 \cdot Al_2O_3 \cdot Na_2O + 2\,SiO_2 \cdot 3\,(Mg,Fe)O \cdot 2\,H_2O,$$
$$\qquad\quad 10\,Ab \qquad\qquad\qquad 5\,(Ant + Fe\text{-}Ant)$$
$$\qquad\quad 2\,Ab \qquad\qquad\qquad\ 1\,(Ant + Fe\text{-}Ant)$$

$$4\,SiO_2 \cdot 3\,MgO \cdot 2\,H_2O = 2\,SiO_2 \cdot 3\,MgO \cdot 2\,H_2O + 2\,SiO_2,$$
$$\quad 7\,Tc \qquad\qquad\qquad 5\,Ant \qquad\qquad 2\,Q$$

$$SiO_2 \cdot Al_2O_3 \cdot FeO \cdot H_2O + FeO \cdot CO_2\,(+\,H_2O) = SiO_2 \cdot Al_2O_3 \cdot 2\,FeO \cdot 2\,H_2O + CO_2,$$
$$\quad 4\,Ot \qquad\qquad\qquad 1\,Sid \quad (1\,W) \qquad\qquad 5\,Fe\text{-}At$$

$$3\,(SiO_2 \cdot Al_2O_3 \cdot FeO \cdot H_2O) + 2\,SiO_2 \cdot 3\,FeO \cdot 2\,H_2O\,(+\,H_2O)$$
$$\qquad 12\,Ot \qquad\qquad\qquad\quad 5\,Fe\text{-}Ant \qquad\quad (1\,W)$$
$$\qquad\qquad = 3\,(SiO_2 \cdot Al_2O_3 \cdot 2\,FeO \cdot 2\,H_2O) + 2\,SiO_2,$$
$$\qquad\qquad\quad 15\,Fe\text{-}At \qquad\qquad\qquad\quad 2\,Q$$

$$3(3\,SiO_2 \cdot Al_2O_3 \cdot 3\,FeO) + 8\,H_2O = 2\,SiO_2 \cdot 3\,FeO \cdot 2\,H_2O$$
$$24\,Alm \qquad (8\,W) \qquad 5\,Fe\text{-}Ant$$
$$+ 3(SiO_2 \cdot Al_2O_3 \cdot 2\,FeO \cdot 2\,H_2O) + 4\,SiO_2,$$
$$15\,Fe\text{-}At \qquad 4\,Q$$

218
$$3(3\,SiO_2 \cdot Al_2O_3 \cdot 3\,FeO) + (4\,H_2O) = 2(2\,SiO_2 \cdot 3\,FeO \cdot 2\,H_2O)$$
$$24\,Alm \qquad (4\,W) \qquad 10\,Fe\text{-}Ant$$
$$+ 3(Al_2O_3 \cdot FeO) + 5\,SiO_2,$$
$$9\,Hz \qquad 5\,Q$$

$$6\,SiO_2 \cdot 2\,Al_2O_3 \cdot Fe_2O_3 \cdot 4\,CaO \cdot H_2O\,(+\,2\,CO_2) = Fe_2O_3 + 2(Al_2O_3 \cdot CaO)$$
$$16\,Pi \qquad (2\,CO_2) \qquad 2\,Hm \qquad 6\,Cal$$
$$+ 2(CaO \cdot CO_2) + 6\,SiO_2\,(+\,H_2O),$$
$$2\,Cc \qquad 6\,Q \qquad (1\,W)$$

$$3(6\,SiO_2 \cdot 2\,Al_2O_3 \cdot Fe_2O_3 \cdot 4\,CaO \cdot H_2O)\,(+\,4\,CO_2) = 3\,Fe_2O_3$$
$$48\,Pi \qquad (4\,CO_2) \qquad 6\,Hm$$
$$+ 2(6\,SiO_2 \cdot 3\,Al_2O_3 \cdot 4\,CaO \cdot H_2O) + 4(CaO \cdot CO_2) + 6\,SiO_2\,(+\,H_2O)$$
$$32\,Zo \qquad 4\,Cc \qquad 6\,Q \qquad (1\,W)$$

or $\qquad 24\,Pi\,(+\,2\,CO_2) = 3\,Hm + 16\,Zo + 2\,Cc + 3\,Q\,(+\,^1/_2\,W)$,

or, because $2(CaO \cdot CO_2) + SiO_2 = SiO_2 \cdot 2\,CaO\,(+\,2\,CO_2)$,
$$2\,Cc \qquad 1\,Q \qquad 3\,Cs \qquad (2\,CO_2)$$
$$24\,Pi = 3\,Hm + 16\,Zo + 3\,Cs + 2\,Q\,(+\,^1/_2\,W),$$

$$2(6\,SiO_2 \cdot 3\,Al_2O_3 \cdot 4\,CaO \cdot H_2O) + Fe_2O_3 = 6\,SiO_2 \cdot 2\,Al_2O_3 \cdot Fe_2O_3 \cdot 4\,CaO \cdot H_2O$$
$$32\,Zo \qquad 2\,Hm \qquad 16\,Pi$$
$$+ 4(Al_2O_3 \cdot CaO) + 6\,SiO_2\,(+\,H_2O),$$
$$12\,Cal \qquad 6\,Q \qquad (1\,W)$$

or $\qquad 16\,Zo + 1\,Hm = 8\,Pi + 6\,Cal + 3\,Q\,(+\,^1/_2\,W)$,

$$3(6\,SiO_2 \cdot 3\,Al_2O_3 \cdot 4\,CaO \cdot H_2O) + 16(2\,SiO_2 \cdot 3\,MgO \cdot 2\,H_2O) + 7\,SiO_2$$
$$48\,Zo \qquad 80\,Ant \qquad 7\,Q$$
$$= 6(8\,SiO_2 \cdot 5\,MgO \cdot 2\,CaO \cdot H_2O) + 9(SiO_2 \cdot Al_2O_3 \cdot 2\,MgO \cdot 2\,H_2O)\,(+\,11\,H_2O),$$
$$90\,Gram \qquad 45\,At \qquad (11\,W)$$

$$3(8\,SiO_2 \cdot 5\,MgO \cdot 2\,CaO \cdot H_2O) + (6\,CO_2 + 7\,H_2O)$$
$$45\,Gram \qquad (6\,CO_2 + 7\,W)$$
$$= 5(2\,SiO_2 \cdot 3\,MgO \cdot 2\,H_2O) + 14\,SiO_2 + 6(CaO \cdot CO_2),$$
$$25\,Ant \qquad 14\,Q \qquad 6\,Cc$$

$$3(SiO_2 \cdot Al_2O_3 \cdot FeO \cdot H_2O) + 2\,SiO_2 \cdot 3\,FeO \cdot 2\,H_2O\,(+\,H_2O)$$
$$12\,Ot \qquad\qquad\qquad 5\,Fe\text{-}Ant \qquad\qquad (1\,W)$$
$$=3(SiO_2 \cdot Al_2O_3 \cdot 2\,FeO \cdot 2\,H_2O) + 2\,SiO_2\,,$$
$$15\,Fe\text{-}At \qquad\qquad\qquad 2\,Q$$

219 $3(3\,SiO_2 \cdot Al_2O_3 \cdot 3\,FeO)\,(+\,7\,H_2O) = 3(SiO_2 \cdot Al_2O_3 \cdot FeO \cdot H_2O)$
$$24\,Alm \qquad\quad (7\,W) \qquad\qquad\qquad 12\,Ot$$
$$+\,2(2\,SiO_2 \cdot 3\,FeO \cdot 2\,H_2O) + 2\,SiO_2\,,$$
$$10\,Fe\text{-}Ant \qquad\qquad\qquad 2\,Q$$

$$3(6\,SiO_2 \cdot 3\,Al_2O_3 \cdot 4\,CaO \cdot H_2O) + 6(CaO \cdot CO_2) + 2(2\,SiO_2 \cdot 3\,FeO \cdot 2\,H_2O)$$
$$48\,Zo \qquad\qquad\qquad\qquad 6\,Cc \qquad\qquad 10\,Fe\text{-}Ant$$
$$=6(3\,SiO_2 \cdot Al_2O_3 \cdot 3\,CaO) + 3(SiO_2 \cdot Al_2O_3 \cdot 2\,FeO \cdot 2\,H_2O)$$
$$48\,Gro \qquad\qquad\qquad 15\,Fe\text{-}At$$
$$+\,SiO_2\,(+\,6\,CO_2 + H_2O)\,,$$
$$1\,Q \quad (6\,CO_2 + 1\,W)$$

$$3\,SiO_2 \cdot Al_2O_3 \cdot 3\,FeO + 2\,Al_2O_3\,(+\,3\,H_2O) = 3(SiO_2 \cdot Al_2O_3 \cdot FeO \cdot H_2O)\,,$$
$$8\,Alm \qquad\qquad 4\,C \quad\;(3\,W) \qquad\qquad\qquad 12\,Ot$$

$$3(6\,SiO_2 \cdot 3\,Al_2O_3 \cdot 4\,CaO \cdot H_2O) + 5(2\,SiO_2 \cdot 3(Mg,\,Fe)O \cdot 2\,H_2O)$$
$$48\,Zo \qquad\qquad\qquad\qquad 25\,(Ant + Fe\text{-}Ant)$$
$$=9(3\,SiO_2 \cdot Al_2O_3 \cdot 3[(Mg,\,Fe,\,Ca)O] + SiO_2\,(+\,13\,H_2O)\,,$$
$$72\,Garnet \qquad\qquad\qquad 1\,Q \quad (13\,W)$$

$$2(7\,SiO_2 \cdot 2\,Al_2O_3 \cdot (Mg,\,Fe)O \cdot K_2O \cdot H_2O) = 2(2\,SiO_2 \cdot Al_2O_3 \cdot K_2O)$$
$$28\,Ph = 28\,Ms_2 \qquad\qquad\qquad\qquad 12\,Kp$$
$$+\,2[Al_2O_3 \cdot (Mg,\,Fe)O] + 10\,SiO_2\,(+\,2\,H_2O)\,,$$
$$6(Sp + Hz) \qquad\qquad 10\,Q \quad\;\;(2\,W)$$

$$2(6\,SiO_2 \cdot 3\,Al_2O_3 \cdot Na_2O \cdot 2\,H_2O) = 2(2\,SiO_2 \cdot Al_2O_3 \cdot Na_2O) + 4\,Al_2O_3$$
$$28\,Pg \qquad\qquad\qquad\qquad 12\,Ne \qquad\qquad\quad 8\,C$$
$$+\,8\,SiO_2\,(+\,4\,H_2O)\,.$$
$$8\,Q \qquad (4\,W)$$

β) **The calculation of a standardized epinorm.** The same reasons given in favor of a standardized katanorm also appear to justify the calculation of a standard epinorm which will correspond as closely as possible to the modus. If the calculations are referred to the basis, the following procedure is suggested (C. Burri and P. Niggli, 1945):

1. After the determination of the amount of the CO_2 present (this should be disregarded in the reaction equations, as well as H_2O), the following is formed:
   a) *Cc* from *Cs*, according to:

$$SiO_2 \cdot 2\,CaO\,(+\,2\,CO_2) = 2(CaO \cdot CO_2) + SiO_2,$$
$$3\,Cs \qquad (2\,CO_2) \qquad 2\,Cc \qquad 1\,Q$$

b) If some CO_2 remains, *Mgs* should be formed from *Fo* according to:

$$SiO_2 \cdot 2\,MgO\,(+\,2\,CO_2) = 2(MgO \cdot CO_2) + SiO_2,$$
$$3\,Fo \qquad (2\,CO_2) \qquad 2\,Mgs \qquad 1\,Q$$

c) Should some CO_2 still remain, *Sid* should be formed from *Fa* as follows:

$$SiO_2 \cdot 2\,FeO\,(+\,2\,CO_2) = 2(FeO \cdot CO_2) + SiO_2.$$
$$3\,Fa \qquad (2\,CO_2) \qquad 2\,Sid \qquad 1\,Q$$

d) If at this stage CO_2 is still available, after using up all available *Cs, Fo* and *Fa*, carbonates should be formed in the following order, from *Cal, Sp, Hz*, according to the relationships:

$$Al_2O_3 \cdot CaO\,(+\,CO_2) = CaO \cdot CO_2 + Al_2O_3,$$
$$3\,Cal \qquad (CO_2) \qquad 1\,Cc \qquad 2\,C$$
$$Al_2O_3 \cdot MgO\,(+\,CO_2) = MgO \cdot CO_2 + Al_2O_3,$$
$$3\,Sp \qquad (CO_2) \qquad 1\,Mgs \qquad 2\,C$$
$$Al_2O_3 \cdot FeO\,(+\,CO_2) = FeO \cdot CO_2 + Al_2O_3.$$
$$3\,Hz \qquad (CO_2) \qquad 1\,Sid \qquad 2\,C$$

e) If, after completing all these operations, an excess of CO_2 is still present, it should be designated as "CO_2-excess" after the completion of the epinorm calculation.

2. *Cp* and *Ru* should be introduced, unchanged, from the basis.

3. *Hm* is formed from *Fs* as follows:

$$SiO_2 \cdot Fe_2O_3 = Fe_2O_3 + SiO_2.$$
$$3\,Fs \qquad 2\,Hm \qquad 1\,Q$$

4. Albite is formed from *Ne* according to:

$$2\,SiO_2 \cdot Al_2O_3 \cdot Na_2O + 4\,SiO_2 = 6\,SiO_2 \cdot Al_2O_3 \cdot Na_2O.$$
$$6\,Ne \qquad\qquad 4\,Q \qquad\qquad 10\,Ab$$
$$\text{or} \qquad 3\,Ne \qquad\qquad 2\,Q \qquad\qquad 5\,Ab$$

5. *Ms* is formed from *Kp* according to:

$$2\,SiO_2 \cdot Al_2O_3 \cdot K_2O + 2\,Al_2O_3 + 4\,SiO_2\,(+\,2\,H_2O)$$
$$6\,Kp \qquad\qquad 4\,C \qquad 4\,Q \qquad (2\,W)$$
$$= 6\,SiO_2 \cdot 3\,Al_2O_3 \cdot K_2O \cdot 2\,H_2O.$$
$$14\,Ms$$

If the *C* gained in 1 d does not suffice for the last-mentioned operation, the following components, in the given order, should be used: *Hz, Sp* and *Cal*, according to the equations:

$$2\,SiO_2 \cdot Al_2O_3 \cdot K_2O + 2(Al_2O_3 \cdot FeO) + 5\,SiO_2(+\,2\,H_2O)$$
$$\underset{6\,Kp}{} \qquad \underset{6\,Hz}{} \qquad \underset{5\,Q}{} \quad \underset{(2\,W)}{}$$
$$= 6\,SiO_2 \cdot 3\,Al_2O_3 \cdot K_2O \cdot 2\,H_2O + SiO_2 \cdot 2\,FeO,$$
$$\underset{14\,Ms}{} \qquad \underset{3\,Fa}{}$$

$$2\,SiO_2 \cdot Al_2O_3 \cdot K_2O + 2(Al_2O_3 \cdot MgO) + 5\,SiO_2(+\,2\,H_2O)$$
$$\underset{6\,Kp}{} \qquad \underset{6\,Sp}{} \qquad \underset{5\,Q}{} \quad \underset{(2\,W)}{}$$
$$= 6\,SiO_2 \cdot 3\,Al_2O_3 \cdot K_2O \cdot 2\,H_2O + SiO_2 \cdot 2\,MgO,$$
$$\underset{14\,Ms}{} \qquad \underset{3\,Fo}{}$$

221 $$2\,SiO_2 \cdot Al_2O_3 \cdot K_2O + 2(Al_2O_3 \cdot CaO) + 5\,SiO_2(+\,2\,H_2O)$$
$$\underset{6\,Kp}{} \qquad \underset{6\,Cal}{} \qquad \underset{5\,Q}{} \quad \underset{(2\,W)}{}$$
$$= 6\,SiO_2 \cdot 3\,Al_2O_3 \cdot K_2O \cdot 2\,H_2O + SiO_2 \cdot 2\,CaO.$$
$$\underset{14\,Ms}{} \qquad \underset{3\,Cs}{}$$

If, even after resorting to this operation, insufficient C is liberated for the transformation of all the available Kp into Ms, the excess of Kp should be converted to orthoclase according to: $3\,Kp + 2\,Q = 5\,Or$.

After carrying out operations 1-5, one of the 5 following remainder combinations will be present, depending on the initial composition:

α)	β)	γ)	δ)	ε)
Or	Cal	Cal	Cal	Cal
Cs	Cs	Sp	Sp	Sp
Fo	Fo	Fo	Hz	Hz
Fa	Fa	Fa	Fa	C
Q	Q	Q	Q	Q.

The following procedures should be applied, respectively, to the various combinations:

α) *Gram* is formed from *Cs* and *Fo*:

$$2(SiO_2 \cdot 2\,CaO) + 5(SiO_2 \cdot 2\,MgO) + 9\,SiO_2(+\,2\,H_2O)$$
$$\underset{6\,Cs}{} \qquad \underset{15\,Fo}{} \qquad \underset{9\,Q}{} \quad \underset{(2\,W)}{}$$
$$= 2(8\,SiO_2 \cdot 5\,MgO \cdot 2\,CaO \cdot H_2O).$$
$$\underset{30\,Gram}{}$$

If some *Cs* remains, *Fe-Akt* should be formed in an analogous manner from *Fa*:

$$6\,Cs + 15\,Fa + 9\,Q(+\,2\,W) = 30\,\text{Fe-}Akt.$$

Should some *Cs* still remain, *Xon* should be formed:

$$3(SiO_2 \cdot 2\,CaO) + 3\,SiO_2(+\,1\,H_2O) = 6\,SiO_2 \cdot 6\,CaO \cdot H_2O.$$
$$\underset{9\,Cs}{} \qquad \underset{3\,Q}{} \quad \underset{(1\,W)}{} \qquad \underset{12\,Xon}{}$$
$$\text{or}\quad \underset{3\,Cs}{} \qquad \underset{1\,Q}{} \quad \underset{(^1/_3\,W)}{} \qquad \underset{4\,Xon}{}$$

If, however, *Fo* and (or) *Fa* remain, serpentine should be formed according to:

$$3(SiO_2 \cdot 2\,MgO) + SiO_2 + (4\,H_2O) = 2(2\,SiO_2 \cdot 3\,MgO \cdot 2\,H_2O).$$
$$9\,Fo 1\,Q \quad (4\,W) 10\,Ant$$
$$\text{or } 9\,Fa 1\,Q \quad (4\,W) 10\,Fe\text{-}Ant$$

β) Initially, zoisite is formed from *Cal* and *Cs*:

$$6(Al_2O_3 \cdot CaO) + SiO_2 \cdot 2\,CaO + 11\,SiO_2(+\,2\,H_2O)$$
$$18\,Cal 3\,Cs 11\,Q \quad (2\,W)$$
$$= 2(6\,SiO_2 \cdot 3\,Al_2O_3 \cdot 4\,CaO \cdot H_2O).$$
$$32\,Zo$$

If *Cs* remains, *Gram*, *Fe-Akt* and perhaps *Xon* should be formed in the given order, in analogy to the procedure given under a), as follows:

$$6\,Cs + 15\,Fo + 9\,Q\,(+\,2\,W) = 30\,Gram$$
$$6\,Cs + 15\,Fa + 9\,Q\,(+\,2\,W) = 30\,Fe\text{-}Akt$$
$$3\,Cs + 1\,Q\,(+\,{}^1/_3\,W) = 4\,Xon,$$

and should an excess of *Fo* + *Fa* be present, *Ant* + *Fe-Ant* should be added.

If, on the other hand, *Cal* is in excess, *Zo* and *At* or *Fe-At* should be formed, *Fo* + *Fa* being used for this purpose, as follows:

$$4(Al_2O_3 \cdot CaO) + SiO_2 \cdot 2\,MgO + 6\,SiO_2(+\,3\,H_2O)$$
$$12\,Cal 3\,Fo 6\,Q \quad (3\,W)$$
$$\text{or } 12\,Cal 3\,Fa 6\,Q \quad (3\,W)$$
$$= 6\,SiO_2 \cdot 3\,Al_2O_3 \cdot 4\,CaO \cdot H_2O + SiO_2 \cdot Al_2O_3 \cdot 2\,MgO \cdot 2\,H_2O.$$
$$16\,Zo 5\,At$$
$$16\,Zo 5\,Fe\text{-}At$$

If, nevertheless, some *Cal* still remains, the next step is the formation of *Zo* + *Ot* from the just-formed *Fe-At*:

$$4(Al_2O_3 \cdot CaO) + SiO_2 \cdot Al_2O_3 \cdot 2\,FeO \cdot 2\,H_2O + 7\,SiO_2\,(+\,H_2O)$$
$$12\,Cal 5\,Fe\text{-}At 7\,Q \quad (1\,W)$$
$$= 6\,SiO_2 \cdot 3\,Al_2O_3 \cdot 4\,CaO \cdot H_2O + 2(SiO_2 \cdot Al_2O_3 \cdot FeO \cdot H_2O).$$
$$16\,Zo 8\,Ot$$

A possible, additional *Cal*-excess should be converted, analogously, in combination with *At*, into *Zo* + *Mg-Ot* as follows:

$$12\,Cal + 5\,At + 7\,Q\,(+\,1\,W) = 16\,Zo + 8\,Mg\text{-}Ot.$$

Additional *Cal* should be incorporated into zoisite and kaolin according to:

$$4(Al_2O_3 \cdot CaO) + 8\,SiO_2(+\,3\,H_2O) = 6\,SiO_2 \cdot 3\,Al_2O_3 \cdot 4\,CaO \cdot H_2O$$
$$12\,Cal 8\,Q \quad (3\,W) 16\,Zo$$
$$+\,2\,SiO_2 \cdot Al_2O_3 \cdot 2\,H_2O.$$
$$4\,Kaol$$

A remainder of *Fo* + *Fa* should be converted into *Ant* + *Fe-Ant*, as shown under a).

γ) Primarily *Cal* is to be eliminated, as discussed under β), *Zo* + *At* or *Fe-At* being formed therefrom:

$$12\,Cal + 3\,Fo + 6\,Q(+3\,W) = 16\,Zo + 5\,At$$
$$12\,Cal + 3\,Fa + 6\,Q(+3\,W) = 16\,Zo + 5\,\text{Fe-At}.$$

If not all of the *Cal* is consumed in these operations, the remainder, together with the *Fe-At* or *At* which was just built, is combined as follows to yield zoisite + chloritoid:

$$12\,Cal + 5\,\text{Fe-At} + 7\,Q(+1\,W) = 16\,Zo + 8\,Ot$$
or
$$12\,Cal + 5\,At\ \ \ \ + 7\,Q(+1\,W) = 16\,Zo + 8\,\text{Mg-}Ot.$$

In the rare cases when a remainder of *Cal* occurs after the formation of chloritoid, margarite should be formed:

$$2(Al_2O_3 \cdot CaO) + 2(Al_2O_3 \cdot MgO) + 5\,SiO_2(+2\,H_2O)$$
$$\ \ \ \ \ 6\,Cal \ \ \ \ \ \ \ \ \ \ \ \ \ \ \ \ \ \ \ 6\,Sp \ \ \ \ \ \ \ \ \ \ \ \ \ \ \ 5\,Q \ \ \ \ \ (2\,W)$$
$$= 2(2\,SiO_2 \cdot 2\,Al_2O_3 \cdot CaO \cdot H_2O) + SiO_2 \cdot 2\,MgO.$$
$$\ \ \ \ \ \ \ \ \ \ \ \ \ \ 14\,Marg \ 3\,Fo$$

If some *Fo* remains, *At* is formed:

$$2(Al_2O_3 \cdot MgO) + SiO_2 \cdot 2\,MgO + SiO_2(+4\,H_2O)$$
$$\ \ \ \ 6\,Sp \ \ \ \ \ \ \ \ \ \ \ \ \ \ \ 3\,Fo \ \ \ \ \ \ \ \ \ \ \ 1\,Q \ \ \ (4\,W)$$
$$= 2(SiO_2 \cdot Al_2O_3 \cdot 2\,MgO \cdot 2\,H_2O).$$
$$\ \ \ \ \ \ \ \ \ \ \ \ \ \ \ \ 10\,At$$

A possible remainder of *Fo* is converted, by silication, into serpentine:

$$3(SiO_2 \cdot 2\,MgO) + SiO_2(+4\,H_2O) = 2(2\,SiO_2 \cdot 3\,MgO \cdot 2\,H_2O).$$
$$\ \ \ 9\,Fo \ \ \ \ \ \ \ \ \ \ \ 1\,Q \ \ \ (4\,W) \ \ \ \ \ \ \ \ \ \ 10\,Ant$$

At + *Fe-At* should be formed if a remainder of *Fa* occurs together with *Sp*:

$$2(Al_2O_3 \cdot MgO) + SiO_2 \cdot 2\,FeO + SiO_2(+4\,H_2O)$$
$$\ \ \ 6\,Sp \ \ \ \ \ \ \ \ \ \ \ \ \ \ 3\,Fa \ \ \ \ \ \ \ \ \ \ \ 1\,Q \ \ \ (4\,W)$$
$$= SiO_2 \cdot Al_2O_3 \cdot 2\,MgO \cdot 2\,H_2O + SiO_2 \cdot Al_2O_3 \cdot 2\,FeO \cdot 2\,H_2O.$$
$$\ \ \ \ \ \ \ \ \ \ \ 5\,At \ 5\,\text{Fe-At}$$

The remaining *Fa* will yield serpentine according to:

$$3(SiO_2 \cdot 2\,FeO) + SiO_2(+4\,H_2O) = 2(2\,SiO_2 \cdot 3\,FeO \cdot 2\,H_2O),$$
$$\ \ \ 9\,Fa \ \ \ \ \ \ \ \ \ \ \ \ 1\,Q \ \ (4\,W) \ \ \ \ \ \ \ \ \ \ 10\,\text{Fe-}Ant$$

whereas the remaining *Sp* should be converted into chloritoid and amesite according to the following equations:

$$2(Al_2O_3 \cdot MgO) + 2 SiO_2 + SiO_2 \cdot Al_2O_3 \cdot 2 FeO \cdot 2 H_2O (+ 2 H_2O)$$
$$\quad 6\,Sp \qquad\qquad 2\,Q \qquad\qquad 5\,Fe\text{-}At \qquad\qquad (2\,W)$$
$$= 2(SiO_2 \cdot Al_2O_3 \cdot FeO \cdot H_2O) + SiO_2 \cdot Al_2O_3 \cdot 2 MgO \cdot 2 H_2O.$$
$$\qquad\qquad 8\,Ot \qquad\qquad\qquad\qquad 5\,At$$

If the available amount of *Fe-At* does not meet the requirements, the remaining *Sp* should be converted into *Mg-Ot*, provided that sufficient *Q* is available, according to:

$$Al_2O_3 \cdot MgO + SiO_2 (+ H_2O) = SiO_2 \cdot Al_2O_3 \cdot MgO \cdot H_2O.$$
$$\quad 3\,Sp \qquad\quad 1\,Q \quad (1\,W) \qquad\qquad 4\,Mg\text{-}Ot$$

δ) One may now proceed with the formation of *Zo*:

$$4(Al_2O_3 \cdot CaO) + SiO_2 \cdot 2 FeO + 6 SiO_2 (+ 3 H_2O)$$
$$\quad 12\,Cal \qquad\qquad 3\,Fa \qquad\quad 6\,Q \qquad (3\,W)$$
$$= 6 SiO_2 \cdot 3 Al_2O_3 \cdot 4 CaO \cdot H_2O + SiO_2 \cdot Al_2O_3 \cdot 2 FeO \cdot 2 H_2O.$$
$$\qquad\qquad 16\,Zo \qquad\qquad\qquad\qquad 5\,Fe\text{-}At$$

Moreover, as much *Fe-At* as possible should be formed from $Hz + Fa$:

$$2(Al_2O_3 \cdot FeO) + SiO_2 \cdot 2 FeO + SiO_2 (+ 4 H_2O) = 2(SiO_2 \cdot Al_2O_3 \cdot 2 FeO \cdot 2 H_2O).$$
$$\quad 6\,Hz \qquad\qquad 3\,Fa \qquad\quad 1\,Q \quad (4\,W) \qquad\qquad 10\,Fe\text{-}At$$

The excess of *Fa* is combined with $Sp + Q$ to form $Fe\text{-}At + At$:

$$SiO_2 \cdot 2 FeO + 2(Al_2O_3 \cdot MgO) + SiO_2 (+ 4 H_2O)$$
$$\quad 3\,Fa \qquad\qquad 6\,Sp \qquad\quad 1\,Q \quad (4\,W)$$
$$= SiO_2 \cdot Al_2O_3 \cdot 2 MgO \cdot 2 H_2O + SiO_2 \cdot Al_2O_3 \cdot 2 FeO \cdot 2 H_2O.$$
$$\qquad\qquad 5\,At \qquad\qquad\qquad\qquad 5\,Fe\text{-}At$$

If some *Fa* remains, one should proceed with the formation of *Fe-Ant*:

$$3(SiO_2 \cdot 2 FeO) + SiO_2 (+ 4 H_2O) = 2(2 SiO_2 \cdot 3 FeO \cdot 2 H_2O).$$
$$\quad 9\,Fa \qquad\qquad 1\,Q \quad (4\,W) \qquad\qquad 10\,Fe\text{-}Ant$$

If, on the other hand, *Hz* remains, chloritoid should be formed as follows:

$$Al_2O_3 \cdot FeO + SiO_2 (+ H_2O) = SiO_2 \cdot Al_2O_3 \cdot FeO \cdot H_2O.$$
$$\quad 3\,Hz \qquad\quad 1\,Q \quad (1\,W) \qquad\qquad 4\,Ot$$

Here again, *Ot* should be formed, using for this purpose the remaining *Sp* combined with the previously formed *Fe-At*:

$$2(Al_2O_3 \cdot MgO) + 2 SiO_2 + SiO_2 \cdot Al_2O_3 \cdot 2 FeO \cdot 2 H_2O (+ 2 H_2O)$$
$$\quad 6\,Sp \qquad\qquad 2\,Q \qquad\qquad 5\,Fe\text{-}At \qquad\qquad (2\,W)$$
$$= 2(SiO_2 \cdot Al_2O_3 \cdot FeO \cdot H_2O) + SiO_2 \cdot Al_2O_3 \cdot 2 MgO \cdot 2 H_2O.$$
$$\qquad\qquad 8\,Ot \qquad\qquad\qquad\qquad 5\,At$$

If the available *Fe-At* is insufficient, Mg-ottrelite should be formed according to:

$$Al_2O_3 \cdot MgO + SiO_2(+ H_2O) = SiO_2 \cdot Al_2O_3 \cdot MgO \cdot H_2O.$$
$$3\,Sp \qquad 1\,Q \quad (1\,W) \qquad\qquad 4\,Mg\text{-}Ot$$

Should there occur an excess of **Fe-At**, it should be presented as such.

ε) Initially, *Cal* is used to form zoisite + kaolin:

$$4\,(Al_2O_3 \cdot CaO) + 8\,SiO_2(+ 3\,H_2O) = (6\,SiO_2 \cdot 3\,Al_2O_3 \cdot 4\,CaO \cdot H_2O)$$
$$12\,Cal \qquad\quad 8\,Q \quad (3\,W) \qquad\qquad\qquad 16\,Zo$$
$$+\; 2\,SiO_2 \cdot Al_2O_3 \cdot 2\,H_2O.$$
$$4\,Kaol$$

Chloritoid is formed from *Hz* and *Sp* as follows:

$$Al_2O_3 \cdot FeO + SiO_2(+ H_2O) = SiO_2 \cdot Al_2O_3 \cdot FeO \cdot H_2O.$$
$$3\,Hz \qquad 1\,Q \quad (1\,W) \qquad\qquad 4\,Ot$$
or $\quad 3\,Sp \qquad 1\,Q \quad (1\,W) \qquad\qquad 4\,Mg\text{-}Ot$

C is used to form kaolin:

$$Al_2O_3 + 2\,SiO_2(+ 2\,H_2O) = 2\,SiO_2 \cdot Al_2O_3 \cdot 2\,H_2O.$$
$$2\,C \qquad 2\,Q \quad (2\,W) \qquad\qquad 4\,Kaol$$
or $\quad 1\,C \qquad 1\,Q \quad (1\,W) \qquad\qquad 2\,Kaol$

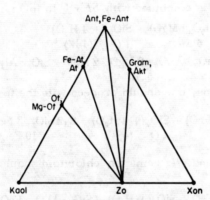

FIGURE 63. Schematic diagram of the *Zo*-bearing mineral combinations of the standard epinorm.

Figure 63 presents a triangular diagrammatic representation of the possible zoisite-bearing combinations: only those components which are located at the corners of one and the same small triangle may occur together.

Thus, if the presence of a sufficient amount of SiO_2 is assumed, the following mineral combinations are possible for the standard epinorm (*Akt* standing for **Gram** + **Fe-Akt**):
Cp, Ru, Hm, Ab, Ms ± carbonates, and in addition:

1.	2.	3.	4.	5.	6.	7.	8.	9.	10.	11.
Or	Or	Or	Zo	Zo	Zo	Zo	Zo	Zo	Zo	Zo
Akt	Akt	Gram	Akt	Akt	Gram	At	At	At	At	Ot
Xon	Fe-Ant	Ant	Xon	Fe-Ant	Ant	Ant	Fe-At	Fe-At	Ot	Mg-Ot
Q	Q	Fe-Ant	Q	Q		Fe-Ant	Fe-Ant	Fe-Ant	Ot	Mg-Ot Kaol
		Q				Q	Q	Q	Q	Q

Examples:

Firstly, the quartz-diorite from Spanish Peak and the cordierite-andesite from Hoyazo will be taken as examples for illustrating the procedure of the standard-epinorm calculation. Subsequently, the same two rocks of extreme composition, which were already used for the calculation of the standard katanorm, will be dealt with. Constant reference to the previously calculated basis is thus made possible.

1. Quartz-diorite from Spanish Peak, California.

The basis yields:

Q	Kp	Ne	Cal	Cs	Fs	Fa	Fo	Ru	Cp	Total
42.3	4.7	21.0	15.6	1.7	3.0	3.2	7.4	0.5	0.6	100.0

One should commence with the formation of:

$$3.0\,Fs = 2.0\,Hm + 1.0\,Q$$
$$21.0\,Ne + 14.0\,Q = 35.0\,Ab$$
$$4.7\,Kp + 4.7\,Cal + 3.9\,Q = 11.0\,Ms + 2.3\,Cs.$$

The following remain:

Cal	Cs	Fo	Fa	Q, corresponding to case β).
10.9	4.0	7.4	3.2	25.4

$$10.9\,Cal + 1.8\,Cs + 6.7\,Q = 19.4\,Zo$$
$$2.2\,Cs + 5.5\,Fo + 3.3\,Q = 11.0\,Gram$$
$$1.9\,Fo + 0.2\,Q = 2.1\,Ant$$
$$3.2\,Fa + 0.4\,Q = 3.6\,Fe\text{-}Ant.$$

The required standard epinorm thus yields:

Q	Ab	Ms	Zo	Gram	Ant	Fe-Ant	Hm	Ru	Cp	Total
14.8	35.0	11.0	19.4	11.0	2.1	3.6	2.0	0.5	0.6	100.0

and corresponds to a grammatite-sericite- and antigorite-bearing albite-zoisite schist. Actually, the grammatite could be somewhat richer in Fe, and the antigorite — in Mg. This may be easily taken into account in the calculation.

2. Cordierite-andesite from Hoyazo, near Nijar, Spain.

The basis yields:

Q	Kp	Ne	Cal	Sp	Fs	Fa	Fo	Ru	Cp	Total
49.4	12.3	9.9	9.3	9.3	3.5	3.9	1.5	0.6	0.3	100.0.

The following components are formed first:

$$3.5\,Fs = 2.3\,Hm + 1.2\,Q$$
$$9.9\,Ne + 6.6\,Q = 16.5\,Ab$$
$$9.3\,Kp + 9.3\,Sp + 7.8\,Q = 21.8\,Ms + 4.6\,Fo$$
$$3.0\,Kp + 3.0\,Cal + 2.5\,Q = 7.0\,Ms + 1.5\,Cs.$$

There remain: Cal Cs Fa Fo Q, corresponding to case β).
6.3 1.5 3.9 6.1 33.7

$$6.3\,Cal + 1.1\,Cs + 3.8\,Q = 11.2\,Zo$$
$$0.4\,Cs + 1.0\,Fo + 0.6\,Q = 2.0\,Gram$$
$$5.1\,Fo + 0.6\,Q = 5.7\,Ant$$
$$3.9\,Fa + 0.4\,Q = 4.3\,Fe\text{-}Ant.$$

The required **standard epinorm** thus yields:

Q	Ab	Ms	Zo	Gram	Ant	Fe-Ant	Hm	Ru	Cp	Total
28.3	16.5	28.8	11.2	2.0	5.7	4.3	2.3	0.6	0.3	100.0

Accordingly, this standard epinorm corresponds to a **serpentine-grammatite and zoisite-bearing albite-sericite schist**. It is important to note that both the pyroxene-diorite from Spanish Peak with $al < (alk + c)$ and the garnet-bearing cordierite-andesite from Hoyazo with $t = al - (alk + c) = 11,3$, qualitatively show the same mineral composition in the standard epinorm. Whereas the alumina excess of the second example is reflected in the standard katanorm by a high cordierite content, in the epinorm this fact is only expressed by a large amount of sericite.

3. **Corundum-cordierite-spinel hornfels** from Tom a'Mhinn, Glen Lednock, Perthshire, Scotland (C. E. Tilley, 1924).

Basis:	Q	Kp	Ne	Cal	Sp	Hz	C	Ru	Cp	Total
	28.2	13.7	11.4	3.3	14.7	20.4	5.2	1.4	1.7	100.0

The calculation is started as follows:

$$11.4\,Ne + 7.6\,Q = 19.0\,Ab$$
$$7.8\,Kp + 5.2\,C + 5.2\,Q = 18.2\,Ms$$
$$5.9\,Kp + 5.9\,Hz + 4.9\,Q = 13.8\,Ms + 2.9\,Fa.$$

There remains the following: Cal Sp Hz Fa Q corresponding to case δ).
3.3 14.7 14.5 2.9 10.5

$$3.3\,Cal + 0.8\,Fa + 1.7\,Q = 4.4\,Zo + 1.4\,Fe\text{-}At$$
$$4.2\,Hz + 2.1\,Fa + 0.7\,Q = 7.0\,Fe\text{-}At$$
$$10.3\,Hz + 3.4\,Q = 13.7\,Ot$$
$$10.1\,Sp + 8.4\,Fe\text{-}At + 3.4\,Q = 13.5\,Ot + 8.4\,At$$
$$4.6\,Sp + 1.5\,Q = 6.1\,Mg\text{-}Ot.$$

The required **standard epinorm** thus yields:

Ab	Ms	Zo	Ot	Mg-Ot	At	Ru	Cp	Q	Total
19.0	32.0	4.4	27.2	6.1	8.4	1.4	1.7	−0.2	100.0

A Q-deficiency of −0.2 is present.

If it is desired to calculate a variant which will correspond to a possible mineral composition, amesite is to be eliminated, since, due to its low stability, its modal presence is doubtful. The formation of chloritoid and serpentine is ruled out due to the paucity of Q. Magnesite may, however, be formed, according to:

$$SiO_2 \cdot Al_2O_3 \cdot 2 MgO \cdot 2 H_2O \, (+ CO_2) = SiO_2 \cdot Al_2O_3 \cdot MgO \cdot H_2O$$
$$5 \, At \qquad\qquad (CO_2) \qquad\qquad 4 \, Mg\text{-}Ot$$
$$8.4 \, At \qquad\qquad\qquad\qquad\qquad 6.7 \, Mg\text{-}Ot$$

$$+ MgO \cdot CO_2 \, (+ H_2O).$$
$$1 \, Mgs \quad (1 \, W)$$
$$1.7 \, Mgs$$

This variant will yield:

Ab	Ms	Zo	Ot	Mg-Ot	Mgs	Ru	Cp	Q	Total
19.0	32.0	4.4	27.2	12.8	1.7	1.4	1.7	−0.2	100.0.

A rock of such a composition should be designated as a chloritoid-sericite-albite schist, or, probably, fels. The chloritoid content of 40 % causes this rock to be of an unusual type. The last remark may also be referred to the corundum-cordierite-spinel hornfels, from which rock the calculation was started.

4. Cordierite-spinel-hornfels from Creag near Iolaire, Glen Lednock, Perthshire, Scotland (C. E. Tilley, 1924).

Basis:	Q	Kp	Ne	Cal	Sp	Hz	Fs	Fa	Ru	Total
	37.6	8.3	10.0	4.1	15.0	18.9	1.8	3.8	0.5	100.0.

The following are formed: $1.8 \, Fs = 1.2 \, Hm + 0.6 \, Q$
$10.0 \, Ne + 6.7 \, Q = 16.7 \, Ab$
$8.3 \, Kp + 8.3 \, Hz + 6.9 \, Q = 19.4 \, Ms + 4.1 \, Fa$.

Thus, the following remain:

	Cal	Sp	Hz	Fa	Q	corresponding to case δ).
	4.1	15.0	10.6	7.9	24.6	

$$4.1 \, Cal + 1.0 \, Fa + 2.1 \, Q = 5.5 \, Zo + 1.7 \, Fe\text{-}At$$
$$10.6 \, Hz + 5.3 \, Fa + 1.8 \, Q = 17.7 \, Fe\text{-}At$$
$$3.2 \, Sp + 1.6 \, Fa + 0.5 \, Q = 2.7 \, At + 2.6 \, Fe\text{-}At$$
$$11.8 \, Sp + 3.9 \, Q + 9.8 \, Fe\text{-}At = 15.7 \, Ot + 9.8 \, At.$$

The required standard epinorm yields:

Q	Ab	Ms	Zo	Fe-At	At	Ot	Hm	Ru	Total
16.3	16.7	19.4	5.5	12.2	12.5	15.7	1.2	0.5	100.0.

This mineral combination, possessing about 1/4 amesite content, should actually occur in nature only very rarely due to the general instability of this mineral. The elimination of this mineral and its substitution by chloritoid and serpentine, in contrast to the previous example, does not give rise to any difficulties, since sufficient free Q is present:

$$3\,(SiO_2 \cdot Al_2O_3 \cdot 2\,MgO \cdot 2\,H_2O) + 2\,SiO_2 = 3\,(SiO_2 \cdot Al_2O_3 \cdot MgO \cdot H_2O)$$
$$15\,At 2\,Q 12\,Mg\text{-}Ot$$
or $15\,Fe\text{-}At 2\,Q 12\,Ot$

$$+ 2\,SiO_2 \cdot 3\,MgO \cdot 2\,H_2O + H_2O.$$
$$5\,Ant (1\,W)$$
$$5\,Fe\text{-}Ant (1\,W)$$

If, correspondingly, the following are formed:

$$12.2\,Fe\text{-}At + 1.6\,Q = 9.8\,Ot + 4.0\,Fe\text{-}Ant$$
$$6.0\,At + 0.8\,Q = 4.8\,Mg\text{-}Ot + 2.0\,Ant,$$

then an average chlorite (clinochlore) with an $At : Serp$ ratio of about $1 : 1$, is obtained instead of the amesite, the chloritoid containing about $1/6$ Mg-Ot. This clinochlore variant yields:

Q	Ab	Ms	Zo	At	Ant	Fe-Ant	Ot	Mg-Ot	Hm	Ru	Total
13.9	16.7	19.4	5.5	6.5	2.0	4.0	25.5	4.8	1.2	0.5	100.0

 Klchl 12.5 Chloritoid 30.3

Such a rock should be designated as a **chlorite-sericite-chloritoid phyllite**. In this context it should be noted that C. E. Tilley (1924) has already indicated that the cordierite-spinel hornfels dealt with here shows a very close chemical resemblance to a **chloritoid phyllite** from Curaglia, Graubünden, which was described by P. Niggli (1912).

The **clinochlore variants** of the epinorm for chloritoid schists from the Gotthard massif were provided by E. Niggli (1944, 250). These should be referred to for the sake of comparison.

5. **Calc-silicate fels** (light variety) from Castione near Bellinzona, Tessin (A. Mittelholzer, 1936).

Basis: Q	Kp	Ne	Cal	Cs	Fa	Fo	Ru	Total
32.1	2.6	4.9	10.8	37.9	7.5	4.0	0.2	100.0.

The analysis indicates the presence of 14.57 % CO_2, corresponding to an equivalent percentage content of 22.4 Cc, as was shown when the calculation of the standard katanorm was discussed. Thus, an equivalent amount of Cs should be desilicated prior to subsequent operations:

$$33.6\,Cs = 22.4\,Cc + 11.2\,Q.$$

As usual, the following should also be formed:

$$4.9\,Ne + 3.3\,Q = 8.2\,Ab.$$
$$2.6\,Kp + 2.6\,Cal + 2.2\,Q = 6.1\,Ms + 1.3\,Cs.$$

The following remain: Cal Cs Fo Fa Q corresponding to case β).
 8.2 5.6 4.0 7.5 37.8

$$8.2\,Cal + 5.0\,Q + 1.4\,Cs = 14.6\,Zo$$
$$1.6\,Cs + 4.0\,Fo + 2.4\,Q = 8.0\,Gram$$
$$2.6\,Cs + 6.5\,Fa + 3.9\,Q = 13.0\,Fe\text{-}Akt$$
$$1.0\,Fa + 0.1\,Q = 1.1\,Fe\text{-}Ant.$$

The required standard epinorm thus yields:

Q	Cc	Ab	Ms	Zo	Gram	Fe-Akt	Fe-Ant	Ru	Total
26.4	22.4	8.2	6.1	14.6	8.0	13.0	1.1	0.2	100.0.

Such a rock should be designated as a zoisite- and actinolite-bearing calc-mica schist.

γ) **Variants of the calculated modus of the standard epinorm.** The calculation of the standard epinorm may also be considered with regard to the distribution of Al_2O_3 among the remaining oxides. Thus, one may also proceed by incorporating the total Al_2O_3, excluding that which is bound to K and Na in *Kp* and *Ne*, into 2 *C*. Orthosilicates such as *Cs, Fo, Fa*, should thus be formed according to the following relationships:

$$2\,(Al_2O_3 \cdot CaO) + SiO_2 = SiO_2 \cdot 2\,CaO + 2\,Al_2O_3,$$
$$6\,Cal \qquad\quad 1\,Q \qquad 3\,Cs \qquad\quad 4\,C$$

$$2\,(Al_2O_3 \cdot MgO) + SiO_2 = SiO_2 \cdot 2\,MgO + 2\,Al_2O_3,$$
$$6\,Sp \qquad\quad 1\,Q \qquad 3\,Fo \qquad\quad 4\,C$$

$$2\,(Al_2O_3 \cdot FeO) + SiO_2 = SiO_2 \cdot 2\,FeO + 2\,Al_2O_3.$$
$$6\,Hz \qquad\quad 1\,Q \qquad 3\,Fa \qquad\quad 4\,C$$

If, in addition, *Ne* is silicated to *Ab* and *Fs* is desilicated to *Hm* + *Q*, the calculation may be started from the following combination:

Ru, Cp, Hm, Ab, Kp, Cs, Fo, Fa, C, Q.

Proceeding on the basis of this combination, *C* should be distributed, as far as possible, among the other basis components according to the following equations:

1. $2SiO_2 \cdot Al_2O_3 \cdot K_2O + 4SiO_2 + 2Al_2O_3 (+ 2H_2O) = 6SiO_2 \cdot 3Al_2O_3 \cdot K_2O \cdot 2H_2O$.
   $\quad 6\,Kp \qquad\qquad\qquad 4\,Q \qquad 4\,C \quad (2\,W) \qquad\qquad 14\,Ms$
   or $\;\; 3\,Kp \qquad\qquad\qquad 2\,Q \qquad 2\,C \quad (1\,W) \qquad\qquad 7\,Ms$

2. $2\,(SiO_2 \cdot 2\,CaO) + 4\,SiO_2 + 3\,Al_2O_3 (+ H_2O) = 6\,SiO_2 \cdot 3\,Al_2O_3 \cdot 4\,CaO \cdot H_2O$.
   $\quad 6\,Cs \qquad\qquad\quad 4\,Q \qquad 6\,C \quad (1\,W) \qquad\qquad 16\,Zo$
   or $\;\; 3\,Cs \qquad\qquad\quad 2\,Q \qquad 3\,C \quad (\tfrac{1}{2}\,W) \qquad\qquad 8\,Zo$

If insufficient *C* is available for the formation of *Zo*, the remaining *Cs* should be converted into *Gram* according to: $6\,Cs + 15\,Fo + 9\,Q\,(+ 2\,W) = 30\,Gram$, as previously shown under a), and one should possibly proceed according to the instructions provided under a).

3. $SiO_2 \cdot 2\,MgO + Al_2O_3 (+ 2\,H_2O) = SiO_2 \cdot Al_2O_3 \cdot 2\,MgO \cdot 2\,H_2O$.
   $\quad 3\,Fo \qquad\quad 2\,C \quad (2\,W) \qquad\qquad 5\,At$

If *C* does not suffice for *At*-formation, the remainder of *Fo* should be used, i.e., $9\,Fo + 1\,Q\,(+ 4\,W) = 10\,Ant$, after which the procedure outlined under a) should once more be followed.

4. $SiO_2 \cdot 2\,FeO + Al_2O_3\,(+\,2\,H_2O) = SiO_2 \cdot Al_2O_3 \cdot 2\,FeO \cdot 2\,H_2O$.
       3 Fa 2 C (2 W)   5 Fe-At

If there is insufficient C for Fe-At-formation, the remaining Fa should be used according to: $9\,Fa + 1\,Q\,(+\,4\,W) = 10$-Fe-$Ant$.

5. If, however, an excess of C remains after the completion of these operations, chloritoid should be formed as follows, by the recalculation of Fe-At or At which were just formed:

$$SiO_2 \cdot Al_2O_3 \cdot 2\,FeO \cdot 2\,H_2O + Al_2O_3 + SiO_2 = 2\,(SiO_2 \cdot Al_2O_3 \cdot FeO \cdot H_2O),$$
          5 Fe-At 2 C 1 Q 8 Ot

$$SiO_2 \cdot Al_2O_3 \cdot 2\,MgO \cdot 2\,H_2O + Al_2O_3 + SiO_2 = 2\,(SiO_2 \cdot Al_2O_3 \cdot MgO \cdot H_2O).$$
          5 At 2 C 1 Q 8 Mg-Ot

6. If some C still remains, it is to be converted into $Kaol$ as follows:

$$Al_2O_3 + 2\,SiO_2\,(+\,2\,H_2O) = 2\,SiO_2 \cdot Al_2O_3 \cdot 2\,H_2O.$$
      2 C 2 Q (2 W)   4 $Kaol$

If too much Q was consumed during the operations 1–6, the Q-deficit may be eliminated in the following way:

a) If Fe-Ant is present together with Hm, Mt should be formed according to:

$$2\,SiO_2 \cdot 3\,FeO \cdot 2\,H_2O + 3\,Fe_2O_3 = 3\,(FeO \cdot Fe_2O_3) + 2\,SiO_2\,(+\,2\,H_2O),$$
      5 Fe-Ant 6 Hm 9 Mt 2 Q (2 W)

Thereby Q is freed and either Hm or Fe-Ant is left together with Mt.

b) If this operation does not suffice, one should proceed with brucite-formation from Ant as follows:

$$2\,SiO_2 \cdot 3\,MgO \cdot 2\,H_2O + H_2O = 3\,(MgO \cdot H_2O) + 2\,SiO_2.$$
      5 Ant (1 W)   3 Bru 2 Q

c) If no Ant is available, or a Q-deficiency remains after the conversion of Ant, additional Q is gained from At as follows:

$$2\,(SiO_2 \cdot Al_2O_3 \cdot 2\,MgO \cdot 2\,H_2O) = 2\,(MgO \cdot H_2O) + 2\,(Al_2O_3 \cdot MgO)$$
            10 At 2 Bru 6 Sp
or   5 At 1 Bru 3 Sp

                                          $+\,2\,SiO_2\,(+\,2\,H_2O)$.
                                               2 Q (2 W)
                                               1 Q (1 W)

d) In order to gain additional Q, Mg-Ot may be broken down according to:

$$SiO_2 \cdot Al_2O_3 \cdot MgO \cdot H_2O = Al_2O_3 \cdot MgO + SiO_2\,(+\,H_2O).$$
       4 Mg-Ot 3 Sp 1 Q (1 W)

e) If necessary, when *Zo* + *Kaol* are present, calc-mica may be formed:

$$6\,SiO_2 \cdot 3\,Al_2O_3 \cdot 4\,CaO \cdot H_2O + 5\,(2\,SiO_2 \cdot Al_2O_3 \cdot 2\,H_2O)$$

   16 *Zo*   20 *Kaol*
or   4 *Zo*   5 *Kaol*

$$= 4\,(2\,SiO_2 \cdot 2\,Al_2O_3 \cdot CaO \cdot H_2O) + 8\,SiO_2\,(+ 7\,H_2O).$$

      28 *Marg*   8 *Q* (7 *W*)
      7 *Marg*   2 *Q* ($^7/_4$ *W*)

f) Finally, the probably available *Xon* may also be converted according to:

$$6\,SiO_2 \cdot 6\,CaO \cdot H_2O\,(+ 6\,CO_2) = 6\,(CaO \cdot CO_2) + 6\,SiO_2\,(+ H_2O)$$

   12 *Xon*   (6 CO_2)   6 *Cc*   6 *Q* (1 *W*).

Alternatively, by the addition of CO_2, *Gram* or *Fe-Akt* may be converted as follows:

$$3\,[8\,SiO_2 \cdot 5\,(Mg, Fe)O \cdot 2\,CaO \cdot H_2O]\,(+ 7\,H_2O + 6\,CO_2)$$

     45 *Akt*   (7 *W* + 6 CO_2)

$$= 5\,[2\,SiO_2 \cdot 3\,(Mg, Fe)O \cdot 2\,H_2O] + 6\,(CaO \cdot CO_2) + 14\,SiO_2.$$

     25 *Serp*   6 *Cc*   14 *Q*

If necessary, the *Ant* may also be converted to *Bru*, as explained under b).

Examples:

1. The application of these variants to the calculation of the standard epinorm, with regard to the example of the corundum-cordierite-spinel hornfels from Tom a'Mhinn, Perthshire (C. E. Tilley, 1924), is demonstrated in the following:

Basis:	*Q*	*Kp*	*Ne*	*Cal*	*Sp*	*Hz*	*C*	*Ru*	*Cp*	Total
	28.2	13.7	11.4	3.3	14.7	20.4	5.2	1.4	1.7	100.0

Initially, *Cal*, *Sp* and *Hz* are converted:

$$3.3\,Cal + 0.5\,Q = 1.6\,Cs + 2.2\,C$$
$$14.7\,Sp + 2.5\,Q = 7.4\,Fo + 9.8\,C$$
$$20.4\,Hz + 3.4\,Q = 10.2\,Fa + 13.6\,C.$$

If, in addition, *Ne* is silicated to *Ab* according to:

$$11.4\,Ne + 7.6\,Q = 19.0\,Ab,$$

the following may serve as the initial combination:

Ru	*Cp*	*Ab*	*Kp*	*Cs*	*Fo*	*Fa*	*C*	*Q*	Total
1.4	1.7	19.0	13.7	1.6	7.4	10.2	30.8	14.2	100.0

According to the given scheme, the following components should be formed in the given order:

(1) $\qquad 13.7\,Kp + 9.1\,Q + 9.1\,C = 31.9\,Ms,$
(2) $\qquad 1.6\,Cs + 1.1\,Q + 1.6\,C = 4.3\,Zo,$
(3) $\qquad 7.4\,Fo + 4.9\,C = 12.3\,At,$
(4) $\qquad 10.2\,Fa + 6.8\,C = 17.0\,Fe\text{-}At.$

Since an amount of $30.8 - (9.1 + 1.6 + 4.9 + 6.8) = 8.4\,C$ remains after these operations, chloritoid is formed as follows, using the $Fe\text{-}At + At$ which were just now formed:

(5) $\qquad 17.0\,Fe\text{-}At + 6.8\,C + 3.4\,Q = 27.2\,Ot,$
$\qquad\ \ 4.0\,At \phantom{\text{-}Xy} + 1.6\,C + 0.8\,Q = 6.4\,Mg\text{-}Ot.$

Since an amount of $9.1 + 1.1 + 3.4 + 0.8 = 14.4\,Q$ was consumed until now, whereas only $14.2\,Q$ was available, there exists a Q-deficit of -0.2. The required standard epinorm yields (the previously-obtained values are given in brackets for comparison):

Q	Ab	Ms	Zo	At	Ot	Mg-Ot	Ru	Cp	Total
−0.2	19.0	31.9	4.3	8.3	27.2	6.4	1.4	1.7	100.0
(−0.2)	(19.0)	(32.0)	(4.4)	(8.4)	(27.2)	(6.1)	(1.4)	(1.7)	(100.0)

2. The calculation of the **cordierite-spinel hornfels** from Creag near Iolaire, Glen Lednock, Perthshire, Scotland (C. E. Tilley, 1924) is carried out as follows:

Basis:	Q	Kp	Ne	Cal	Sp	Hz	Fs	Fa	Ru	Total
	37.6	8.3	10.0	4.1	15.0	18.9	1.8	3.8	0.5	100.0

Since, in contrast to the previous example, Fs is present, it should first be desilicated into Hm:

$$1.8\,Fs = 1.2\,Hm + 0.6\,Q.$$

In addition, Cal, Sp and Hz are converted in the following way:

$$4.1\,Cal + 0.7\,Q = 2.1\,Cs + 2.7\,C$$
$$15.0\,Sp + 2.5\,Q = 7.5\,Fo + 10.0\,C$$
$$18.9\,Hz + 3.1\,Q = 9.4\,Fa + 12.6\,C,$$

and Ne is silicated to Ab:

$$10.0\,Ne + 6.7\,Q = 16.7\,Ab.$$

Thus the starting combination is obtained:

Ru	Hm	Ab	Kp	Cs	Fo	Fa	C	Q	Total
0.5	1.2	16.7	8.3	2.1	7.5	13.2	25.3	25.2	100.0

Subsequently, the following components should be formed in accordance with the given regulations:

(1) $8.3\,Kp + 5.5\,Q + 5.5\,C = 19.3\,Ms$,
(2) $2.1\,Cs + 1.4\,Q + 2.1\,C = 5.6\,Zo$,
(3) $7.5\,Fo + 5.0\,C = 12.5\,At$,
(4) $13.2\,Fa + 8.8\,C = 22.0\,\text{Fe-}At$,
(5) $9.8\,\text{Fe-}At + 3.9\,C + 1.9\,Q = 15.6\,Ot$.

Thus, the required standard epinorm yields (the values which were obtained previously are given in brackets for comparison):

Q	Ab	Ms	Zo	At	Fe-At	Ot	Hm	Ru	Total
16.4	16.7	19.3	5.6	12.5	12.2	15.6	1.2	0.5	100.0
(16.3)	(16.7)	(19.4)	(5.5)	(12.5)	(12.2)	(15.7)	(1.2)	(0.5)	(100.0)

3. For the calc-silicate fels from Castione near Bellinzona, Tessin, the following calculation is made:

Basis:	Q	Kp	Ne	Cal	Cs	Fa	Fo	Ru	Total
	32.1	2.6	4.9	10.8	37.9	7.5	4.0	0.2	100.0

By the conversion of *Cal*, according to:

$$10.8\,Cal + 1.8\,Q = 5.4\,Cs + 7.2\,C$$

and the silication of *Ne*, according to:

$$4.9\,Ne + 3.3\,Q = 8.2\,Ab,$$

the following starting combination is obtained:

Ru	Ab	Kp	Cs	Fo	Fa	C	Q	Total
0.2	8.2	2.6	43.3	4.0	7.5	7.2	27.0	100.0

In addition, the following components are formed, according to the given scheme:

(1) $2.6\,Kp + 1.7\,Q + 1.7\,C = 6.0\,Ms$.

If one refers to $7.2 - 1.7 = 5.5\,C$, one further obtains:

(2) $5.5\,Cs + 3.7\,Q + 5.5\,C = 14.7\,Zo$.

As much *Gram* as possible should be formed, on the basis of $4.0\,Fo$, from the remaining $43.3 - 5.5 = 37.8\,Cs$:

$$1.6\,Cs + 4.0\,Fo + 2.4\,Q = 8.0\,Gram.$$

In addition, **Fe-Akt** is analogously formed on the basis of $7.5\,Fa$:

$$3.0\,Cs + 7.5\,Fa + 4.5\,Q = 15.0\,\text{Fe-}Akt.$$

From the remaining $37.8 - (1.6 + 3.0) = 33.2\,Cs$, *Xon* should be formed according to:

$$33.2\,Cs + 11.1\,Q = 44.3\,Xon,$$

and after the completion of these operations, a Q-remainder of $27.0 - (1.7 + 3.7 + 2.4 + 4.5 + 11.1) = 3.6\,Q$ is left, the required standard epinorm being:

Q	Ab	Ms	Zo	Gram	Fe-Akt	Xon	Ru	Total
3.6	8.2	6.0	14.7	8.0	15.0	44.3	0.2	100.0

Since the analysis contained 14.57 wt. % CO_2, which corresponds to 22.4 equivalent % Cc, Xon may be eliminated by Cc-formation, as discussed previously under f):

$$44.3\,Xon = 22.1\,Cc + 22.2\,Q.$$

In order to make up the deficiency of $22.4 - 22.1 = 0.3\,Cc$, an additional Ca-bearing component must be converted, e.g., Fe-Akt. This is carried out in the following manner:

$$3(8\,SiO_2 \cdot 5\,FeO \cdot 2\,CaO \cdot H_2O) \;(+\;6\,CO_2 + 7\,H_2O) = 6(CaO \cdot CO_2)$$

$$\begin{array}{lll} 45\,Fe\text{-}Akt & (6\,CO_2 + 7\,W) & 6\,Cc \\ 2.2\,Fe\text{-}Akt & & 0.3\,Cc \end{array}$$

$$+\;5(2\,SiO_2 \cdot 3\,FeO \cdot 2\,H_2O) + 14\,SiO_2,$$

$$\begin{array}{ll} 25\,Fe\text{-}Ant & 14\,Q \\ 1.2\,Fe\text{-}Ant & 0.7\,Q \end{array}$$

and the **standard epinorm** is as follows (the previously-obtained values are again provided in brackets):

Q	Cc	Ab	Ms	Zo	Gram	Fe-Akt	Fe-Ant	Ru	Total
26.5	22.4	8.2	6.0	14.7	8.0	12.8	1.2	0.2	100.0
(26.4)	(22.4)	(8.2)	(6.1)	(14.6)	(8.0)	(13.0)	(1.1)	(0.2)	(100.0)

δ) **The calculation of epinorm variants adapted to the modus.** Often, the previously-calculated epinorm will correspond satisfactorily to the modus. In other cases, however, large differences may be found between the calculated normative composition and the observed mineral composition. This fact was previously indicated during the calculation of the examples. If considerable differences occur, one should attempt to approach the actual composition by forming variants, as was done in the case of the katanorm. The following points are of importance in this regard.

For example, rather frequently it is found that the calculation yields very antigorite- or amesite-rich **chlorites**. Since the commonly occurring chlorites, e.g., clinochlore, are generally mid-members between antigorite and amesite, the amount of $(Ant + Fe\text{-}Ant)$ or $(At + Fe\text{-}At)$, which is in excess of the ratio $1:1$, enters into other components. If, for example, much Q is present, together with Ant, SiO_2-rich $Talc$ is formed according to:

$$2\,SiO_2 \cdot 3\,MgO \cdot 2\,H_2O + 2\,SiO_2 = 4\,SiO_2 \cdot 3\,MgO \cdot 2\,H_2O.$$

$$\begin{array}{lll} 5\,Ant & 2\,Q & 7\,Tc \end{array}$$

If a large amount of **Ab** is present, together with **Ant**, glaucophane is often formed. The following equation presents the ideal form of this relationship:

$$(6\,SiO_2 \cdot Al_2O_3 \cdot Na_2O) + [2\,SiO_2 \cdot 3(Mg,Fe)O \cdot 2\,H_2O]$$
$$10\,Ab \qquad\qquad 5(Ant + Fe\text{-}Ant)$$
$$= [8\,SiO_2 \cdot Al_2O_3 \cdot 3(Mg,Fe)O \cdot Na_2O \cdot H_2O] + 1\,H_2O.$$
$$15\,Glph \qquad\qquad (1\,W)$$

In the presence of a high **Fe-At** content, garnet may also be formed, e.g., according to the following equation:

$$3(SiO_2 \cdot Al_2O_3 \cdot 2\,FeO \cdot 2\,H_2O) + (2\,SiO_2 \cdot 3\,FeO \cdot 2\,H_2O) + 4\,SiO_2$$
$$15\,Fe\text{-}At \qquad\qquad 5\,Fe\text{-}Ant \qquad\qquad 4\,Q$$
$$= 3(3\,SiO_2 \cdot Al_2O_3 \cdot 3\,FeO)\,(+ 8\,H_2O)$$
$$24\,Alm \qquad\qquad (8\,W).$$

Alternatively, chloritoid is formed as follows:

$$3(SiO_2 \cdot Al_2O_3 \cdot 2\,FeO \cdot 2\,H_2O) + 2\,SiO_2 = 3(SiO_2 \cdot Al_2O_3 \cdot FeO \cdot H_2O)$$
$$15\,Fe\text{-}At \qquad\qquad 2\,Q \qquad\qquad 12\,Ot$$
$$+ 2\,SiO_2 \cdot 3\,FeO \cdot 2\,H_2O\,(+ H_2O).$$
$$5\,Fe\text{-}Ant \qquad (1\,W)$$

According to the given regulations, **Zo** should be formed prior to **At** in the standard-epinorm calculation. However, the equation:

$$3(6\,SiO_2 \cdot 3\,Al_2O_3 \cdot 4\,CaO \cdot H_2O) + 16(2\,SiO_2 \cdot 3\,MgO \cdot 2\,H_2O) + 7\,SiO_2$$
$$48\,Zo \qquad\qquad 80\,Ant \qquad\qquad 7\,Q$$
$$= 6(8\,SiO_2 \cdot 5\,MgO \cdot 2\,CaO \cdot H_2O) + 9(SiO_2 \cdot Al_2O_3 \cdot 2\,MgO \cdot 2\,H_2O)\,(+ 11\,H_2O)$$
$$90\,Gram \qquad\qquad 45\,At \qquad\qquad (11\,W)$$

shows that **Zo + Ant** may also be substituted for by **Gram** and **At**, which fact should be taken into account under certain circumstances.

The formation of epidote, itself a common epi-mineral, requires special consideration. This mineral was not taken into account in the introduction to the calculation of the standard epinorm. The reason for this lies in the fact that the determination of the amount of epidote is dependent on the often arbitrary nature of the degree of oxidation of the iron.

If desired, however, the Fe-end member **Pi** (pistacite), among others, may be formed as follows:

$$6\,SiO_2 + Fe_2O_3 + 2(Al_2O_3 \cdot CaO) + 2(CaO \cdot CO_2)\,(+ H_2O)$$
$$6\,Q \quad 2\,Hm \qquad 6\,Cal \qquad\qquad 2\,Cc \qquad (1\,W)$$
$$= 6\,SiO_2 \cdot 2\,Al_2O_3 \cdot Fe_2O_3 \cdot 4\,CaO \cdot H_2O\,(+ 2\,CO_2),$$
$$16\,Pi \qquad\qquad\qquad (2\,CO_2)$$

$$3\,Fe_2O_3 + 2(6\,SiO_2 \cdot 3\,Al_2O_3 \cdot 4\,CaO \cdot H_2O) + 4(CaO \cdot CO_2) + 6\,SiO_2\,(+ H_2O)$$
$$6\,Hm \qquad\qquad 32\,Zo \qquad\qquad 4\,Cc \qquad 6\,Q \quad (1\,W)$$
$$\text{or} \quad 3\,Hm \qquad\qquad 16\,Zo \qquad\qquad 2\,Cc \qquad 3\,Q \quad (\tfrac{1}{2}W)$$

$$= 3(6\,SiO_2 \cdot 2\,Al_2O_3 \cdot Fe_2O_3 \cdot 4\,CaO \cdot H_2O)\ (+\,4\,CO_2),$$
$$48\ Pi \qquad (4\,CO_2)$$
$$24\ Pi \qquad (2\,CO_2)$$

238
$$3\,Fe_2O_3 + 2(6\,SiO_2 \cdot 3\,Al_2O_3 \cdot 4\,CaO \cdot H_2O) + 2(SiO_2 \cdot 2\,CaO) + 4\,SiO_2\,(+\,H_2O)$$
$$6\ Hm \qquad 32\ Zo \qquad\qquad 6\ Cs \qquad 4\ Q \quad (1\ W)$$
or $\ 3\ Hm \qquad 16\ Zo \qquad\qquad 3\ Cs \qquad 2\ Q \quad (^1/_2 W)$
$$= 3(6\,SiO_2 \cdot 2\,Al_2O_3 \cdot Fe_2O_3 \cdot 4\,CaO \cdot H_2O),$$
$$48\ Pi$$
$$24\ Pi$$

$$Fe_2O_3 + 2(6\,SiO_2 \cdot 3\,Al_2O_3 \cdot 4\,CaO \cdot H_2O) = 6\,SiO_2 \cdot 2\,Al_2O_3 \cdot Fe_2O_3 \cdot 4\,CaO \cdot H_2O$$
$$2\ Hm \qquad 32\ Zo \qquad\qquad\qquad 16\ Pi$$
or $\ 1\ Hm \qquad 16\ Zo \qquad\qquad\qquad 8\ Pi$
$$+\ 6\,SiO_2 + 4(Al_2O_3 \cdot CaO)\ (+\,H_2O).$$
$$6\ Q \qquad 12\ Cal \qquad (1\ W)$$
$$3\ Q \qquad 6\ Cal \qquad (^1/_2 W)$$

According to the two last equations, for example, either $Cs + Q$ are consumed during the formation of Pi from $Hm + Zo$, or $Cal + Q$ are set free, allowing for the elimination of Xon or Cc according to:

$$6\,SiO_2 \cdot 6\,CaO \cdot H_2O = 3(SiO_2 \cdot 2\,CaO) + 3\,SiO_2\,(+\,H_2O),$$
$$12\ Xon \qquad\qquad 9\ Cs \qquad 3\ Q \quad (1\ W)$$

$$2(CaO \cdot CO_2) + SiO_2 = 2\,CaO \cdot SiO_2\,(+\,2\,CO_2).$$
$$2\ Cc \qquad 1\ Q \qquad 3\ Cs \qquad (2\,CO_2)$$

$Cal + Cs$ again yields Zo. Alternatively, Cs combined with Ru yields Tit according to:

$$SiO_2 \cdot 2\,CaO + 2\,TiO_2 + SiO_2 = 2(SiO_2 \cdot TiO_2 \cdot CaO).$$
$$3\ Cs \qquad 2\ Ru \qquad 1\ Q \qquad 6\ Tit$$

d) **Additional selected examples for the calculation of metamorphic rocks:**

α) **The mineral composition of the prasinites:** The prasinites comprise a characteristic group among the initial magmatites of the Mediterranean mountain chains (ophiolites, Pietre verdi). The relations between the chemism and the mineral composition of these rocks lend themselves to ready representation and discussion by means of the methods dealt with here. The **prasinites**, with a gabbroidal chemism, show the following mineral composition: albite, epidote-clinozoisite and chlorite, with a typical poikiloblastic development of the albite. Thus, the prasinite-type is defined both mineralogically and texturally (A. Gansser, 1937, 449). In addition, members of the grammatite-actinolite group, and more rarely also glaucophane, biotite and calcite, may be present, and should be expressed in the name of the rock. Hornblende-prasinites lose their poikiloblastic texture through the enrichment of hornblende, thus trending toward albite-amphibolites. The prasinites belong to the epi-facies of the calc-aluminosilicate rocks, in the classification of Grubenmann-Niggli, or to the greenschist facies of Eskola. The **prasinite-type** is found in surprisingly constant development in the Pennine zone of the Alps, from the Cottiennes and Graie Alps, through the Pennines, the Adula and the region of the Oberhalbstein-Engadine, up to the Hohe Tauern, as well as in the Toscanian Appenines (Apuane Alps) and at Elba (C. Burri and P. Niggli, 1948).

Figure 64 illustrates 25 prasinite analyses from the previously-mentioned localities, arbitrarily selected from the literature, projected on a QLM-triangle. They originate from works by A. Bianchi, Gb. Dal Piaz, E. A. Diehl, St. Bonatti, E. Grill, P. Rossoni, H. Ph. Roothaan and P. Niggli, Fr. de Quervain and R. U. Winterhalder (1930).

Regarding the high content of members of the epidote-clinozoisite series, which is always present as required by definition, no agreement should be expected, a priori, with the standard epinorm calculated according to the given regulations. This is exemplified by a series of 5 selected examples, from various parts of the Alps.

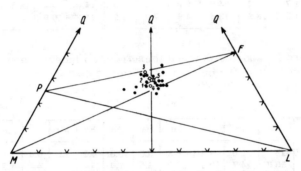

FIGURE 64. QLM-projection of 25 arbitrarily selected prasinites

Table 10

	SiO_2	Al_2O_3	Fe_2O_3	FeO	MnO	MgO	CaO	Na_2O	K_2O	TiO_2	P_2O_5	CO_2	H_2O+	H_2O-	Total
1.	49.01	16.12	1.20	6.43	—	8.02	10.82	3.21	0.34	2.98	0.20	—	2.11	0.31	100.75
2.	49.60	14.55	4.39	5.93	0.15	6.92	9.61	3.25	0.60	1.83	0.02	1.00	1.90	0.07	99.82
3.	47.59	15.99	4.38	5.25	0.14	7.70	11.43	1.45	0.26	1.99	0.16	0.16	2.70	0.02	99.22
4.	47.88	15.41	4.91	5.69	0.18	6.87	10.71	3.67	0.49	1.67	—	—	2.74	0.04	100.26
5.	48.82	14.59	7.28	5.93	0.29	5.25	7.98	3.36	1.13	3.05	0.40	Sp.	1.37	0.28	99.89*)

*) Incl. Cl 0.03, S 0.18, BaO 0.02, — 0.07, for O.

1. **Prasinite**, Punta Segnale, Valle della Germanasca, Cottiennes Alps. E. Grill, in "Alla Memoria del Prof. Ernesto Manasse (1875-1922)", Florence (1924). Analysis by author.
2. **Epidote-hornblende-albite schist** (prasinite), Glacier, Val d'Ollomont, Prov. Aosta, Italy. E. A. Diehl, Schweiz. Min. Petr. Mitt. 18 (1938) 346. Analysis by author.

3. Hornblende-chlorite-albite-epidote schist. Basispartie, Glacier, Val d'Ollomont, Prov. Aosta, Italy. E. A. Diehl, loc. cit. 346. Analysis by author.

4. Greenschist (prasinite), N Leisalp-Oberstaffel, Vals, Graubünden. H. Ph. Roothaan, Jahrb. Natf. Ges. Graubünden, Chur, 59 (1919) 63. Analysis by author.

5. Biotite-amphibole prasinite, Bocchetta del Vento di Dentro, Alto Adige Orientale, Italy. A. Bianchi, Mem. Ist. Geol. R. Univ. Padova 10 (1934) 131. Analysis by author.

From these analyses the basis values of Table 11 are calculated.

Table 11

	Q	Kp	Ne	Cal	Cs	Fs	Fa	Fo	Cp	Ru	Total
1.	28.8	1.2	17.4	17.4	7.2	1.3	7.5	16.7	0.4	2.1	100.0
2.	29.4	2.3	18.0	14.5	7.5	4.8	7.3	14.9	—	1.3	100.0
3.	32.0	0.9	8.1	23.0	6.1	4.8	6.6	16.7	0.3	1.5	100.0
4.	26.2	1.7	20.2	14.9	8.9	5.3	7.0	14.6	—	1.2	100.0
5.	28.9	4.2	18.8	13.4	4.9	7.9	7.5	11.3	0.9	2.2	100.0

The standard katanorms derived therefrom do indeed show that rocks of gabbroidal chemism are being dealt with (Table 12).

Table 12

	Q	Or	Ab	An	Wo	En	Hy	Fo	Fa	Mt	Cp	Ru	Total
1.	—	2.0	29.0	29.0	9.6	8.3	3.6	10.5	4.2	1.3	0.4	2.1	100.0
2.	—	3.8	30.0	24.2	10.0	18.0	6.5	1.4	—	4.8	—	1.3	100.0
3.	4.1	1.5	13.5	38.3	8.1	22.3	5.6	—	—	4.8	0.3	1.5	100.0
4.	—	2.8	33.7	24.8	11.9	—	5.2	14.6	0.5	5.3	—	1.2	100.0
5.	2.0	7.0	31.3	22.3	6.5	15.1	4.8	—	—	7.9	0.9	2.2	100.0

The standard epinorms, calculated according to the given regulations, yield:

Table 13

	Q	Ab	Ms	Zo	Gram	Fe-Akt	Ant	Fe-Ant	Hm	Cp	Ru	Cc	Total
1.	—2.2	29.0	2.8	28.8	25.5	—	4.4	8.3	0.9	0.4	2.1	—	100.0
2.	2.5	30.0	5.4	21.7	22.5	—	4.0	8.1	3.2	—	1.3	1.3	100.0
3.	8.4	13.5	2.1	39.2	13.0	—	11.3	7.3	3.2	0.3	1.5	0.2	100.0
4.	—6.6	33.6	4.0	23.5	29.2	8.5	—	3.1	3.5	—	1.2	—	100.0
5.	1.0	31.3	16.3	9.8	22.6	5.0	—	5.6	5.3	0.9	2.2	—	100.0

A comparison with the previously-provided mineral composition shows that indeed no agreement exists between the modus and the standard epinorm. As is well known, the norm calculation only reckons the Fe^{3+}-free end member **Zo** of the clinozoisite-epidote series. Thus, due to the Fe^{3+}-content, a considerable amount is converted to **Hm**. In addition, Al-free serpentine occurs instead of clinochlore. Thus the question arises whether, by some other means, an epinorm variant may be obtained, which is better adapted to the modus.

As was shown by E.A. Diehl (1938) in connection with his investigations of greenschists of the Combin-zone in the Val d'Ollomont area (Prov. Aosta, Italy), it is possible to achieve this by application of the "remainder triangle" (compare section B II 10 c). According to this procedure, it is possible to eliminate by calculation both the components *Cc, Cp, Tit* and *Ms (Ser)* which occur only in very small amounts, if at all, and the principal component *Ab*, the composition of which very closely approaches the ideal. The residual remainder *R* is then projected on a *QLM*-triangle (Figure 65). The three remaining principal components, hornblende (*Ho*), epidote (*Ep*) and clinochlore (*Kl*), are also plotted in the same triangle, using suitably analyzed examples of naturally-occurring specimens instead of the idealized compositions. If the projection point of *R* falls within the triangle *Ho–Ep–Kl*, this may be considered as an oblique-angled concentration triangle. The distribution of *R* among the three components *Ho, Ep* and *Kl* may then be determined graphically and be recalculated to the total composition. If the projection point of *R* does not fall within the triangle, special assumptions must be made, as will be shown hereafter.

The following analyzed examples, suggested by E.A. Diehl, will serve to represent the composition of the three principal components *Ho, Ep* and *Kl*:

	1. Epidote	2. Grammatite	3. Clinochlore
SiO_2	37.99	57.62	31.18
Al_2O_3	29.53	1.41	18.28
Fe_2O_3	5.67	0.12	2.00
FeO	0.53	1.03	4.85
MnO	0.21	—	0.03
MgO	—	23.48	31.11
CaO	23.85	13.94	—
Na_2O	—	0.26	—
K_2O	—	0.14	—
TiO_2	—	0.04	0.10
H_2O+	2.04	2.29	12.62
	99.82	100.33	100.17

1. Epidote, Huntington, Mass., USA (average of 2 analyses). Anal. E.A. Forbes, in E.A. Forbes, Z. Kristallogr. 26 (1896) 1939.
2. Grammatite, Monte Spinosa, Campigliese, Toscana, Italy. Anal. Fr. Rodolico, in Fr. Rodolico, Rendic. R. Accad. Lincei, Cl. sc. fis. mat. e. nat. (6) 13.1° sem. (1931) 706.
3. Clinochlore from chlorite schists, Besafotra River, Madagascar. Anal. J. Orcel, in J. Orcel, Thèse Fac. Sc. Paris (1927) 193.

[241]

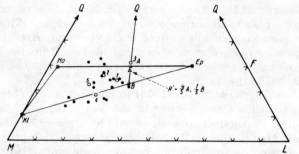

FIGURE 65. Analyses of remainders of 25 prasinites (partly according to E. A. Diehl, 1938).

From these analyses the following cation percentages are calculated:

	Si^{4+}	Al^{3+}	Fe^{3+}	$Fe^{2+}+Mn^{2+}$	Mg^{2+}	Ca^{2+}	Na^+	K^+	Ti^{4+}	Total
1. Ep	36.8	33.8	4.1	0.6	—	24.7	—	—	—	100.0
2. Ho	52.1	1.5	$Sp.$	0.8	31.5	13.5	0.4	0.2	$Sp.$	100.0
3. Kl	29.8	20.6	1.4	3.9	44.3	—	—	—	$Sp.$	100.0

The following basic values were obtained:

	1. Ep		2. Ho		3. Kl	
Q	30.7	Q 30.7	28.7	Q 28.7	10.2	Q 10.2
Kp	—		0.5		—	
Ne	—	L 50.7	1.3	L 3.1	—	L —
Cal	50.7		1.3		—	
Sp	—		—		30.6	
Cs	11.5		19.6		—	
Fs	6.2	M 18.6	0.1	M 68.2	2.2	M 89.8
Fo	—		47.4		51.1	
Fa	0.9		1.1		5.9	
	100.0	100.0	100.0	100.0	100.0	100.0

For the **prasinite analysis No. 1** (Punta Segnale, Valle della Germanasca) the following calculation is provided, on the basis of the cation percentages, and, as will be the procedure hereafter, Mn^{2+} is added to Fe^{2+}:

	Cation %	Tit	Cp	Ab	Ms	Remainder R*
Si^{4+}	45.9	2.1		17.4	1.2	25.2
Al^{3+}	17.8			5.8	1.2	10.8
Fe^{3+}	0.8					0.8
Fe^{2+}	5.0					5.0
Mg^{2+}	11.2					11.2
Ca^{2+}	10.9	2.1	0.2			8.6
Na^+	5.8			5.8		61.6
K^+	0.4				0.4	38.4
Ti^{4+}	2.1	2.1				100.0
P^{5+}	0.1		0.1			
	100.0	6.3	0.3	29.0	2.8	Total 38.4

* [R denotes remainder in all following cases].

243   Initially, according to the regulations of section B II 6, *Tit, Cp, Ab, Ms* (and, if CO_2 is present, also *Cc*) are formed, their sum amounting to 38.4. The remainder (totalling 61.6) is then recalculated to basis components in the following way:

	R	Cal	Cs	Fs	Fa	Fo	Q
Si^{4+}	25.2		1.6	0.4	2.5	5.6	15.1
Al^{3+}	10.8	10.8					
Fe^{3+}	0.8			0.8			
Fe^{2+}	5.0				5.0		
Mg^{2+}	11.2					11.2	
Ca^{2+}	8.6	5.4	3.2				
	61.6	16.2	4.8	1.2	7.5	16.8	15.1 Total 61.6.

On recalculating to the sum 100 one obtains:

26.3   7.8   1.9   12.2   27.3   24.5   Total 100.0

L $M_R = 49.2$ Q_R

According to the group values Q_R, L_R, M_R (thus designated in order to distinguish them from the group values Q, L and M, which were intended to characterize the total chemism of the rock), R may now be introduced in the QLM-triangle. The projection point does indeed fall within the remainder triangle Ho–Ep–Kl, as shown in Figure 65, and it should thus be possible to express the remainder R by these three minerals. The following values are read from the figure:

        Ho 22.2%   Ep 50.0%   Kl 27.8%   Total 100.0.

The reading may be carried out either by drawing parallels to the three sides of the triangle, or by means of the previously-mentioned vector method.

On recalculation from the remainder sum of 61.6, one thus obtains: Ho 13.7, Ep 30.8, Kl 17.1, and the required variant of the epinorm yields:

Ab	Ep	Kl	Ho	Ms	Tit	Cp	Total
29.0	30.8	17.1	13.7	2.8	6.3	0.3	100.0.

One should now check whether the chemism of R can indeed be expressed by the three accepted mineral compositions. This check is deemed necessary, mainly because Q, L and M represent composite components consisting of groups of isomorphically-substituting atom species or components. It is preferable to start from the cation percentages, using the previously-provided atom-percentage compositions of epidote, hornblende and clinochlore.

	13.7 Ho	30.8 Ep	17.1 Kl	Total S	R	$R-S$	Balance
Si^{4+}	7.2	11.3	5.1	23.6	25.2	+1.6	+1.6 ⎫
Al^{3+}	0.2	10.4	3.5	14.1	10.8	−3.3	⎧ −1.6 ⎬
							⎩ −1.7 ⎭
Fe^{3+}	Sp.	1.3	0.2	1.5	0.8	−0.7 ⎫	⎧ +1.7 ⎫
Fe^{2+}	0.1	0.2	0.7	1.0	5.0	+4.0 ⎭	⎩ +1.6 ⎭
Mg^{2+}	4.3	—	7.6	11.9	11.2	−0.7	−0.7
Ca^{2+}	1.8	7.6	—	9.4	8.6	−0.8	−0.8
Na^+ K^+ ⎫⎬⎭	0.1	—	—	0.1	—	−0.1	−0.1
	13.7	30.8	17.1	61.6	61.6		

As may be expected, small differences do show up, which may, however, be eliminated by the introduction of minute corrections of the isomorphic substitute. If, for example, it is assumed that the +1.6 Si^{4+} of R substitutes for an equal amount of Al^{3+}, then the Al^{3+} deficiency is reduced to −1.7. In order to compensate for this amount with Fe^{3+}, the oxidation ratio of the iron must be changed, so that +1.6 Fe^{2+} will still remain, corresponding to an equal number of $Mg^{2+} + Ca^{2+} + Na^+ + K^+$. This leads, as already indicated by E. A. Diehl, to an insignificant substitution of Ca^{2+} and Fe^{2+}. It should be noted that the fact that the equalization has proved successful, in itself does not constitute a proof of the correctness of the assumptions made, because for $R = S$ it always holds true that $\sum (R_i - S_i) = 0$. The attempt at calculating an epinorm variant which is adapted to the modus thus leads to a satisfactory result.

For analysis 2 (prasinite, Glacier, Val d'Ollomont) one should start, for a change, from the basis:

		Cc	Tit	Ab	Ms	R		To the sum of 100
Q	29.4	−0.6	0.6	12.0	1.9	15.5	15.5	$Q_R = 26.1$
Kp	2.3				2.3			
Ne	18.0			18.0				
Cal	14.5				2.3	12.2	12.2	$L_R = 20.5$
Cs	7.5	1.9	1.9		−1.1	4.8 ⎫		
Fs	4.8					4.8 ⎬		
Fa	7.3					7.3 ⎬	31.8	$M_R = 53.4$
Fo	14.9					14.9 ⎭		100.0
Ru	1.3		1.3			59.5 = R		
$(CO_2$	1.3)	(1.3)				40.5		
						100.0		
	100.0	1.3	3.8	30.0	5.4	Total 40.5		

The corresponding reaction equations yield:

$$1.9\, Cs = 1.3\, Cc + 0.6\, Q$$
$$1.9\, Cs + 1.3\, Ru + 0.6\, Q = 3.8\, Tit$$
$$18.0\, Ne + 12.0\, Q = 30.0\, Ab$$
$$2.3\, Kp + 2.3\, Cal + 1.9\, Q = 5.4\, Ms + 1.1\, Cs.$$

The projection point of R again falls within the triangle $Ho-Ep-Kl$. The

following values may be read off:
Ho 46.2, Ep 37.1, Kl 16.7. Recalculated to R = 59.5 this yields: Ho 27.5, Ep 22.1, Kl 9.9. The required **variant of the epinorm** is thus obtained:

Ab	Ep	Kl	Ho	Ms	Tit	Cp	Cc	Total
30.0	22.1	9.9	27.5	5.4	3.8	Sp.	1.3	100.0 ·

Here again, one should check, by reverse calculations, if R may indeed be expressed by the three selected mineral compositions. In order to do this, R should first be recalculated to cation percentages, according to section B II 4 c:

$$Si^{4+} = 1/_3 (Cs + Fs + Fo + Fa) + Q = 26.1$$
$$Al^{3+} = 2/_3 Cal = 8.1$$
$$Fe^{3+} = 2/_3 Fs = 3.2$$
$$Fe^{2+} = 2/_3 Fa = 4.9$$
$$Mg^{2+} = 2/_3 Fo = 9.9$$
$$Ca^{2+} = 2/_3 Cs + 1/_3 Cal = 7.3$$
$$\overline{ 59.5}$$

By using the previously-provided compositions of Ho, Ep and Kl, R may again be calculated as follows:

	27.5 Ho	22.1 Ep	9.9 Kl	Total S	R	R–S	Balance
Si^{4+}	14.3	8.1	2.9	25.3	26.1	+ 0.8	+ 0.8 ⎫
Al^{3+}	0.4	7.5	2.1	10.0	8.1	– 1.9	– 0.8 ⎬ –1.1
Fe^{3+}	Sp.	0.9	0.1	1.0	3.2	+ 2.2	+1.1 ⎭
Fe^{2+}	0.2	0.1	0.4	0.7	4.9	+ 4.2	+5.3
Mg^{2+}	8.7	—	4.4	13.1	9.9	– 3.2	
Ca^{2+}	3.7	5.5	—	9.2	7.3	– 1.9	– 5.3
Na^+ / K^+	0.2	—	—	0.2	—	– 0.2	
	27.5	22.1	9.9	59.5	59.5		

For this example also, a satisfactory agreement was found. It is seen that the epidote is somewhat richer in Fe^{3+} than was assumed, that slightly more Mg^{2+} is substituted for by Fe^{2+}, and that here also some Ca^{2+} may probably enter in place of $Mg^{2+} + Fe^{2+}$.

The calculation of **analysis 5** (prasinite, Bocchetta del Vento di Dentro) takes the following form, once again starting from the cation percentages:

	Cp	Tit	Ab	Ms	R	Cal	Cs	Fs	Fa	Fo	Q
Si^{4+}	47.2	2.2	18.9	4.2	21.9		1.4	2.6	2.5	3.7	11.7
Al^{3+}	16.6		6.3	4.2	6.1	6.1					
Fe^{3+}	5.3				5.3			5.3			
Fe^{2+}	5.0				5.0				5.0		
Mg^{2+}	7.5				7.5					7.5	
Ca^{2+}	8.3	0.3	2.2		5.8	3.0	2.8				
Na^+	6.3		6.3		51.6	9.1	4.2	7.9	7.5	11.2	11.7
K^+	1.4			1.4	48.4				Total 51.6		
Ti^{4+}	2.2	2.2			100.0	calculated to the sum of 100					
P^{5+}	0.2	0.2				17.6	8.2	15.3	14.5	21.7	22.7
	100.0	0.5	6.6	31.5	9.8 Total 48.4 L_R		$M_R = 59.7$				Q_R

The projection point of R again falls within the triangle Ho-Ep-Kl, and the following values may be read off: Ho 32.5, Ep 32.5, Kl 35.0. By recalculation to the sum of 51.6 = R, it follows that: Ho 16.8, Ep 16.8, Kl 18.0.

Thus, the required **epinorm variant** yields:

Ab	Ep	Kl	Ho	Ms	Tit	Cp	Total
31.5	16.8	18.0	16.8	9.8	6.6	0.5	100.0.

The reverse calculation provides:

	16.8 Ho	16.8 Ep	18.0 Kl	Total S	R	$R-S$	Balance
Si^{4+}	8.8	6.2	5.4	20.4	21.9	+ 1.5	+ 1.5
Al^{3+}	0.3	5.7	3.7	9.7	6.1	− 3.6	−1.5 / −2.1
Fe^{3+}	—	0.7	0.3	1.0	5.3	+ 4.3	+ 2.1
Fe^{2+}	0.1	0.1	0.7	0.9	5.0	+ 4.1	+ 6.3
Mg^{2+}	5.2	—	7.9	13.1	7.5	− 5.6	
Ca^{2+}	2.3	4.1	—	6.4	5.8	− 0.6	− 6.3
Na^+ / K^+	0.1	—	—	0.1	—	− 0.1	
	16.8	16.8	18.0	51.6	51.6		

The agreement is again satisfactory. A higher substitution for Mg^{2+} by Fe^{2+} is apparent, and the epidote is more Fe^{3+}-rich.

The following calculation is given for **analysis No. 4** (greenschist, Leisalp-Oberstaffel, Vals):

	Tit	Ab	Ms	R	Cal	Cs	Fs	Fa	Fo	Q
Si^{4+}	45.5	1.2	20.1	1.8	22.4		2.7	1.7	2.3	4.8	10.9
Al^{3+}	17.2		6.7	1.8	8.7	8.7				
Fe^{3+}	3.5				3.5			3.5		
Fe^{2+}	4.7				4.7				4.7	
Mg^{2+}	9.7				9.7					9.7
Ca^{2+}	10.9	1.2			9.7	4.3	5.4			
Na^+	6.7		6.7		58.7	13.0	8.1	5.2	7.0	14.5	10.9
K^+	0.6			0.6	41.3					Total 58.7
Ti^{4+}	1.2	1.2			100.0	Recalculated to the sum of 100				
	100.0	3.6	33.5	4.2	Total	22.2	13.8	8.8	11.9	24.7	18.6
					41.3					
						L_R		$M_R = 59.2$			Q_R

In this example, the projection point of R (Figure 65) falls outside the remainder triangle, i. e., slightly below the side Kl−Ep. If, by approximation, the point is considered to be located exactly on the tie line Kl−Ep, the composition of R is given by Ep 42.8, Kl 57.2, or, taking R = 58.7, one obtains Ep = 25.1 and Kl 33.6 respectively.

The required epinorm variant thus yields:

Ab	Ep	Kl	Ms	Tit	Total
33.5	25.1	33.6	4.2	3.6	100.0

On reversing the calculation, one obtains:

	25.1 Ep	33.6 Kl	Total S	R	$R-S$	Balance
Si^{4+}	9.2	10.0	19.2	22.4	$+3.2$	$+3.2 \brace -3.2$
Al^{3+}	8.5	6.9	15.4	8.7	-6.7	$\{-3.5$
Fe^{3+}	1.0	0.5	1.5	3.5	$+2.0\}$	$+3.5\}$
Fe^{2+}	0.2	1.3	1.5	4.7	$+3.2\}$	$+1.7\}$
Mg^{2+}	—	14.9	14.9	9.7	$-5.2\}$	$-1.7\}$
Ca^{2+}	6.2	—	6.2	9.7	$+3.5\}$	
	25.1	33.6	58.7	58.7		

A good agreement was also found here. In addition to a slightly higher Fe^{3+}-content, the substitution for Ca by Mg + Fe was also noted, but in contrast to the cases dealt with up to now, more Ca is present in R than in the graphically determined amount of Ep.

In the examples hitherto dealt with (analyses 1, 2, 4 and 5) the projection points of the remainder R always all within the triangle $Ho-Ep-Kl$ or upon a side of the triangle. Thus, a mathematical distribution of R among these three components, or among two of them, was immediately possible. A reverse calculation provides a means of control for checking whether the isomorphous mixture ratios, assumed to represent the components of these examples, do indeed correspond to the actual rock's chemism. This was essentially the case in the treated examples, and the balance did not show any dubious results. In other cases this will occur less frequently, and other compositions must be introduced into the calculation of the remainder components.

If, however, the projection points of R do not fall within the triangle $Ho-Ep-Kl$, a different procedure should be employed. One should then distinguish whether the remainder point falls above or below the triangle. An example of the first case is provided by analysis No. 3 of a prasinite (originally designated as a hornblende-chlorite-albite-epidote schist) from Glacier (Basispartie), Val d'Ollomont, which was also discussed by E. A. Diehl. Below is given a recalculation taking into consideration the small content of Cp and Cc.

Analysis No. 3: Hornblende-chlorite-albite-epidote schist, Glacier (Basispartie), Val d'Ollomont.

	Cp	Tit	Ab	Ms	Cc	R	Cal	Cs	Fs	Fa	Fo	Q
Si^{4+}	46.4	1.5	8.1	0.9		35.9		1.3	1.6	2.2	5.6	25.2
Al^{3+}	18.3		2.7	0.9		14.7	14.7					
Fe^{3+}	3.2					3.2			3.2			
Fe^{2+}	4.4					4.4				4.4		
Mg^{2+}	11.2					11.2					11.2	
Ca^{2+}	11.9	0.2	1.5		0.2	10.0	7.3	2.7				
Na^+	2.7		2.7			79.4	22.0	4.0	4.8	6.6	16.8	25.2
K^+	0.3			0.3		20.6					Total	79.4
Ti^{4+}	1.5	1.5				100.0						
P^{5+}	0.1	0.1					Recalculated to the sum of 100
CO_2	(0.2)				(0.2)		27.7	5.0	6.1	8.3	21.2	31.7
	100.0	0.3	4.5	13.5	2.1	0.2	L_R	$M_R = 40.6$			Q_R	

Total 20.6

As shown in Figure 65, the projection point of R falls above the triangle $Ho-Ep-Kl$. If a certain amount of Q is subtracted from R, this point is displaced into the [small] triangle along a line which issues from the Q-apex, coming to rest between A and B. The following compositions may be read off for these two points from the figure:

	A			B	
Ho	46.4		Kl	38.1	
Ep	53.6		Ep	61.9	
	100.0			100.0	

If, as an experiment, a composition of $^4/_5\,A$ and $^1/_5\,B$ is considered, the projection point will come to rest at R'. From the equations for the center of gravity (or by reading off triangular coordinations), it is found that this displacement corresponds to a subtraction of 5 % Q. From the previously calculated R, the composition of R' is thus found to be 79.4 - 5.0 = 77.4, yielding:

Cal	Cs	Fs	Fa	Fo	Q	Total
22.0	4.0	4.8	6.6	16.8	20.2	74.4,

or in cation percentages:

$$Si^{4+} = {}^1/_3(Cs + Fs + Fa + Fo) + Q = 30.9$$
$$Al^{3+} = {}^2/_3\,Cal = 14.7$$
$$Fe^{3+} = {}^2/_3\,Fs = 3.2$$
$$Fe^{2+} = {}^2/_3\,Fa = 4.4$$
$$Mg^{2+} = {}^2/_3\,Fo = 11.2$$
$$Ca^{2+} = {}^2/_3\,Cs + {}^1/_3\,Cal = 10.0$$
$$\phantom{Ca^{2+} = {}^2/_3\,Cs + {}^1/_3\,Cal = {}}\overline{74.4.}$$

The compositions of A and B, calculated on the basis of the new remainder sum $R' = 74.4$, is given by:

	A			B	
Ho	46.4	34.5	Kl	38.1	28.3
Ep	53.6	39.9	Ep	61.9	46.1
	100.0	74.4		100.0	74.4,

or, expressed in cation percentages:

	A			B		
	34.5 Ho	39.9 Ep	Total S_A	28.3 Kl	46.1 Ep	Total S_B
Si^{4+}	17.9	14.7	32.6	8.4	16.9	25.3
Al^{3+}	0.5	13.5	14.0	5.8	15.6	21.4
Fe^{3+}	Sp.	1.6	1.6	0.4	1.9	2.3
Fe^{2+}	0.3	0.2	0.5	1.1	0.3	1.4
Mg^{2+}	10.9	—	10.9	12.6	—	12.6
Ca^{2+}	4.7	9.9	14.6	—	11.4	11.4
Na^+ } K^+ }	0.2	—	0.2	—	—	—
	34.5	39.9	74.4	28.3	46.1	74.4.

The calculation based on $^4/_5 A + ^1/_5 B$ takes the following form:

	S_A	S_B	$(^4/_5 A + ^1/_5 B) = M$	R'	$R' - M$	Balance	
Si^{4+}	32.6	25.3	31.1	30.9	−0.2	−1.0	
Al^{3+}	14.0	21.4	15.5	14.7	−0.8		
Fe^{3+}	1.6	2.3	1.8	3.2	+1.4	+1.0	
						+0.4	+4.1
Fe^{2+}	0.5	1.4	0.7	4.4	+3.7	+3.7	
Mg^{2+}	10.9	12.6	11.2	11.2	—		
Ca^{2+}	14.6	11.4	14.0	10.0	−4.0	−4.2	
$Na^+ + K^+$	0.2	—	0.2	—	−0.2		
	74,4	74,4	74,5	74,4			

The agreement may be considered as satisfactory. Here again, as in the other examples, the substitution for $Fe^{2+} + Mg^{2+}$ by Ca^{2+} is confirmed.

The composition $^4/_5 A + ^1/_5 B$ of R', which was taken into calculation, corresponds to 27.6 Ho, 41.1 Ep, 5.7 Kl. Thus the required **epinorm variant** takes the following form:

Q	Ab	Ep	Ho	Kl	Ms	Tit	Cp	Cc	Total
5.0	13.5	41.1	27.6	5.7	2.1	4.5	0.3	0.2	100.0.

If the projection point of R happens to fall below the triangle Ho–Ep–Kl, different possibilities exist for displacing it into the triangle. This was already discussed in greater detail in section B II 10 c. In the case of the considered rocks, the following possibilities are of prime interest:

1. **Displacement from the direction of the feldspar point F**, i.e., the subtraction of An, resulting in the formation of albite-oligoclase instead of albite.

2. **Displacement from the direction of the oxide point.** Depending on the chemism, this may entail the formation of Hm, Mt or additional Cc. In nature these possibilities may occur in combination, and a calculation on such a basis is only of value if modal relationships are reckoned with.

3. **The possibility of Na-sericite formation** or the consideration of the possible existence of an Na-content in the sericite which up to now was assumed to be a pure K-component in the calculation. This was already indicated by L. A. Diehl. According to statistical investigations of P. Niggli (1937), Na-contents of 1/8 to 1/5 of the total alkalis may be attributed to muscovites and sericites. One should thus investigate in what manner this fact affects the norm calculation or the position of the projection points in the QLM-triangle.

The formation of Na-sericite (Na-Sc) at the expense of albite is carried out in a manner analogous to muscovite formation:

$$6\,SiO_2 \cdot Al_2O_3 \cdot K_2O + 2\,Al_2O_3 (+ 2\,H_2O) = 6\,SiO_2 \cdot 3\,Al_2O_3 \cdot K_2O \cdot 2\,H_2O,$$
$$10\,Or \qquad\qquad 4\,C \quad (2\,W) \qquad\qquad 14\,Ms\,(Sc)$$

according to the relationship:

$$6\,SiO_2 \cdot Al_2O_3 \cdot Na_2O + 2\,Al_2O_3 (+ 2\,H_2O) = 6\,SiO_2 \cdot 3\,Al_2O_3 \cdot Na_2O \cdot 2\,H_2O. \quad (a)$$
$$10\,Ab \qquad\qquad 4\,C \quad (2\,W) \qquad\qquad 14\,Na\text{-}Sc$$

The required amount of Al₂O₃ is obtained by breaking up *Cal*. At the same time Q is consumed for the formation of *Cs*:

$$2(Al_2O_3 \cdot CaO) + SiO_2 = 2\,CaO \cdot SiO_2 + 2\,Al_2O_3. \qquad \text{(b)}$$
$$6\,Cal 1\,Q 3\,Cs 4\,C$$

By addition of (a) and (b), $4C$ is eliminated on both sides of the equation and one obtains:

$$6\,SiO_2 \cdot Al_2O_3 \cdot Na_2O + 2(Al_2O_3 \cdot CaO) + SiO_2(+\,2\,H_2O)$$
$$10\,Ab 6\,Cal 1\,Q (2\,W)$$
$$= 6\,SiO_2 \cdot 3\,Al_2O_3 \cdot Na_2O \cdot 2\,H_2O + SiO_2 \cdot 2\,CaO\,.$$
$$14\,Na\text{-}Sc 3\,Cs$$

The formation of **Na-Sc** thus requires $Cal + Q$, and Cs is liberated. The remainder R is therefore not simply displaced along the lines drawn through the *Cal*-pole from outside the triangle (i. e., the L-apex of the triangle). Instead, these lines must originate from the projection point of the complex

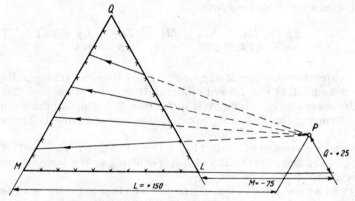

FIGURE 66. Displacement of the remainder point for prasinite analysis, considering a Na-content of the sericite. After E. A. Diehl (1938).

$(6\,Cal + 1\,Q - 3\,Cs)$. The corresponding coordinates, expressed in Q, L and M, are found as follows:

$$\left.\begin{array}{l} Q = x \\ L = 6x \\ M = -3x \end{array}\right\} \text{ corresponding to } 1\,Q + 6\,Cal - 3\,Cs.$$

Since, however, $Q + L + M = x + 6x - 3x = 100$, it is found that $x = 25$, and thus $Q = +25.\ L = +150.\ M = -75$.

As follows from Figure 66, the formation of **Na-Sc** may lead to the same result as the displacement of a point located below the triangle $Ho-Ep-Kl$.

β) Sillimanite-biotite-cordierite-garnet fels, from La Vieille, Comba di Vessona, Valpelline (Aosta province, Italy).

aa) Calculation of the standard katanorm. The basis is calculated from the chemical analysis (anal. R. Masson) of the example which was investigated in detail by R. Masson in connection with his description of the rocks from the Valpelline-series (R. Masson, 1938, 158), as follows:

	Weight %	Atomic equivalent numbers	Ru	Kp	Ne	Cal	Sp	Hz	Fa	Fs	Q
SiO_2	48.66	727			30	13			38.5	22	623.5
Al_2O_3	28.42	557		30	13	20	292	202			
Fe_2O_3	3.45	44							44		
FeO	12.74	177 ⎫ 178						101	77		
MnO	0.07	1 ⎭									
MgO	5.88	146					146				
CaO	0.53	10				10					
Na_2O	0.39	13			13						
K_2O	1.40	30		30							
TiO_2	2.78	35	35								
P_2O_5	0.00	1740	35	90	39	30	438	303	115.5	66	623.5
H_2O+	0.80										
H_2O-	0.00										
	100.12										

Q	Kp	Ne	Cal	Sp	Hz	Fa	Fs	Ru	Total
35.9	5.2	2.2	1.7	25.2	17.4	6.6	3.8	2.0	100.0.

The standard katanorm is given as follows:

$3.8\,Fs + 1.9\,Fa = 3.8\,Mt + 1.9\,Q$ Q-Balance:

$1.7\,Cal + 1.1\,Q = 2.8\,An$ Used for

$5.2\,Kp + 3.5\,Q = 8.7\,Or$ An   1.1   On hand: 35.9

$2.2\,Ne + 1.5\,Q = 3.7\,Ab$ Or   3.5   From Mt-form.:   1.9

$25.2\,Sp + 21.0\,Q = 46.2\,Cord$ Ab   1.5   Available: 37.8

$17.4\,Hz + 14.5\,Q = 31.9\,Fe\text{-}Cord$ Cord   21.0

$4.7\,Fa + 1.6\,Q = 6.3\,Hy$ Fe-Cord   14.5

Hy   1.6

Total consumption: 43.2
Available: 37.8
Q-deficit: −5.4 Q

The deficit could be eliminated by desilication of Hy and $Ab + Or$ purely by calculation. However, as this operation would introduce the components Ne and Lc, which do not occur in metamorphic kata-rocks, the introduction of $-Q$ should be preferred. The standard katanorm thus yields:

Or	Ab	An	Cord	Fe-Cord	Hy	Mt	Ru	Q	Total
8.7	3.7	2.8	46.2	31.9	6.3	3.8	2.0	−5.4	100.0
	Plag 6.5		Cord* 78.1						

A rock of such a composition should be designated as a **hypersthene-bearing feldspar-cordierite fels**. The high Fe-content of the cordierite should be noted:

$\beta\beta$) Calculation of a biotite variant. A **biotite variant** can be calculated, if the Or is eliminated according to:

$$10\,Or + 9(Fo + Fa) = 16\,Bi + 3\,Q.$$

The required amount of $(Fo + Fa)$ is gained by desilication of $Cord$ or Fe-$Cord$ according to:

$$11(Cord + \text{Fe-}Cord) = 3(Fo + Fa) + 6\,Sil + 2\,Q,$$

and also by desilication of Hy according to:

$$4\,Hy = 3\,Fa + 1\,Q.$$

Since the normative $Cord^*$ ($= Cord + $ Fe-$Cord$) is very Fe-rich, only Fe-Bi should be formed at first:

$$
\begin{aligned}
11.4\ \text{Fe-}Cord &= 3.1\,Fa + 6.2\,Sil + 2.1\,Q \\
6.3\,Hy &= 4.7\,Fa + 1.6\,Q \\
8.7\,Or + 7.8\,Fa &= 13.9\ \text{Fe-}Bi + 2.6\,Q.
\end{aligned}
$$

Because $2.1 + 1.6 + 2.6 = 6.3\,Q$ are set free during these conversions, the Q-deficit of $-5.4\,Q$ is converted to an excess of $0.9\,Q$, so that the **biotite-variant** yields:

Q	An	Ab	Fe-Bi	Cord	Fe-Cord	Sil	Mt	Ru	Total
0.9	2.8	3.7	13.9	46.2	20.5	6.2	3.8	2.0	100.0.

Since the ratio of Fe:Mg in the cordierite or biotite can only be determined arbitrarily, one may also write:

Q	An	Ab	Bi	Cord*	Sil	Mt	Ru	Total
0.9	2.8	3.7	13.9	66.7	6.2	3.8	2.0	100.0.

$\gamma\gamma$) An attempt at calculating a **modus**. According to R. Masson's estimates, the modus yields:

$$Bi\ 20,\ Gr\ 25,\ Sil\ 25,\ Cord\ 20,\ Q\ 10.$$

Thus, this corresponds to a feldspar-free biotite-sillimanite-garnet variant of the kata-standardnorm. For this purpose both Kp and Ne should be introduced into the biotite, and An or Cal be used for garnet formation. Firstly, one should carry out the calculation of a alkali-feldspar cordierite-biotite-sillimanite variant, without garnet. The required amount of $(Fo + Fa)$ for Bi-formation should be gained exclusively from $Cord + Fe$-$Cord$, omitting the available Hy. In order to convert the 8.7 Or + 3.7 Ab into Bi, 11.2 $(Fo + Fa)$ are required. The latter are gained by the conversion of 41.1 $(Cord + Fe$-$Cord)$.

$$41.1\,(Cord + Fe\text{-}Cord) = 11.2\,(Fo + Fa) + 22.4\,Sil + 7.5\,Q$$
$$12.4\,(Or + Ab) + 11.2\,(Fo + Fa) = 19.9\,Bi + 3.7\,Q.$$

The required **garnet-free cordierite-biotite-sillimanite variant** is thus given by:

Q	An	Cord*	Bi	Hy	Sil	Mt	Ru	Total
5.8	2.8	37.0	19.9	6.3	22.4	3.8	2.0	100.0.

The **formation of garnet** follows according to the relations:

$$2\,(3\,SiO_2 \cdot Al_2O_3 \cdot 3\,MgO) = 3\,(SiO_2 \cdot 2\,MgO) + 2\,(SiO_2 \cdot Al_2O_3) + SiO_2,$$
$$16\,Pyp 9\,Fo 6\,Sil 1\,Q$$

$$2\,(3\,SiO_2 \cdot Al_2O_3 \cdot 3\,FeO) = 3\,(SiO_2 \cdot 2\,FeO) + 2\,(SiO_2 \cdot Al_2O_3) + SiO_2,$$
$$16\,Alm \qquad\qquad 9\,Fa \qquad\qquad 6\,Sil \qquad 1\,Q$$

$$2\,(3\,SiO_2 \cdot Fe_2O_3 \cdot 3\,CaO) = 3\,(SiO_2 \cdot 2\,CaO) + 2\,(SiO_2 \cdot Fe_2O_3) + SiO_2,$$
$$16\,Andr \qquad\qquad 9\,Cs \qquad\qquad 6\,Fs \qquad 1\,Q$$

or $\quad 16\,Gran\,(Pyp + Alm + Andr) \quad 9\,(Fo + Fa + Cs) + 6\,(Sil + Fs) + 1\,Q.$

Modally, 25 $Gran$ (by estimation) are present. According to R. Masson, however, only 20 units of a garnet, the composition of which corresponds to the observed refractive index and density, can be formed. Thus one obtains:

$$20\,Gran = 11.3\,(Fo + Fa + Cs) + 7.5\,(Sil + Fs) + 1.2\,Q.$$

Cs is gained by conversion of An:

$$2\,(2\,SiO_2 \cdot Al_2O_3 \cdot CaO) = SiO_2 \cdot 2\,CaO + 2\,(SiO_2 \cdot Al_2O_3) + SiO_2.$$
$$10\,An \qquad\qquad 3\,Cs \qquad\qquad 6\,Sil \qquad 1\,Q$$
$$2.8\,An \qquad\qquad 0.8\,Cs \qquad\qquad 1.7\,Sil \qquad 0.3\,Q$$

In andradite the ratio of $Cs:Fs = 3:2$. Thus, $^2/_3\,0.8\,C = 0.5\,Fs$, which are obtained by converting a corresponding amount of Mt, correspond to 0.8 Cs:

$$0.5\,Mt + 0.3\,Q = 0.5\,Fs + 0.3\,Fa,$$

from which one obtains:

$$0.8\,Cs + 0.5\,Fs + 0.1\,Q = 1.4\,Andr.$$

The *Fa* required for the formation of *Alm* is obtained from *Hy*:

$$6.3\,Hy = 4.7\,Fa + 1.6\,Q.$$

Adding to this the $0.3\,Fa$ gained from the conversion of *Mt*, one obtains:

$$5.0\,Fa + 3.3\,Sil + 0.6\,Q = 8.9\,Alm.$$

Since altogether 20 *Gran* should be formed, 20 − (1.4 + 8.9) = 9.7 *Pyp* remain to be formed. The required amount of *Fo* is gained from *Cord*:

$$5\,SiO_2 \cdot 2\,Al_2O_3 \cdot 2\,MgO = SiO_2 \cdot 2\,MgO + 2\,(SiO_2 \cdot Al_2O_3) + 2\,SiO_2,$$

| 11 Cord | 3 Fo | 6 Sil | 2 Q |
| 20.2 Cord | 5.5 Fo | 11.0 Sil | 3.7 Q |

$$5.5\,Fo + 3.6\,Sil + 0.6\,Q = 9.7\,Pyp.$$

The balances for the various norm minerals are provided in the following:

Q-balance

present	5.8	consumed for *Fs*	0.3
from *An*-conver-	0.3	,, ,, *Andr*	0.1
from *Hy*- ,,sion	1.6	,, ,, *Alm*	0.6
from *Cord*- ,,	3.7	,, ,, *Pyp*	0.6
available	11.4	Total consumption	1.6
	1.6		
Q-excess	9.8 *Q*		

Sil-balance:

present	22.4
gained from *An*	1.7
gained from *Cord*	11.0
available	35.1
	6.9
Sil-excess	28.2

consumed for *Alm*	3.3
,, ,, *Pyp*	3.6
Total consumption of *Sil*	6.9

Mt-balance

present	3.8
converted to *Fs*	0.5
Mt-excess B	3.3 *Mt*

Thus, the garnet-biotite variant of the standard katanorm shows a satisfactory agreement with the estimated modus provided hereafter, taking into consideration the relatively coarse-grained character of the rock:

| Q | Bi | Gran | Cord* | Sil | Mt | Ru | Total |
| 9.8 | 19.9 | 20.0 | 16.8 | 28.2 | 3.3 | 2.0 | 100.0. |

Here again, the distribution of Mg and Fe between *Bi* and *Cord** was left unfixed.

δδ) **Calculation of the standard epinorm from the basis.** It is interesting to find out which mineral composition of the mentioned Al-rich rock would have occurred under epi-building conditions. One of the various possibilities is provided by the standard epinorm, from which others may be obtained by variant construction.

Basis:	Q	Kp	Ne	Cal	Sp	Hz	Fa	Fs	Ru	Total
	35.9	5.2	2.2	1.7	25.2	17.4	6.6	3.8	2.0	100.0.

One commences by forming Hm, Ab and Ms according to:

$$3.8\,Fs = 2.5\,Hm + 1.3\,Q$$
$$2.2\,Ne + 1.5\,Q = 3.7\,Ab$$
$$5.2\,Kp + 5.2\,Hz + 4.3\,Q = 12.1\,Ms + 2.6\,Fa.$$

After these preparatory operations there remain:

Cal	Sp	Hz	Fa	Q	Ru
1.7	25.2	12.2	9.2	31.4	2.0,

corresponding to case δ) provided in the introduction to the calculation of the standard epinorm. Thus Zo is formed from Cal and Fa according to:

$$1.7\,Cal + 0.4\,Fa + 0.9\,Q = 2.3\,Zo + 0.7\,Fe\text{-}At.$$

Together with Hz the remaining Fa again yields, as far as possible, Fe-At, or otherwise Fe-Ant according to:

$$12.2\,Hz + 6.1\,Fa + 2.0\,Q = 20.3\,Fe\text{-}At$$
$$2.7\,Fa + 0.3\,Q = 3.0\,Fe\text{-}Ant.$$

Together with **Fe-At** and Q, Sp yields Ot and At; in this particular case it so happens that exactly the required equivalent amounts of Sp and **Fe-At** are available:

$$25.2\,Sp + 8.4\,Q + 21.0\,Fe\text{-}At = 33.6\,Ot + 21.0\,At,$$

the required **standard epinorm** thus yielding:

Q	Ab	Ms	Zo	At	Fe-Ant	Ot	Hm	Ru	Total
19.8	3.7	12.1	2.3	21.0	3.0	33.6	2.5	2.0	100.0.
				24.0 Chl					

A rock of such a composition should be designated as a **muscovite- (or sericite) chlorite-chloritoid phyllite** with Al-rich chlorite.

εε) **Garnet variant of the standard epinorm.** Among others, a garnet variant may be calculated from the katanorm, in that the garnet which was formed is subsequently preserved. Hereby Sil may be converted to $Dist$, whereas the Bi, $Cord$* and Mt should be converted to the corresponding basis components:

$$3.3\,Mt + 0.6\,Q = 2.2\,Hm + 1.7\,Fa$$
$$19.9\,Bi = 7.5\,(Kp + Ne) + 11.2\,(Fo + Fa) + 1.2\,Q$$
$$16.8\,Cord^* = 9.2\,(Sp + Hz) + 7.6\,Q.$$

The following components are present:

Q	Kp	Ne	Sp+Hz	Fo+Fa	Hm	Ru	Gran	Dist	Total
18.0	5.3	2.2	9.2	12.9	2.2	2.0	20.0	28.2	100.0

Primarily, *Ab* and *Ms* should be formed according to:

$$2.2\,Ne + 1.5\,Q = 3.7\,Ab$$
$$5.3\,Kp + 5.3\,(Sp + Hz) + 4.4\,Q = 12.4\,Ms + 2.6\,(Fo + Fa).$$

Thus the following remain:

	Sp + Hz	Fo + Fa	Q
	9.2	12.9	18.0
	−5.3	+2.6	−5.9
	3.9	15.5	12.1

wherefrom chlorite should be formed according to the following:

$$3.9\,(Sp + Hz) + 1.9\,(Fo + Fa) + 0.7\,Q = 6.5\,(At + Fe\text{-}At)$$
$$13.6\,(Fo + Fa) + 1.5\,Q = 15.1\,(Ant + Fe\text{-}Ant).$$

Thus the **garnet variant of the epinorm** yields:

Q	Ab	Ms	Dist	At+Fe-At	Ant+Fe-Ant	Gran	Hm	Ru	Total
9.9	3.7	12.4	28.2	6.5	15.1	20.0	2.2	2.0	100.0

21.6 Chlorite (bracketing At+Fe-At and Ant+Fe-Ant)

A rock of such a composition should be designated as a **kyanite-garnet-sericite-chlorite phyllite**.

ξξ) **Calculation of a mesonorm.** A mesofacies mineral composition may be calculated in an easy way from the katanorm, by substituting for cordierite and expressing sillimanite as kyanite.

The **staurolite-formation** is carried out as follows:

$$5\,SiO_2 \cdot 2\,Al_2O_3 \cdot 2\,(Mg, FeO) + 2\,(Al_2O_3 \cdot SiO_2)\ (+\ 2\,H_2O)$$

11 Cord*	6 Sil	(2 W)
16.8 Cord*	9.2 Sil	

$$= 2\,[2\,SiO_2 \cdot 2\,Al_2O_3 \cdot (Fe, Mg)O \cdot H_2O] + 3\,SiO_2,$$

	14 Staur	3 Q
	21.4 Staur	4.6 Q

and the **mesonorm** (staurolite variant) yields:

Q	Bi	Gran	Staur	Dist	Mt	Ru	Total
14.4	19.9	20.0	21.4	19.0	3.3	2.0	100.0

which corresponds to the mineral composition of a **garnet-biotite-kyanite-staurolite schist**.

η) The garnet-free variant of the mesonorm. The garnet which was introduced from the standard katanorm, may be broken up as follows:

$$1.4\, Andr = 0.8\, Cs + 0.5\, Fs + 0.1\, Q$$
$$8.9\, Alm = 5.0\, Fa + 3.3\, Sil + 0.6\, Q$$
$$9.7\, Pyp = 5.5\, Fo + 3.6\, Sil + 0.6\, Q.$$

Mt is formed from Fs:

$$0.5\, Fs + 0.2\, Fa = 0.5\, Mt + 0.2\, Q,$$

and from $(Fo + Fa)$ and Cs, $(Gram + Fe\text{-}Akt) = Akt$ is built:

$$2\, Cs + 5\,(Fo + Fa) + 3\, Q\,(+\,^2/_3\, W) = 10\, Akt$$
$$0.8\, Cs + 2.0\,(Fo + Fa) + 1.2\, Q = 4.0\, Akt.$$

The remaining $10.5 - (0.2 + 2.0) = 8.3\,(Fo + Fa)$ may be converted to chlorite. The following equations are given for the formation of an average chlorite (clinochlore Kl) having the following ratios: $(Ant + Fe\text{-}Ant):(At + Fe\text{-}At) \sim 1:1$:

$$4\,[SiO_2 \cdot 2(Mg, Fe)O] + 2\,[Al_2O_3 \cdot (Mg, Fe)O] + 2\, SiO_2\,(+\, 8\, H_2O)$$
$$12\,(Fo + Fa) \qquad 6\,(Sp + Hz) \qquad 2\,Q \quad (8\,W)$$
$$= 2\,[SiO_2 \cdot Al_2O_3 \cdot 2(Mg, Fe)O \cdot 2\, H_2O] + 2\,[2\, SiO_2 \cdot 3(Mg, Fe)O \cdot 2\, H_2O].$$
$$10\,(At + Fe\text{-}At) \qquad\qquad 10\,(Ant + Fe\text{-}Ant)$$
$$\underbrace{\hspace{6cm}}_{20\, Kl}$$

Another possibility is given by:

$$SiO_2 \cdot 2(Mg, Fe)O + 2\,(SiO_2 \cdot Al_2O_3) = 2\,[Al_2O_3 \cdot (Mg, Fe)O] + 3\, SiO_2.$$
$$3\,(Fo + Fa) \qquad 6\, Sil\,(Dist) \qquad 6\,(Sp + Hz) \qquad 3\, Q$$

By addition of the two equations, $6\,(Sp + Hz)$ are eliminated and one obtains:

$$15\,(Fo + Fa) + 6\, Sil\,(Dist) = 20\, Kl + 1\, Q$$
$$8.3\,(Fo + Fa) + 3.3\, Sil\,(Dist) = 11.0\, Kl + 0.6\, Q.$$

The garnet-free variant of the mesonorm thus yields:

Q	Bi	Staur	Dist	Akt	Kl	Mt	Ru	Total
15.3	19.9	21.4	22.6	4.0	11.0	3.8	2.0	100.0

which would correspond to the mineral composition of a biotite-staurolite-kyanite-chloritoid schist.

C. THE SUGGESTIONS OF T.F.W. BARTH AND P. ESKOLA FOR THE CONSIDERATION OF ANIONS IN PETROCHEMICAL CALCULATIONS, AND THEIR RELATIONSHIP TO NIGGLI'S EQUIVALENT NORM.

I. General

The calculation methods dealt with hitherto were mainly based on the cations. Only in a very few cases were the anions also considered, e.g., in the Niggli-values by the introduction of the values p, co_2, cl_2, f_2, s, so_3, etc, or in the equivalent norm by the formation of normative components such as *Ap, Fr, Pr, Sod, Hau, Nos, Canc, Cc, Sid, Mgs, Nc, Hl, Th, Ma, Me,* etc. This was possible provided that the considered anions were determined analytically. The most important and frequently occurring anion, oxygen, was considered either as a constituent of the oxide (molecular- or atomic-equivalent) formula units, or, as in the case of the calculation of the cation percentages, were not considered at all. In the majority of cases such a procedure may be applied immediately. From the discussion on the role of the oxygen in the lithosphere, provided in Section A I, where the possibility of presenting the rock's chemism in the so-called "oxide form" was explained, it is clearly seen that in relation to the other available cations, the amount of oxygen may vary only within very narrow limits. Thus, by giving the amounts of the cations, one also, in a manner of speaking, determines the amount of oxygen.

However, the abundance of oxygen in the composition of the upper lithosphere is 46.6 % by weight, 60.5 % by atom percentages, and, assuming exclusively heteropolar bonds, 91.8 % by volume. Thus, oxygen is primarily responsible for the volume of the rock-forming minerals and thus also for the rocks themselves. It appears justified to seek a method suitable for considering these facts. The incorporation of the anions, and primarily oxygen, has opened entirely new possibilities, as will be shown later, for the treatment of isovolumetric, metasomatic processes.

II. The Calculation of Rock Analyses with Consideration of the Anions

The calculation of the rock analyses with consideration of the anions was explained by T. F. W. Barth (1948, 1952, 1955) and P. Eskola (1954). It will be exemplified here, as based on the suggestions of the mentioned authors, using the three examples provided in the previous sections. The method closely resembles the calculation of the cation percentages, which was discussed in Section B I 1 b. Here again the calculation may be started either from the weight percentages or from the Niggli-values.

1. Calculation Starting from the Weight Percentages

Quartz-diorite, Spanish Peak, California

	Weight %	atomic equivalent numbers × 1000	Cation	%	Number of O	
SiO_2	59.68	994	Si^{4+}	56.0	$56.0 \cdot 2 =$	112.0
Al_2O_3	17.09	336	Al^{3+}	18.9	$18.9 \cdot 3/2 =$	28.3
Fe_2O_3	2.85	36	Fe^{3+}	2.0	$2.0 \cdot 3/2 =$	3.0
FeO	2.75	38	Fe^{2+}	2.1	$2.1 \cdot 1 =$	2.1
MgO	3.54	88	Mg^{2+}	5.0	$5.0 \cdot 1 =$	5.0
CaO	6.62	118	Ca^{2+}	6.7	$6.7 \cdot 1 =$	6.7
Na_2O	3.87	124	Na^+	7.0	$7.0 \cdot 1/2 =$	3.5
K_2O	1.31	28	K^+	1.6	$1.6 \cdot 1/2 =$	0.8
TiO_2	0.65	8	Ti^{4+}	0.5	$0.5 \cdot 2 =$	1.0
P_2O_5	0.25	4	P^{5+}	0.2	$0.2 \cdot 5/2 =$	0.5
H_2O+	1.00	(56)*		(3.2)		
H_2O-	0.15					
incl.	0.27					
	100.03	1774		100.0		162.9 O-ions
						−3.2 for OH
						159.7 O-ions
						+6.4 OH-ions
						166.1 (O + OH)-ions

* In contrast to the equivalent numbers of the metal atoms, the term (56), which was obtained as 1.00:18.016, is the molecular equivalent number of H_2O, and thus corresponds to the relative number of H_2O-molecules and not to the H-ions.

The number of O per 100 cations is obtained by multiplying the cation percentages by the number of O per cation. H_2O- is neglected since it expresses the hygroscopic water. However, considering the actual role of H_2O+, namely, that H does not occur as a cation and that the H_2O-content of the rock-forming minerals (excluding the zeolitic H_2O which occurs only seldom) is always present in the hydroxyl form, H_2O+ should be totally represented in the calculation as $(OH)^-$. Since 1 H_2O is equivalent to $2(OH)^-$, an additional 1 oxygen is required per H_2O for hydroxyl formation. This should be subtracted from the calculated total number of the O-ions. 6.4 $(OH)^-$ thus corresponds to the 3.2 H_2O in the previous example, and an additional 3.2 O are required. There remain thus $162.9 - 3.2 =$ $= 159.7$ O-ions, yielding a total of 159.7 $O^{2-} + 6.4$ $(OH)^- = 166.1$ anions per 100 cations.

The method here outlined corresponds to the suggestion of P. Eskola (1954). The same result can be arrived at in a somewhat different way by the method of T. F. W. Barth (1955). In this case, the weight-percentage content of the H_2O+ is not divided by $H_2O = 18.016$, but by $1/2$ $H_2O = 9.008$. By this procedure the atomic-equivalent number of H^+- ions is found to be 112. The corresponding %-number, 6.4, at the same time represents the number of (OH)⁻. For its formation one requires an additional half of the amount, i.e., 3.2, of oxygen, which should be thus deducted from the calculated sum of O:

Number of O-ions	162.9
Number of OH-ions	6.4
Sum of (O + OH)-ions	169.3
subtraction of the O-required for OH-building	3.2
Total of anions	166.1

2. Calculation Starting from the Niggli-Values

The following Niggli-values are given for the **cordierite-andesite** from Hoyazo near Nijar, Spain:

si	al	fm	c	alk	k	mg	ti	p	w	h
229	38.2	34.9	12.7	14.2	0.55	0.45	2.3	0.2	0.48	33.0

		Cation %	Number of O per 100 cations
$si = Si^{4+}$	= 229.0	59.6	$59.6 \cdot 2 = 119.2$
$2\,al = Al^{3+}$	= 76.4	19.9	$19.9 \cdot {}^3/_2 = 29.8$
$w(1-mg)/m = Fe^{3+}$	= 9.2	2.4	$2.4 \cdot {}^3/_2 = 3.6$
$(1-w)(1-mg)/m = Fe^{2+}$	= 10.0	2.6	$2.6 \cdot 1 = 2.6$
$mg \cdot /m = Mg^{2+}$	= 15.7	4.1	$4.1 \cdot 1 = 4.1$
$c = Ca^{2+}$	= 12.7	3.3	$3.3 \cdot 1 = 3.3$
$2(1-k)\,alk = Na^+$	= 12.8	3.3	$3.3 \cdot {}^1/_2 = 1.65$
$2k \cdot alk = K^+$	= 15.6	4.1	$4.1 \cdot {}^1/_2 = 2.05$
$ti = Ti^{4+}$	= 2.3	0.6	$0.6 \cdot 2 = 1.2$
$2p = P^{5+}$	= 0.4	0.1	$0.1 \cdot {}^5/_2 = 0.2$
$2h = H^+$	(66.0)	(17.2)	
	384.1	100.0	167.7
			−8.6 for OH
			159.1 O-ions
			+ 17.2 OH-ions
		Total	176.3 (O + OH)-ions

263   In this case the number of H⁺-ions was calculated as 17.2. It also corresponds to the number of the OH-ions which are formed, assuming that all of the hydrogen is always introduced into the hydroxyl radical and does not occur as a cation. In order to form these OH-ions, however, an additional 8.6 O are required, which should be subtracted from the oxygen calculated to the sum of 100 cations.

3. Calculation Involving the Presence of Additional Anions

The arfvedsonite-lujaurite from Lille-Elv, Kangerdluarsuk, Greenland, will serve as an example of a rock which contains Cl in addition to O and OH-ions. Here again one should start from the weight percentages.

	Weight %	Atomic equivalent numbers × 1000	Cation %		Number of O per 100 cations	
SiO_2	53.01	882	Si^{4+}	48.4	$48.4 \cdot 2$	$= 96.8$
Al_2O_3	15.33	300	Al^{3+}	16.5	$16.5 \cdot 3/2$	$= 24.7$
Fe_2O_3	9.14	114	Fe^{3+}	6.3	$6.3 \cdot 3/2$	$= 9.5$
FeO	4.44 }	63	Fe^{2+}	3.5	$3.5 \cdot 1$	$= 3.5$
MnO	0.13 }					
MgO	0.10	2	Mg^{2+}	0.1	$0.1 \cdot 1$	$= 0.1$
CaO	0.67	12	Ca^{2+}	0.7	$0.7 \cdot 1$	$= 0.7$
Na_2O	11.86	383	Na^+	21.0	$21.0 \cdot 1/2$	$= 10.5$
K_2O	2.60	55	K^+	3.0	$3.0 \cdot 1/2$	$= 1.5$
TiO_2	0.33	4	Ti^{4+}	0.2	$0.2 \cdot 2$	$= 0.4$
ZrO_2	0.65	5	Zr^{4+}	0.3	$0.3 \cdot 2$	$= 0.6$
P_2O_5	*Sp.*	*Sp.*	P^{5+}	*Sp.*		148.3
Cl	0.23	(6)		(0.3)		−5.7 for OH
H_2O+	1.88	(208)		(11.4)		142.6 O-ions
		1820		100.0		+11.4 OH-ions
						+ 0.3 Cl-ions
					Total	154.3 anions

III. Barth's Standard Cell

The calculated example, as well as general experience, show that on the average rocks contain about 160 (O + OH) per 100 cations. For weathered and epi-facies rocks, this number of O is somewhat larger, and in addition more OH is present. For the kata-facies the number of O is somewhat smaller, and OH also plays a somewhat less important role. According to T. F. W. Barth, geological oxidation prevails in the upper levels of the lithosphere, leading to the removal of cations from the crystal lattices and their substitution by H. This results in a relative O-enrichment. At greater depths, in contrast, geological reduction prevails, causing the removal of H_2O and its substitution by cations. This results in the decrease of the amount of O. This not only applies to rocks, but also to the typomorphic rock-forming minerals, as can be shown by handling a computable equivalent unit of Niggli, based on one cation.

The fact that the number of (O + OH)-ions, based on 100 cations, varies only very slightly can now be used for arriving at equal-volume calculation units, according to the suggestions of T. F. W. Barth (1948, 1952). These units can be used for the building of material balances for metasomatic and weathering transformations without volume changes. This calculation unit, based on 160 (O + OH), is designated as Barth's standard cell.

The procedure for the calculation of Barth's standard cell will again be illustrated by recourse to a previously-provided example. The calculation may either be started from the cation percentages, the weight percentages, or the Niggli-values.

1. Quartz-diorite, Spanish Peak, California (Calculation from the Cation Percentages)

From the cation percentages or anion numbers of the previous section:

Si^{4+}	Al^{3+}	Fe^{3+}	Fe^{2+}	Mg^{2+}	Ca^{2+}	Na^+	K^+	Ti^{4+}	P^{5+}	O^{2-}	OH^-
56.0	18.9	2.0	2.1	5.0	6.7	7.0	1.6	0.5	0.2	159.7	6.4

By simple ratio calculations, e.g., $166.1 : 100 = 160 : x$, the number x of the cations for 160 anions = 96.3. If the cation percentages are recalculated to this new sum (using to advantage a logarithmic slide rule), the following is obtained:

Si^{4+}	Al^{3+}	Fe^{3+}	Fe^{2+}	Mg^{2+}	Ca^{2+}	Na^+	K^+	Ti^{4+}	P^{5+}	Total
53.9	18.2	1.9	2.0	4.8	6.5	6.8	1.5	0.5	0.2	96.3

The distribution for O and OH is obtained from the ratios:

$$166.1 : 160 = 159.7 : x \quad x = \text{Number of O} = 153.8$$
$$166.1 : 160 = 6.4 : y \quad y = \text{Number of OH} = 6.2$$
$$x + y = 160.0.$$

The standard cell may thus be formulated as follows:

$K_{1.5}\ Na_{6.8}\ Ca_{6.5}\ Mg_{4.8}\ Fe_{3.9}\ Al_{18.2}\ Si_{53.9}\ Ti_{0.5}\ P_{0.2}\ O_{153.8}\ OH_{6.2}]$.

2. Quartz-diorite, Spanish Peak, California (Calculation from the Weight Percentages)

In order to calculate the standard cell from weight percentages, the number of the O is first recalculated to the sum of 160. On the basis of this recalculation the distribution of the cations is carried out.

	Weight %	Atomic equivalent numbers × 1000	Number of O	Total O = 160		Number of cations based on 1600
SiO_2	59.68	994	1988	107.9	Si^{4+}	53.9
Al_2O_3	17.09	336	504	27.3	Al^{3+}	18.2
Fe_2O_3	2.85	36	54	2.9	Fe^{3+}	1.9
FeO	2.75	38	38	2.1	Fe^{2+}	2.1
MgO	3.54	88	88	4.8	Mg^{2+}	4.8
CaO	6.62	118	118	6.4	Ca^{2+}	6.4
Na_2O	3.87	124	62	3.4	Na^+	6.8
K_2O	1.31	28	14	0.8	K^+	1.6
TiO_2	0.65	8	16	0.9	Ti^{4+}	0.4
P_2O_5	0.25	4	10	0.5	P^{5+}	0.2
H_2O+	1.00	(56)*	56	3.0	H^+	(6.0)
		1774	2948	160.0		96.3
				−6.0 10r OH		
				154.0 O		

* Molecular equivalent number of H_2O.

The reduction of the number of O to 160 is most simply carried out by multiplication by the factor 160/2948 = 0.054 using a logarithmic slide rule. The number of cations is obtained by multiplication by various factors: 1/2 for Si and Ti, 2/3 for Al and Fe^{3+}, 1 for Fe^{2+}, Mg and Ca, 2 for Na and K, and 2/5 for P. The 6H which, by convention, are not calculated as cations, require an equal amount of O for hydroxyl formation, so that 154 O + 6 OH result as anions.

The result:

$$K_{1.6}\ Na_{6.8}\ Ca_{6.4}\ Mg_{4.8}\ Fe_{4.0}\ Al_{18.2}\ Si_{53.9}\ Ti_{0.4}\ P_{0.2}\ [O_{154.0}\ OH_{6.0}]$$

is in satisfactory agreement with the results obtained from the cation percentages.

3. Quartz-diorite, Spanish Peak, California (Calculation from the Niggli-Values)

Under certain circumstances, e. g., when interpolating values from a variation diagram, one may desire to calculate the standard cell directly from the Niggli-values. For the **quartz-diorite** from Spanish Peak the following are obtained:

si	al	fm	c	alk	k	mg	ti	p	h
190	32.1	30.9	22.5	14.5	0.18	0.54	1.5	0.4	10.7

On the basis of the relationship between the Niggli-values and atomic-equivalent numbers one obtains*:

		Number of O	Total O = 160	Number of cations based on 160 O
$si = Si^{4+} =$	190.0	380.0	106.6	53.3
$2\,al = Al^{3+} =$	64.2	96.3	27.0	18.0
$(1-mg)\,fm = Fe^{2+} + Fe^{3+} =$	14.2	14.2	4.0	4.0
$mg \cdot fm = Mg^{2+} =$	16.7	16.7	4.7	4.7
$c = Ca^{2+} =$	22.5	22.5	6.3	6.3
$(1-k)\,2\,alk = Na^+ =$	23.8	11.9	3.3	6.6
$k \cdot 2\,alk = K^+ =$	5.2	2.6	0.7	1.4
$ti = Ti^{4+} =$	1.5	3.0	0.8	0.4
$2\,p = P^{5+} =$	0.8	2.0	0.6	0.2
$2\,h = H^+ =$	21.4	21.4	6.0	(6.0)
	360.3	570.6	160.0	
			−6.0 for OH	
			154.0	

* The degree of oxidation of the Fe may be taken into account if the value w is known. It is obtained as follows: $Fe^{3+} = w\,(1-mg)\,fm$ or $Fe^{2+} = (1-w)\,(1-mg)\,fm$.

The following standard cell is obtained and shows good agreement with the previous results:

$K_{1.4}$ $Na_{6.6}$ $Ca_{6.3}$ $Mg_{4.7}$ $Fe_{4.0}$ $Al_{18.0}$ $Si_{53.3}$ $Ti_{0.4}$ $P_{0.2}$ $[O_{154.0}$ $OH_{6.0}]$.

The previous example was calculated on the basis of the assumption that the h-value is given. Otherwise, only a simplified standard cell, based on O = 160, can be calculated, neglecting the OH.

4. The Norm of the Standard Cell

Using the cations based on 160 (O + OH), a norm can be calculated according to the regulations given in Section B II 6, which will provide information on the possible mineral composition. This norm will differ somewhat from the standard norm, with the exception of the case when exactly 100 cations are present for each 160 (O + OH). In many cases, however, the deviation may be ignored. If required, the recalculation can be carried out to the sum of 100 by means of a logarithmic slide rule without much time-consumption, i.e., the common standard norm is calculated.

The following relationships are obtained for the quartz-diorite from Spanish Peak, California:

	Katanorm of the standard cell	Standard katanorm (Total 100)	Standard katanorm (calculated directly from analysis)
Q	11.7	12.1	12.6
Or	8.0	8.3	7.8
Ab	34.0	35.3	35.0
An	24.5	25.5	26.0
Wo	2.4	2.5	2.3
En	9.6	10.0	9.9
Hy	2.4	2.5	2.3
Mt	2.8	2.9	3.0
Cp	0.5	0.5	0.6
Ru	0.4	0.4	0.5
	96.3	100.0	100.0

The differences between the norm of the standard cell and the standard katanorm, obtained by recalculation to the sum of 100, are indeed very small in the mentioned example. In addition, the agreement with the standard katanorm, directly calculated from the weight percentages via the basis, is for most purposes entirely sufficient.

5. The Application of Barth's Standard Cell for the Representation of the Material Balance of Isovolumetric Metasomatic Processes

Since metasomatic processes, as shown by numerous field observations, very often do not involve any change of volume, Barth's standard cell, which, due to the equal number of (O + OH)-ions, makes use of equal volumes, is very suitable for the representation of material balances. This should be illustrated by a few examples.

a) The injection metamorphism in the Stavanger Region (Norway)

The example of the conversion of **quartz-muscovite-chlorite phyllite** into **albite-porphyroblastic schists** in the Stavanger region (V. M. Goldschmidt, 1920) was previously discussed by T. F. W. Barth (1948). The weight-percentage compositions of both rocks (average sample) yield (Anal. O. Röer):

	a)	b)	
SiO_2	58.32	64.70	
Al_2O_3	20.00	15.45	
Fe_2O_3	2.01	1.36	
FeO	4.98	4.18	
MnO	0.22	0.05	
MgO	1.85	1.48	
CaO	0.66	2.92	
BaO	0.10	0.10	a) Quartz-muscovite-chlorite phyllite
Na_2O	1.26	3.09	
K_2O	4.49	3.46	b) Albite-porphyroblastic schist
TiO_2	0.98	0.59	
P_2O_5	0.15	0.19	
CO_2	0.43	0.85	
S	0.06	0.05	
H_2O+	4.05	1.82	
H_2O-	0.05	0.15	
incl.	0.41		
	100.02	100.44	

Barth's standard cells are calculated as follows:

(a) $K_{5.2}\ Na_{2.2}\ Ca_{0.6}\ Mg_{2.4}\ Fe_{5.1}\ Al_{20.6}\ Si_{51.0}\ Ti_{0.7}\ P_{0.1}\ C_{0.5}\ [O_{136.4}\ OH_{23.6}]$,

(b) $K_{4.0}\ Na_{5.2}\ Ca_{2.7}\ Mg_{2.0}\ Fe_{4.0}\ Al_{16.0}\ Si_{56.8}\ Ti_{0.4}\ P_{0.2}\ C_{1.0}\ [O_{149.4}\ OH_{10.6}]$.

For the conversion of the phyllite to the albite schist, the following material transformations are required, based on 160 (O + OH):

Addition:	Subtraction:
3.0 Na$^+$	1.2 K$^+$
2.1 Ca^{2+}	0.4 Mg^{2+}
5.8 Si^{4+}	1.1 Fe $\{$ 0.4 Fe^{2+} ; 0.7 Fe^{3+} $\}$
0.5 C^{4+}	
0.1 P^{5+}	4.6 Al^{3+}
11.5 metal ions	0.3 Ti^{4+}
(32.9 valencies)	13.0 OH$^-$
	7.6 metal- + 13.0 OH-ions
	(32.9 valences)

The material balance here calculated does not match the one which was given at the time by V. M. Goldschmidt. Since the latter, however, called for a volume change of about 25 %, a fact unsupported by field observations, Barth's solution is in better correspondence with the facts.

b) The augite-granite in the granodiorite massif of Boraja (West Serbia)

The near-surface, solidified Tertiary granodiorite mass of Boraja in West Serbia is composed of quartz-monzonites in the region of the Ramnaja River. Near the contact of these rocks with skarns and marbles the hornblende is substituted by diopside, the alkali-feldspar content increasing simultaneously, so that a transition to augite-granite takes place. All the observations in the field, as well as in thin sections, clearly show that the augite-granite originated from the hornblende-quartz monzonites (St. Karamata, 1957). Since not only the Ca-content, but also that of the alkali-feldspars, increases, this cannot be considered as a simple case of endomorphism by calc-addition, and one is obliged to conclude that complicated metasomatic processes have occurred. St. Karamata has attempted to calculate these, assuming the calc-content to have remained unchanged. According to this reasoning, the increase of the percentage calc-content was relative, and was caused by the outward migration of the elements Fe, Mn, and Mg, and to a lesser degree also of Na, Si, and Al. The application of the method of Barth's standard cell provides a solution which, in contrast to the above-mentioned solution, does not exclude any migration of the important metals. No discussion of which of the two possibilities is more plausible will be entered upon. The chemical composition of the two rocks is illustrated by the following analyses (St. Karamata, loc. cit., 57, Anal. V. Knežević):

	a)	b)	
SiO_2	62.01	61.52	
Al_2O_3	16.21	17.56	
Fe_2O_3	2.21	0.61	
FeO	2.68	1.55	
MnO	0.11	0.08	
MgO	3.67	2.25	a) Hornblende-quartz monzonite
CaO	4.68	6.32	
Na_2O	3.31	2.86	b) Augite-granite
K_2O	4.63	6.37	
TiO_2	0.36	0.39	
P_2O_5	0.21	0.20	
H_2O+	0.29	0.24	
H_2O-	0.05	0.21	
	100.42	100.16	

Barth's standard cells, based on 160 (O + OH), are calculated as follows:

(a) $K_{5.4}\ Na_{5.8}\ Ca_{4.6}\ Mg_{5.0}\ Fe_{3.7}\ Al_{17.3}\ Si_{56.3}\ Ti_{0.2}\ P_{0.2}\ [O_{158.2}\ OH_{1.8}]$,

(b) $K_{7.4}\ Na_{5.0}\ Ca_{6.2}\ Mg_{3.1}\ Fe_{1.7}\ Al_{18.9}\ Si_{56.3}\ Ti_{0.2}\ P_{0.2}\ [O_{158.6}\ OH_{1.4}]$.

For the conversion of quartz-monzonite to augite-granite the following material balance is required:

Addition:	Subtraction:
2.0 K^+	0.8 Na^+
1.6 Ca^{2+}	1.9 Mg^{2+}
1.6 Al^{3+}	2.0 Fe { 1.0 Fe^{2+} / 1.0 Fe^{3+} }
5.2 metal ions	0.4 OH^-
(10.0 valencies)	5.1 metal- + 0.4 OH-ions
	(10.0 valencies)

According to the view expressed in these calculations, namely the addition of Ca connected with the outward migration of Mg + Fe, it is well understood that the hornblende with (Mg + Fe):Ca ~ 5:3 is replaced by diopside with (Mg + Fe) : Ca = 1 : 1.

c) The relationship between the lavas of Young-Somma and Vesuvius.

As Rittmann was able to show (A. Rittmann, 1933, 1944), the lavas of the Vesbic volcano (Somma-Vesuvius) during their development have provided an example of progressive desilication, which has found its expression in an increasing leucite content. This development could be well explained by a carbonate-assimilation for which there exists unequivocal geological evidence, this being connected with an alkali-metasomatism.

The material balances associated herewith lend themselves to examination by utilizing the methods which were just dealt with, as will be shown for the development of the recent Vesuvius lavas from those of the Young-Somma. According to Rittmann (1933, 90), these may be characterized by the following average values, as expressed in Niggli-values:

	si	al	fm	c	alk	k	mg	ti	p
Young-Somma	121	26.5	35	23.8	15	0.62	0.60	2.0	0.2
Vesuvius	113.8	25	33.4	24.4	17.2	0.64	0.47	1.5	0.5

Since neither w nor h are known, the calculation of the standard cell may only be carried out by resorting to simplification, namely, all of the Fe is considered to be bivalent, and the formation of OH-ions is disregarded, i.e., the calculation is based on 160 O. According to this method one obtains:

Young-Somma $K_{7.4}$ $Na_{4.6}$ $Ca_{9.5}$ $Mg_{8.4}$ $Fe_{5.6}$ $Al_{21.0}$ $Si_{48.5}$ $Ti_{0.8}$ $P_{0.2}$ $[O_{160.0}]$,

Vesuvius $K_{9.2}$ $Na_{5.2}$ $Ca_{10.2}$ $Mg_{6.6}$ $Fe_{7.4}$ $Al_{21.0}$ $Si_{47.5}$ $Ti_{0.6}$ $P_{0.4}$ $[O_{160.0}]$.

Therefore, the following material balance is obtained:

Addition:

1.8 K^+
0.6 Na^+
0.7 Ca^{2+}
1.8 Fe^{2+}
0.2 P^{5+}
―――
5.1 metal ions
(8.4 valencies)

Subtraction:

1.8 Mg^{2+}
1.0 Si^{4+}
0.2 Ti^{4+}
―――
3.0 metal ions
(8.4 valencies)

Basing the calculation on 160 O, a slight addition of alkalis, Fe and Ca takes place. Whereas the increase of Ca may be explained by assimilation of calcic sediments, in the case of the alkalis this increase must assumably be attributed to metasomatic addition from the depths of the magmatic source. The bulk of the additional Fe should also be attributed to metasomatism, whereas for a smaller part assimilation should also be considered. As far as outward migration is concerned, it is of interest to note that the ratio Si : Mg is closely related to the ratio 1 : 2, which may suggest that the gravitational separation of olivine played an important role. This is in good agreement with the fact that segregated peridoticic rocks are quite frequent in the vicinity of the Vesbic volcano. The CO_2 evolved during the assimilation of the calc is not introduced into the calculation, this representing an advantage of the used method.

D. APPENDIX

I. Alphabetical Index of Used Basis- and Equivalent-Normative Components

Symbol	Designation	Formula Unit	Equivalent Weight
A	Anhydrite	$1/2\ (CaO \cdot SO_3)$	68.1
Ab	Albite (Na-Feldspar)	$1/10\ (6\ SiO_2 \cdot Al_2O_3 \cdot Na_2O)$	52.4
Ac	Acmite (Aegirine)	$1/8\ (4\ SiO_2 \cdot Fe_2O_3 \cdot Na_2O)$	57.7
$K\text{-}Ac$	K-Aegirine	$1/8\ (4\ SiO_2 \cdot Fe_2O_3 \cdot K_2O)$	61.8
Ak	Akermanite	$1/5\ (2\ SiO_2 \cdot MgO \cdot 2\ CaO)$	54.5
$Fe\text{-}Ak$	Fe-Akermanite	$1/5\ (2\ SiO_2 \cdot FeO \cdot 2\ CaO)$	60.8
Akt	Normal Actinolite	$1/30\ (16\ SiO_2 \cdot 8{,}5\ MgO \cdot 1{,}5\ FeO \cdot 4\ CaO \cdot 2\ H_2O)$	55.7
$Fe\text{-}Akt$	Fe-Actinolite	$1/15\ (8\ SiO_2 \cdot 5\ FeO \cdot 2\ CaO \cdot H_2O)$	64.7
$Alkf$	Alkali Feldspars	Mixture of $Ab\text{-}Or$	52.4–55.6
Alm	Almandine	$1/8\ (3\ SiO_2 \cdot Al_2O_3 \cdot 3\ FeO)$	62.2
An	Anorthite (Ca-Feldspar)	$1/5\ (2\ SiO_2 \cdot Al_2O_3 \cdot CaO)$	55.6
Anc	Analcite	$1/8\ (4\ SiO_2 \cdot Al_2O_3 \cdot Na_2O \cdot 2\ H_2O)$	55.0
$Andr$	Andradite	$1/8\ (3\ SiO_2 \cdot Fe_2O_3 \cdot 3\ CaO)$	63.5
Ant	Antigorite	$1/5\ (2\ SiO_2 \cdot 3\ MgO \cdot 2\ H_2O)$	55.4
$Fe\text{-}Ant$	Fe-Antigorite	$1/5\ (2\ SiO_2 \cdot 3\ FeO \cdot 2\ H_2O)$	74.3
$Anth_1$	Antophyllite$_1$	$1/15\ (8\ SiO_2 \cdot 7\ MgO \cdot H_2O)$	52.0
$Anth_2$	Antophyllite$_2$	$1/30\ (16\ SiO_2 \cdot 7\ MgO \cdot 7\ FeO \cdot 2\ H_2O)$	59.4
Ap	Apatite	$1/16\ (3\ P_2O_5 \cdot 9\ CaO \cdot CaF_2)$	63.0
At	Amesite	$1/5\ (SiO_2 \cdot Al_2O_3 \cdot 2\ MgO \cdot 2\ H_2O)$	55.7
$Fe\text{-}At$	Fe-Amesite	$1/5\ (SiO_2 \cdot Al_2O_3 \cdot 2\ FeO \cdot 2\ H_2O)$	68.3
Aug	Augite, in general		
$Mg\text{-}Bi$	Mg-Biotite (Phlogopite)	$1/16\ (6\ SiO_2 \cdot Al_2O_3 \cdot 6\ MgO \cdot K_2O \cdot 2\ H_2O)$	52.1
$Fe\text{-}Bi$	Fe-Biotite	$1/16\ (6\ SiO_2 \cdot Al_2O_3 \cdot 6\ FeO \cdot K_2O \cdot 2\ H_2O)$	64.0
Bru	Brucite	$1\ (MgO \cdot H_2O)$	58.3
C	Corundum	$1/2\ (Al_2O_3)$	51.0
Cal	Ca-Aluminate	$1/3\ (Al_2O_3 \cdot CaO)$	52.7
$Canc$	Cancrinite	$1/20\ (6\ SiO_2 \cdot 3\ Al_2O_3 \cdot 3\ Na_2O \cdot 2\ CaO \cdot 2\ CO_2)$	52.6
Cc	Calcite	$1\ (CaO \cdot CO_2)$	100.1
$Cord$	Cordierite	$1/11\ (5\ SiO_2 \cdot 2\ Al_2O_3 \cdot 2\ MgO)$	53.2
$Fe\text{-}Cord$	Fe-Cordierite	$1/11\ (5\ SiO_2 \cdot 2\ Al_2O_3 \cdot 2\ FeO)$	58.9
$Cord^*$	Mixed crystals of	$Cord + Fe\text{-}Cord$	
Cp	Ca-Phosphate	$1/5\ (P_2O_5 \cdot 3\ CaO)$	62.0
Cm	Chromite	$1/3\ (Cr_2O_3 \cdot FeO)$	74.6
Cs	Ca-Orthosilicate (Larnite)	$1/3\ (SiO_2 \cdot 2\ CaO)$	57.4
Di	Diopside	$1/4\ (2\ SiO_2 \cdot MgO \cdot CaO)$	54.1
En	Enstatite	$1/2\ (SiO_2 \cdot MgO)$	50.2
Ep	Edipote, in general		
Fa	Fayalite	$1/3\ (SiO_2 \cdot 2\ FeO)$	67.9
Fks	K-Ferrisilicate	$1/6\ (2\ SiO_2 \cdot Fe_2O_3 \cdot K_2O)$	62.3
Fns	Na-Ferrisilicate	$1/6\ (2\ SiO_2 \cdot Fe_2O_3 \cdot Na_2O)$	57.0

Symbol	Designation	Formula Unit	Equivalent Weight
Fo	Forsterite	$\frac{1}{3}(SiO_2 \cdot 2 MgO)$	46.9
Fr	Fluorite	$1 (CaF_2)$	78.1
Fs	Ferrisilicate	$\frac{1}{3}(SiO_2 \cdot Fe_2O_3)$	73.2
Ge	Gehlenite	$\frac{1}{5}(SiO_2 \cdot Al_2O_3 \cdot 2 CaO)$	54.8
Fe-Ge	Fe-Gehlenite	$\frac{1}{10}(2 SiO_2 \cdot Al_2O_3 \cdot Fe_2O_3 \cdot 4 CaO)$	60.6
Na-Ge	Na-Gehlenite	$\frac{1}{10}(4 SiO_2 \cdot Al_2O_3 \cdot 2 CaO \cdot Na_2O)$	51.6
Glph	Glaucophane	$\frac{1}{15}(8 SiO_2 \cdot Al_2O_3 \cdot 3 (Mg, Fe) O \cdot Na_2O \cdot H_2O)$	52.2–58.5
Gram	Grammatite	$\frac{1}{15}(8 SiO_2 \cdot 5 MgO \cdot 2 CaO \cdot H_2O)$	54.1
Gro	Grossularite	$\frac{1}{8}(3 SiO_2 \cdot Al_2O_3 \cdot 3 CaO)$	56.3
Hau	Hauyne	$\frac{1}{22}(6 SiO_2 \cdot 3 Al_2O_3 \cdot 3 Na_2O \cdot 2 CaO \cdot 2 SO_3)$	51.2
Hed	Hedenbergite	$\frac{1}{4}(2 SiO_2 \cdot FeO \cdot CaO)$	62.0
Hl	Rock salt (Halite)	$1 (NaCl)$	58.5
Hm	Hematite	$\frac{1}{2}(Fe_2O_3)$	79.8
Ho	Common Hornblene (idealized)	$\frac{1}{15}(7 SiO_2 \cdot Al_2O_3 \cdot 4 (Mg, Fe) O \cdot 2 CaO \cdot H_2O)$	54.3–62.7
Hy	Hypersthene	$\frac{1}{2}(SiO_2 \cdot FeO)$	65.9
Hz	Hercynite	$\frac{1}{3}(Al_2O_3 \cdot FeO)$	57.9
Kl	Clinochlore (average chlorite) with *Ant* : *At* ~ 1 : 1		~ 55.0
Ilm	Ilmenite	$\frac{1}{2}(TiO_2 \cdot FeO)$	75.9
Jd	Jadeite	$\frac{1}{8}(4 SiO_2 \cdot Al_2O_3 \cdot Na_2O)$	50.5
Kaol	Kaolin	$\frac{1}{4}(2 SiO_2 \cdot Al_2O_3 \cdot 2 H_2O)$	64.5
Kp	Kaliophylite	$\frac{1}{6}(2 SiO_2 \cdot Al_2O_3 \cdot K_2O)$	52.7
Ks	K-Silicate	$\frac{1}{3}(SiO_2 \cdot K_2O)$	51.4
Lc	Leucite	$\frac{1}{8}(4 SiO_2 \cdot Al_2O_3 \cdot K_2O)$	54.5
Lep	Lepidolite	$\frac{1}{32}(12 SiO_2 \cdot 5 Al_2O_3 \cdot 3 Li_2O \cdot 4 KF \cdot 2 H_2O)$	49.6
Ma$_1$	Chloride-Marialite	$\frac{1}{22}(18 SiO_2 \cdot 3 Al_2O_3 \cdot 3 Na_2O \cdot 2 NaCl)$	52.8
Ma$_2$	Carbonate-Marialite	$\frac{1}{32}(18 SiO_2 \cdot 3 Al_2O_3 \cdot 4 Na_2O \cdot CO_2)$	52.4
Ma$_3$	Sulfate-Marialite	$\frac{1}{33}(18 SiO_2 \cdot 3 Al_2O_3 \cdot 4 Na_2O \cdot SO_3)$	52.0
Marg	Margarite	$\frac{1}{7}(2 SiO_2 \cdot 2 Al_2O_3 \cdot CaO \cdot H_2O)$	57.0
Me$_1$	Chloride-Meionite	$\frac{1}{16}(6 SiO_2 \cdot 3 Al_2O_3 \cdot 3 CaO \cdot CaCl_2)$	59.1
Me$_2$	Carbonate-Meionite	$\frac{1}{16}(6 SiO_2 \cdot 3 Al_2O_3 \cdot 4 CaO \cdot CO_2)$	58.4
Me$_3$	Sulfate-Meionite	$\frac{1}{17}(6 SiO_2 \cdot 3 Al_2O_3 \cdot 4 CaO \cdot SO_3)$	57.1
Mel	Melilite, in general,	mixture of *Ak*, Fe-*Ak*, *Ge*, Fe-*Ge*, Na-*Ge*	
Mgs	Magnesite	$1 (MgO \cdot CO_2)$	84.3
Mont	Monticellite	$\frac{1}{3}(SiO_2 \cdot MgO \cdot CaO)$	52.1
Ms$_1$	Muscovite$_1$	$\frac{1}{14}(6 SiO_2 \cdot 3 Al_2O_3 \cdot K_2O \cdot 2 H_2O)$	56.9
Ms$_2$	Muscovite$_2$ (Phengite)	$\frac{1}{14}(7 SiO_2 \cdot 2 Al_2O_3 \cdot (Mg, Fe) O \cdot K_2O \cdot 2 H_2O)$	56.8–59.0
Mt	Magnetite	$\frac{1}{3}(FeO \cdot Fe_2O_3)$	77.1
Nc	Na-Carbonate	$\frac{1}{2}(Na_2O \cdot CO_2)$	53.0
Ne	Nepheline	$\frac{1}{6}(2 SiO_2 \cdot Al_2O_3 \cdot Na_2O)$	47.3
Nos	Nosean	$\frac{1}{21}(6 SiO_2 \cdot 3 Al_2O_3 \cdot 4 Na_2O \cdot SO_3)$	47.3
Ns	Na-Silicate	$\frac{1}{3}(SiO_2 \cdot Na_2O)$	40.7
Ol	Olivine, in general		
Omph	Omphacite	$\frac{1}{16}(8 SiO_2 \cdot Al_2O_3 \cdot 2 MgO \cdot 2 CaO \cdot Na_2O)$	52.3
Or	Orthoclase (K-Feldspar)	$\frac{1}{10}(6 SiO_2 \cdot Al_2O_3 \cdot K_2O)$	55.6
Ot	Fe-Ottrelite	$\frac{1}{4}(SiO_2 \cdot Al_2O_3 \cdot FeO \cdot H_2O)$	63.0
Mg-Ot	Mg-Ottrelite	$\frac{1}{4}(SiO_2 \cdot Al_2O_3 \cdot MgO \cdot H_2O)$	55.1
Pf	Perovskite	$\frac{1}{2}(TiO_2 \cdot CaO)$	68.0
Pg	Paragonite	$\frac{1}{14}(6 SiO_2 \cdot 3 Al_2O_3 \cdot Na_2O \cdot 2 H_2O)$	54.6
Pi (= Ep)	Pistazite (end member)	$\frac{1}{16}(6 SiO_2 \cdot Fe_2O_3 \cdot 2 Al_2O_3 \cdot 4 CaO \cdot H_2O)$	60.4
Pig	Pigeonite, in general		
Plag	Plagioclase	Mixtures *Ab-An*	52.4–55.6
Pph	Pyrophyllite	$\frac{1}{6}(4 SiO_2 \cdot Al_2O_3 \cdot 2 H_2O)$	63.0

Symbol	Designation	Formula Unit	Equivalent Weight
Pr	Pyrite	$1/3\ (FeS_2)$	40.0
Ps	Periclase	$1\ (MgO)$	40.3
Pyp	Pyrope	$1/8\ (3\ SiO_2 \cdot Al_2O_3 \cdot 3\ MgO)$	50.4
Q	Quartz	$1\ (SiO_2)$	60.1
Rb_1	Riebeckite$_1$	$1/15\ (8\ SiO_2 \cdot Fe_2O_3 \cdot 3\ FeO \cdot Na_2O \cdot H_2O)$	62.3
Rb_2	Riebeckite$_2$	$1/32\ (16\ SiO_2 \cdot Fe_2O_3 \cdot 8\ FeO \cdot 3\ Na_2O \cdot 2\ H_2O)$	59.9
Rb_3	Riebeckite$_3$	$1/32\ (16\ SiO_2 \cdot 2\ Fe_2O_3 \cdot 6\ FeO \cdot 3\ Na_2O \cdot H_2O)$	60.4
Ru	Rutile	$1\ (TiO_2)$	79.9
Sid	Siderite	$1\ (FeO \cdot CO_2)$	115.8
Sil	Sillimanite, Andalusite, Kyanite	$1/3\ (SiO_2 \cdot Al_2O_3)$	54.0
Sod	Sodalite	$1/20\ (6\ SiO_2 \cdot 3\ Al_2O_3 \cdot 3\ Na_2O \cdot 2\ NaCl)$	48.4
Sp	Spinel	$1/3\ (Al_2O_3 \cdot MgO)$	47.4
Spe	Spessartite	$1/8\ (3\ SiO_2 \cdot Al_2O_3 \cdot 3\ MnO)$	61.9
$Spod$	Spodumene	$1/8\ (4\ SiO_2 \cdot Al_2O_3 \cdot Li_2O)$	46.5
$Spur$	Spurrite	$1/7\ (2\ SiO_2 \cdot 5\ CaO \cdot CO_2)$	63.5
$Staur$	Staurolite	$1/7\ (2\ SiO_2 \cdot 2\ Al_2O_3 \cdot FeO \cdot H_2O)$	56.5
Tc	Talc	$1/7\ (4\ SiO_2 \cdot 3\ MgO \cdot 2\ H_2O)$	56.7
$Tephr$	Tephroite	$1/3\ (SiO_2 \cdot 2\ MnO)$	67.3
Th	Thenardite	$1/3\ (Na_2O \cdot SO_3)$	47.3
$Till$	Tilleyite	$1/7\ (2\ SiO_2 \cdot 5\ CaO \cdot 2\ CO_2)$	69.8
Tit	Titanite	$1/3\ (SiO_2 \cdot TiO_2 \cdot CaO)$	65.3
Ts	Tschermak's Molecule	$1/4\ (SiO_2 \cdot Al_2O_3 \cdot CaO)$	54.5
Ves	Vesuvianite	$1/25\ (9\ SiO_2 \cdot 2\ Al_2O_3 \cdot 10\ CaO \cdot 2\ MgO \cdot 2\ H_2O)$	56.9
W	Water	$1\ (H_2O)$	(18.0)
Wo	Wollastonite	$1/2\ (SiO_2 \cdot CaO)$	58.1
Xon	Xonotlite	$1/12\ (6\ SiO_2 \cdot 6\ CaO \cdot H_2O)$	59.6
Z	Zircon	$1/2\ (SiO_2 \cdot ZrO_2)$	91.6
Zo	Zoisite	$1/16\ (6\ SiO_2 \cdot 3\ Al_2O_3 \cdot 4\ CaO \cdot H_2O)$	56.8
Zwd	Zinnwaldite	$1/16\ (6\ SiO_2 \cdot 2\ Al_2O_3 \cdot 2\ FeO \cdot 2\ LiF \cdot 2\ KF)$	54.7

II. Compilation of Important Reaction Relations Between Basis- and Equivalent-Normative Components*

a) Feldsparthoids, including the melilite group

1. $4\ Lc = 3\ Kp + 1\ Q$
2. $8\ Lc = 5\ Or + 3\ Kp$
3. $4\ Anc = 3\ Ne + 1\ Q\ (+ 1\ W)$
4. $10\ Sod = 9\ Ne + 1\ Hl$
5. $11\ Hau = 9\ Ne + 2\ A$
6. $21\ Nos = 18\ Ne + 3\ Th$
7. $10\ Ge = 6\ Cal + 3\ Cs + 1\ Q$
8. $10\ Ge + 3\ Q = 10\ An + 3\ Cs$
9. $10\ Na\text{-}Ge + 3\ Q = 10\ Ab + 3\ Cs$

* The equations are arranged in the order of the most important mineral groups. In order to facilitate the finding of the required relations, a number of equations are repeated in the different mineral groups.

10. $10\,Na\text{-}Ge = 6\,Ne + 3\,Cs + 1\,Q$
11. $5\,Ge = 3\,Cal + 2\,Wo$
12. $10\,Ak = 6\,Cs + 3\,Fo + 1\,Q$
13. $5\,Ak = 3\,Cs + 2\,En$
14. $10\,Ak = 3\,Cs + 3\,Fo + 4\,Wo$

b) Feldspars

1. $5\,Or = 3\,Kp + 2\,Q$
2. $5\,Ab = 3\,Ne + 2\,Q$
3. $5\,An = 3\,Cal + 2\,Q$
4. $5\,An = 4\,Ts + 1\,Q$
5. $5\,An + 3\,Fo = 4\,Ts + 4\,En$
6. $20\,An + 5\,Or(+ 2\,W) = 7\,Ms + 16\,Zo + 2\,Q$
7. $15\,Or(+ 1\,W) = 7\,Ms + 3\,Ks + 5\,Q$
8. $40\,Or + 33\,Cord(+ 8\,W) = 42\,Ms + 16\,Mg\text{-}Bi + 15\,Q$
9. $40\,Or + 33\,Fe\text{-}Cord(+ 8\,W) = 42\,Ms + 16\,Fe\text{-}Bi + 15\,Q$
10. $5\,Or + 2\,C(+ 1\,W) = 7\,Ms$
11. $5\,Or + 3\,Sil(+ 1\,W) = 7\,Ms + 1\,Q$
12. $5\,Or + 5\,An(+ 1\,W) = 7\,Ms + 2\,Wo + 1\,Q$

c) Pyroxenes

1. $4\,En = 3\,Fo + 1\,Q$
2. $4\,Hy = 3\,Fa + 1\,Q$
3. $4\,Wo = 3\,Cs + 1\,Q$
4. $8\,Di = 3\,Fo + 3\,Cs + 2\,Q$
5. $4\,Di = 2\,En + 2\,Wo$
6. $4\,Hed = 2\,Hy + 2\,Wo$
7. $4\,Ts = 3\,Cal + 1\,Q$
8. $8\,Ts = 3\,Cs + 4\,C + 1\,Q$
9. $2\,Ts = 1\,Wo + 1\,C$
10. $4\,Ts + 1\,Q = 5\,An$
11. $4\,Ts + 4\,En = 5\,An + 3\,Fo$
12. $4\,Ts + 4\,Hy = 5\,An + 3\,Fa$
13. $8\,Ac = 3\,Ns + 2\,Hm + 3\,Q$
14. $8\,Ac = 6\,Fns + 2\,Q$
15. $4\,Jd = 3\,Ne + 1\,Q$
16. $16\,Omph = 6\,Ne + 3\,Fo + 3\,Cs + 4\,Q$
17. $16\,Omph = 10\,Ab + 3\,Fo + 3\,Cs$
18. $8\,Di + 6\,En + 1\,Q(+ 1\,W) = 15\,Gram$
19. $8\,Hed + 6\,Hy + 1\,Q(+ 1\,W) = 15\,Fe\text{-}Akt$
20. $2\,En + 3\,Fo(+ 2\,W) = 5\,Ant$
21. $2\,Hy + 3\,Fa(+ 2\,W) = 5\,Fe\text{-}Ant$

d) Amphipoles

1. $10\,Gram = 2\,Cs + 5\,Fo + 3\,Q(+\,^2/_3\,W)$
2. $10\,Fe\text{-}Akt = 2\,Cs + 5\,Fa + 3\,Q(+\,^2/_3\,W)$
3. $15\,Gram = 8\,Di + 6\,En + 1\,Q(+\,1\,W)$
4. $15\,Fe\text{-}Akt = 8\,Hed + 6\,Hy + 1\,Q(+\,1\,W)$
5. $10\,Glph = 4\,Ne + 3\,(Fo + Fa) + 3\,Q + (+\,^2/_3\,W)$
6. $15\,Glph + 1\,Q = 10\,Ab + 6\,(En + Hy)\,(+\,1\,W)$
7. $3\,Glph\,(+\,^1/_5\,W) = 2\,Ab + 1\,(Ant + Fe\text{-}Ant)$
8. $10\,Ho = 2\,Cal + 1\,Cs + 4\,Fo + 3\,Q(+\,^2/_3\,W)$
9. $15\,Ho = 5\,An + 6\,Fo + 2\,Wo + 2\,Q(+\,1\,W)$
10. $15\,Ho = 4\,Ts + 2\,Wo + 6\,Fo + 3\,Q(+\,1\,W)$
11. $10\,Rb_1 = 2\,Fs + 2\,Ns + 3\,Fa + 3\,Q(+\,^2/_3\,W)$
12. $30\,Rb_2 = 16\,Ac + 9\,Fa + 5\,Q(+\,2\,W)$
13. $15\,Anth_2 = 7\,En + 7\,Hy + 1\,Q(+\,1\,W)$

e) Micas

1. $16\,Mg\text{-}Bi = 6\,Kp + 9\,Fo + 1\,Q(+\,2\,W)$
2. $16\,Fe\text{-}Bi = 6\,Kp + 9\,Fa + 1\,Q(+\,2\,W)$
3. $16\,Mg\text{-}Bi + 3\,Q = 10\,Or + 9\,Fo(+\,2\,W)$
4. $16\,Fe\text{-}Bi + 3\,Q = 10\,Or + 9\,Fa(+\,2\,W)$
5. $8\,Mg\text{-}Bi + 3\,Q = 5\,Or + 6\,En(+\,2\,W)$
6. $8\,Fe\text{-}Bi + 3\,Q = 5\,Or + 6\,Hy(+\,2\,W)$
7. $7\,Ms = 3\,Kp + 2\,Q + 2\,C(+\,1\,W)$
8. $7\,Ms + 3\,Ks + 5\,Q = 15\,Or(+\,1\,W)$
9. $14\,Ms + 3\,Cs = 6\,Kp + 6\,Cal + 5\,Q(+\,2\,W)$
10. $14\,Ms + 3\,Fo = 6\,Kp + 6\,Sp + 5\,Q(+\,2\,W)$
11. $14\,Ms + 3\,Fa = 6\,Kp + 6\,Hz + 5\,Q(+\,2\,W)$
12. $7\,Ms + 16\,Zo + 2\,Q = 20\,An + 5\,Or(+\,2\,W)$
13. $7\,Ms + 8\,Alm = 8\,Fe\text{-}Bi + 6\,Sil + 1\,Q$
14. $7\,Ms = 5\,Or + 2\,C(+\,1\,W)$
15. $7\,Ms + 1\,Q = 5\,Or + 3\,Sil(+\,1\,W)$
16. $7\,Ms + 2\,Wo + 1\,Q = 5\,An + 5\,Or(+\,1\,W)$
17. $42\,Ms + 16\,Mg\text{-}Bi + 15\,Q = 40\,Or + 33\,Cord\,(+\,8\,W)$
18. $42\,Ms + 16\,Fe\text{-}Bi + 15\,Q = 40\,Or + 33\,Fe\text{-}Cord\,(+\,8\,W)$
19. $14\,Ph = 6\,Kp + 3\,(Sp + Hz) + 5\,Q(+\,1\,W)$
20. $7\,Marg = 3\,Cal + 2\,C + 2\,Q(+\,1\,W)$
21. $7\,Pg = 3\,Ne + 2\,C + 2\,Q(+\,1\,W)$

f) Al-Silicates

1. $3\,Sil\,(Dist, And) = 2\,C + 1\,Q$
2. $11\,Cord = 6\,Sp + 5\,Q$
3. $11\,Fe\text{-}Cord = 6\,Hz + 5\,Q$

4. $11\,Cord = 4\,En + 6\,Sil + 1\,Q$
5. $11\,Fe\text{-}Cord = 4\,Hy + 6\,Sil + 1\,Q$
6. $11\,Cord = 3\,Fo + 4\,C + 4\,Q$
7. $11\,Fe\text{-}Cord = 3\,Fa + 4\,C + 4\,Q$
8. $11\,Cord\,(+\,2\,W) = 8\,Mg\text{-}Ot + 3\,Q$
9. $11\,Fe\text{-}Cord\,(+\,2\,W) = 8\,Ot + 3\,Q$
10. $11\,Cord + 8\,En = 16\,Pyp + 3\,Q$
11. $11\,Fe\text{-}Cord + 8\,Hy = 16\,Alm + 3\,Q$
12. $22\,Cord + 15\,Or + 8\,En\,(+\,9\,W) = 21\,Ms + 10\,Ant + 5\,At + 9\,Q$
13. $22\,Fe\text{-}Cord + 15\,Or + 8\,Hy\,(+\,9\,W) = 21\,Ms + 10\,Fe\text{-}Ant + 5\,Fe\text{-}At + 9\,Q$
14. $33\,Cord = 16\,Pyp + 12\,Sil + 5\,Q$
15. $33\,Fe\text{-}Cord = 16\,Alm + 12\,Sil + 5\,Q$
16. $33\,Cord + 40\,Or\,(+\,8\,W) = 42\,Ms + 16\,Mg\text{-}Bi + 15\,Q$
17. $33\,Fe\text{-}Cord + 40\,Or\,(+\,8\,W) = 42\,Ms + 16\,Fe\text{-}Bi + 15\,Q$
18. $33\,Cord + 10\,Or\,(+\,2\,W) = 16\,Mg\text{-}Bi + 18\,Sil + 9\,Q$
19. $33\,Fe\text{-}Cord + 10\,Or\,(+\,2\,W) = 16\,Fe\text{-}Bi + 18\,Sil + 9\,Q$
20. $7\,Staur = 3\,Hz + 3\,Sil + 1\,Q\,(+\,1\,W)$
21. $14\,Staur + 3\,Q = 11\,Fe\text{-}Cord + 6\,Sil\,(+\,2\,W)$
22. $21\,Staur + 2\,Q = 8\,Alm + 15\,Sil\,(+\,3\,W)$
23. $7\,Staur = 4\,Ot + 3\,Sil\,(Dist)$
24. $2\,Kaol = 1\,C + 1\,Q\,(+\,1\,W)$
25. $4\,Kaol = 3\,Sil + 1\,Q\,(+\,2\,W)$

g) Garnet group

1. $8\,Gro = 3\,Cal + 3\,Cs + 2\,Q$
2. $8\,Gro = 3\,Cal + 4\,Wo + 1\,Q$
3. $8\,Gro = 5\,An + 3\,Cs$
4. $8\,Gro = 4\,Ts + 4\,Wo$
5. $8\,Pyp = 3\,Sp + 3\,Fo + 2\,Q$
6. $8\,Pyp = 3\,Sp + 4\,En + 1\,Q$
7. $16\,Pyp + 3\,Q = 11\,Cord + 8\,En$
8. $8\,Alm = 3\,Hz + 3\,Fa + 2\,Q$
9. $8\,Alm = 3\,Hz + 4\,Hy + 1\,Q$
10. $16\,Alm + 3\,Q = 11\,Fe\text{-}Cord + 8\,Hy$
11. $16\,Andr = 4\,Hm + 9\,Cs + 3\,Q$
12. $4\,Andr = 1\,Hm + 3\,Wo$
13. $8\,Alm + 7\,Ms = 8\,Fe\text{-}Bi + 6\,Sil + 1\,Q$
14. $16\,Alm + 3\,Q = 8\,Hy + 11\,Fe\text{-}Cord$
15. $24\,Gro\,(+\,5\,CO_2 + 1\,W) = 16\,Zo + 5\,Cc + 3\,Q$
16. $24\,Alm\,(+\,8\,W) = 5\,Fe\text{-}Ant + 15\,Fe\text{-}At + 4\,Q$
17. $24\,Alm\,(+\,2\,W) = 10\,Fe\text{-}Ant + 9\,Hz + 5\,Q$
18. $12\,Alm\,(+\,^7/_2\,W) = 6\,Ot + 5\,Fe\text{-}Ant + 1\,Q$
19. $4\,Alm + 2\,C\,(+\,2\,W) = 6\,Ot$

h) Serpentine, Chlorite, Talc

1. $10\,Ant = 9\,Fo + 1\,Q\,(+\,4\,W)$
2. $10\,\text{Fe-}Ant = 9\,Fa + 1\,Q\,(+\,4\,W)$
3. $10\,At = 6\,Sp + 3\,Fo + 1\,Q\,(+\,4\,W)$
4. $10\,\text{Fe-}At = 6\,Hz + 3\,Fa + 1\,Q\,(+\,4\,W)$
5. $5\,At = 3\,Sp + 2\,En\,(+\,2\,W)$
6. $5\,\text{Fe-}At = 3\,Hz + 2\,Hy\,(+\,2\,W)$
7. $5\,Ant = 3\,Fo + 2\,En\,(+\,2\,W)$
8. $5\,\text{Fe-}Ant = 3\,Fa + 2\,Hy\,(+\,2\,W)$
9. $5\,Ant\,(+\,1\,W) = 3\,Bru + 2\,Q$
10. $14\,Tc = 9\,Fo + 5\,Q\,(+\,4\,W)$
11. $7\,Tc = 6\,En + 1\,Q\,(+\,2\,W)$
12. $7\,Tc\,(+\,1\,W) = 5\,Ant + 2\,Q$
13. $5\,\text{Fe-}At\,(+\,CO_2) = 4\,Ot + 1\,Sid\,(+\,1\,W)$
14. $5\,At\,(+\,CO_2) = 4\,\text{Mg-}Ot + 1\,Mgs\,(+\,1\,W)$
15. $15\,\text{Fe-}At + 2\,Q = 12\,Ot + 5\,\text{Fe-}Ant\,(+\,1\,W)$
16. $15\,At + 2\,Q = 12\,\text{Mg-}Ot + 5\,Ant\,(+\,1\,W)$
17. $10\,Ant + 5\,At + 21\,Ms + 9\,Q = 22\,Cord + 15\,Or + 8\,En\,(+\,9\,W)$
18. $10\,Kl = 6\,(Fo + Fa) + 3\,(Sp + Hz) + 1\,Q\,(+\,4\,W)$

i) Chloritoid

1. $4\,Ot = 3\,Hz + 1\,Q\,(+\,1\,W)$
2. $4\,\text{Mg-}Ot = 3\,Sp + 1\,Q\,(+\,1\,W)$
3. $4\,Ot + 1\,Sid\,(+\,1\,W) = 5\,\text{Fe-}At\,(+\,1\,CO_2)$
4. $4\,\text{Mg-}Ot + 1\,Mgs\,(+\,1\,W) = 5\,At\,(+\,1\,CO_2)$
5. $12\,Ot + 5\,\text{Fe-}Ant\,(+\,1\,W) = 15\,\text{Fe-}At + 2\,Q$
6. $12\,\text{Mg-}Ot + 5\,Ant\,(+\,1\,W) = 15\,At + 2\,Q$
7. $6\,Ot + 5\,\text{Fe-}Ant + 1\,Q = 12\,Alm\,(+\,^7/_2\,W)$
8. $6\,\text{Mg-}Ot + 5\,Ant + 1\,Q = 12\,Pyp\,(+\,^7/_2\,W)$
9. $6\,Ot = 4\,Alm + 2\,C\,(+\,^3/_2\,W)$
10. $6\,\text{Mg-}Ot = 4\,Pyp + 2\,C\,(+\,^3/_2\,W)$
11. $8\,Ot + 3\,Q = 11\,\text{Fe-}Cord\,(+\,2\,W)$
12. $8\,\text{Mg-}Ot + 3\,Q = 11\,Cord\,(+\,2\,W)$
13. $4\,Ot + 3\,Sil = 7\,Staur$

j) Epidote-Zoisite group

1. $32\,Zo = 18\,Cal + 3\,Cs + 11\,Q\,(+\,2\,W)$
2. $32\,Zo + 9\,Fo = 18\,Sp + 12\,Cs + 11\,Q\,(+\,2\,W)$
3. $16\,Zo + 5\,At = 12\,Cal + 3\,Fo + 6\,Q\,(+\,3\,W)$
4. $16\,Zo + 7\,Ms + 2\,Q = 20\,An + 5\,Or\,(+\,2\,W)$
5. $16\,Zo + 5\,Cc + 3\,Q = 24\,Gro\,(+\,5\,CO_2 + 1\,W)$
6. $8\,Zo = 3\,Cs + 3\,C + 2\,Q\,(+\,^1/_2\,W)$

7. $16\,Zo = 9\,Cal + 2\,Wo + 5\,Q\,(+1\,W)$
8. $48\,Zo + 80\,Ant + 7\,Q = 90\,Gram + 45\,At\,(+11\,W)$
9. $16\,Pi = 6\,Cal + 3\,Cs + 5\,Q + 2\,Hm\,(+1\,W)$
10. $8\,Pi\,(+1\,CO_2) = 3\,Q + 1\,Hm + 3\,Cal + 1\,Cc\,(+\,^{1}/_{2}\,W)$
11. $24\,Pi\,(+2\,CO_2) = 3\,Hm + 16\,Zo + 2\,Cc + 3\,Q\,(+\,^{1}/_{2}\,W)$
12. $24\,Pi = 3\,Hm + 16\,Zo + 3\,Cs + 2\,Q\,(+\,^{1}/_{2}\,W)$
13. $8\,Pi + 3\,Q + 6\,Cal\,(+\,^{1}/_{2}\,W) = 1\,Hm + 16\,Zo$

k) Calcium Silicates

1. $4\,Xon = 3\,Cs + 1\,Q\,(+\,^{1}/_{3}\,W)$
2. $7\,Spur = 6\,Cs + 1\,Cc$
3. $7\,Till\,(+\,CO_2) = 3\,Cc + 4\,Wo$
4. $7\,Till = 6\,Cs + 1\,Cc\,(+\,CO_2)$
5. $25\,Ves = 3\,Fo + 15\,Cs + 6\,Sil + 1\,Q$
6. $6\,Mont = 3\,Fo + 3\,Cs$

l) Carbonates

1. $3\,Cs\,(+\,2\,CO_2) = 2\,Cc + 1\,Q$
2. $3\,Fo\,(+\,2\,CO_2) = 2\,Mgs + 1\,Q$
3. $3\,Fa\,(+\,2\,CO_2) = 2\,Sid + 1\,Q$

m) Accessories

1. $6\,Tit = 2\,Ru + 3\,Cs + 1\,Q$
2. $3\,Tit = 1\,Ru + 2\,Wo$
3. $2\,Ru + 3\,Cs = 1\,Q + 4\,Pf$
4. $3\,Tit = 2\,Pf + 1\,Q$
5. $15\,Cp + 1\,Fr = 16\,Ap$
6. $2\,Fs + 1\,Fa = 2\,Mt + 1\,Q$
7. $3\,Fa + 2\,Ru = 1\,Q + 4\,Ilm$
8. $3\,Fs = 2\,Hm + 1\,Q$
9. $2\,Fs + 1\,Fa = 2\,Mt + 1\,Q$

III. Tables of the Molecular and Atomic Equivalent Numbers, Multiplied by 1000, for the Important Rock-Forming Oxides

Recalculated for intervals of 0.1 wt %,
allowing for interpolation of 0.01 wt %.
By W. Oberholzer

SiO$_2$ 60.06

Weight % 0.0—20.9

%	0.0	0.1	0.2	0.3	0.4	0.5	0.6	0.7	0.8	0.9	%
0	0	2	3	5	7	8	10	12	13	15	0
1	17	18	20	22	23	25	27	28	30	32	1
2	33	35	37	38	40	42	43	45	47	48	2
3	50	52	53	55	57	58	60	62	63	65	3
4	67	68	70	72	73	75	76	78	80	82	4
5	83	85	87	88	90	92	93	95	97	98	5
6	100	102	103	105	107	108	110	112	113	115	6
7	117	118	120	122	123	125	127	128	130	132	7
8	133	135	137	138	140	142	143	145	147	148	8
9	150	152	153	155	157	158	160	162	163	165	9
10	166	168	170	171	173	175	176	178	180	181	10
11	183	185	186	188	190	191	193	195	196	198	11
12	200	201	203	205	206	208	210	211	213	215	12
13	216	218	220	221	223	225	226	228	230	231	13
14	233	235	236	238	240	241	243	245	246	248	14
15	250	251	253	255	256	258	260	261	263	265	15
16	266	268	270	271	273	275	276	278	280	281	16
17	283	285	286	288	290	291	293	295	296	298	17
18	300	301	303	305	306	308	310	311	313	315	18
19	316	318	320	321	323	325	326	328	330	331	19
20	333	335	336	338	340	341	343	345	346	348	20
%	0.0	0.1	0.2	0.3	0.4	0.5	0.6	0.7	0.8	0.9	%

1

%	
0.01	0.1
2	0.2
3	0.3
4	0.4
5	0.5
6	0.6
7	0.7
8	0.8
9	0.9

2

%	
0.01	0.2
2	0.4
3	0.6
4	0.8
5	1.0
6	1.2
7	1.4
8	1.6
9	1.8

SiO₂ 60.06

Weight % 20.0—40.9

%	0.0	0.1	0.2	0.3	0.4	0.5	0.6	0.7	0.8	0.9	%
20	333	335	336	338	340	341	343	345	346	348	20
21	350	351	353	355	356	358	360	361	363	365	21
22	366	368	370	371	373	375	376	378	380	381	22
23	383	385	386	388	390	391	393	395	396	398	23
24	400	401	403	405	406	408	410	411	413	415	24
25	416	418	420	421	423	425	426	428	430	431	25
26	433	435	436	438	440	441	443	445	446	448	26
27	450	451	453	455	456	458	460	461	463	465	27
28	466	468	470	471	473	475	476	478	480	481	28
29	483	485	486	488	490	491	493	495	496	498	29
30	500	501	503	504	506	508	509	511	513	514	30
31	516	518	519	521	523	524	526	528	529	531	31
32	533	534	536	538	539	541	543	544	546	548	32
33	549	551	553	554	556	558	559	561	563	564	33
34	566	568	569	571	573	574	576	578	579	581	34
35	583	584	586	588	589	591	593	594	596	598	35
36	599	601	603	604	606	608	609	611	613	614	36
37	616	618	619	621	623	624	626	628	629	631	37
38	633	634	636	638	639	641	643	644	646	648	38
39	649	651	653	654	656	658	659	661	663	664	39
40	666	668	669	671	673	674	676	678	679	681	40
%	0.0	0.1	0.2	0.3	0.4	0.5	0.6	0.7	0.8	0.9	%

1

%	
0.01	0.1
2	0.2
3	0.3
4	0.4
5	0.5
6	0.6
7	0.7
8	0.8
9	0.9

2

%	
0.01	0.2
2	0.4
3	0.6
4	0.8
5	1.0
6	1.2
7	1.4
8	1.6
9	1.8

SiO₂ 60.06

Weight % 40.0—60.9

%	0.0	0.1	0.2	0.3	0.4	0.5	0.6	0.7	0.8	0.9	%
40	666	668	669	671	673	674	676	678	679	681	40
41	683	684	686	688	689	691	693	694	696	698	41
42	699	701	703	704	706	708	709	711	713	714	42
43	716	718	719	721	723	724	726	728	729	731	43
44	733	734	736	738	739	741	743	744	746	748	44
45	749	751	753	754	756	758	759	761	763	764	45
46	766	768	769	771	773	774	776	778	779	781	46
47	783	784	786	788	789	791	793	794	796	798	47
48	799	801	803	804	806	808	809	811	813	814	48
49	816	818	819	821	823	824	826	828	829	831	49
50	832	834	836	837	839	841	842	844	846	847	50
51	849	851	852	854	856	857	859	861	862	864	51
52	866	867	869	871	872	874	876	877	879	881	52
53	882	884	886	887	889	891	892	894	896	897	53
54	899	901	902	904	906	907	909	911	912	914	54
55	916	917	919	921	922	924	926	927	929	931	55
56	932	934	936	937	939	941	942	944	946	947	56
57	949	951	952	954	956	957	959	961	962	964	57
58	966	967	969	971	972	974	976	977	979	981	58
59	982	984	986	987	989	991	992	994	996	997	59
60	999	1001	1002	1004	1006	1007	1009	1011	1012	1014	60
%	0.0	0.1	0.2	0.3	0.4	0.5	0.6	0.7	0.8	0.9	%

1

%	
0.01	0.1
2	0.2
3	0.3
4	0.4
5	0.5
6	0.6
7	0.7
8	0.8
9	0.9

2

%	
0.01	0.2
2	0.4
3	0.6
4	0.8
5	1.0
6	1.2
7	1.4
8	1.6
9	1.8

SiO₂ 60.06

Weight % 60.0—80.9

%	0.0	0.1	0.2	0.3	0.4	0.5	0.6	0.7	0.8	0.9	%
60	999	1001	1002	1004	1006	1007	1009	1011	1012	1014	60
61	1016	1017	1019	1021	1022	1024	1026	1027	1029	1031	61
62	1032	1034	1036	1037	1039	1041	1042	1044	1046	1047	62
63	1049	1051	1052	1054	1056	1057	1059	1061	1062	1064	63
64	1066	1067	1069	1071	1072	1074	1076	1077	1079	1081	64
65	1082	1084	1086	1087	1089	1091	1092	1094	1096	1097	65
66	1099	1101	1102	1104	1106	1107	1109	1111	1112	1114	66
67	1116	1117	1119	1121	1122	1124	1126	1127	1129	1131	67
68	1132	1134	1136	1137	1139	1141	1142	1144	1146	1147	68
69	1149	1151	1152	1154	1156	1157	1159	1161	1162	1164	69
70	1166	1167	1169	1170	1172	1174	1175	1177	1179	1180	70
71	1182	1184	1185	1187	1189	1190	1192	1194	1195	1197	71
72	1199	1200	1202	1204	1205	1207	1209	1210	1212	1214	72
73	1215	1217	1219	1220	1222	1224	1225	1227	1229	1230	73
74	1232	1234	1235	1237	1239	1240	1242	1244	1245	1247	74
75	1249	1250	1252	1254	1255	1257	1259	1260	1262	1264	75
76	1265	1267	1269	1270	1272	1274	1275	1277	1279	1280	76
77	1282	1284	1285	1287	1289	1290	1292	1294	1295	1297	77
78	1299	1300	1302	1304	1305	1307	1309	1310	1312	1314	78
79	1315	1317	1319	1320	1322	1324	1325	1327	1329	1330	79
80	1332	1334	1335	1337	1339	1340	1342	1344	1345	1347	80
%	0.0	0.1	0.2	0.3	0.4	0.5	0.6	0.7	0.8	0.9	%

1 %	
0.01	0.1
2	0.2
3	0.3
4	0.4
5	0.5
6	0.6
7	0.7
8	0.8
9	0.9

2 %	
0.01	0.2
2	0.4
3	0.6
4	0.8
5	1.0
6	1.2
7	1.4
8	1.6
9	1.8

SiO₂ 60.06

Weight %, 80.0 – 99.9

%	0.0	0.1	0.2	0.3	0.4	0.5	0.6	0.7	0.8	0.9	%
80	1332	1334	1335	1337	1339	1340	1342	1344	1345	1347	80
81	1349	1350	1352	1354	1355	1357	1359	1360	1362	1364	81
82	1365	1367	1369	1370	1372	1374	1375	1377	1379	1380	82
83	1382	1384	1385	1387	1389	1390	1392	1394	1395	1397	83
84	1399	1400	1402	1404	1405	1407	1409	1410	1412	1414	84
85	1415	1417	1419	1420	1422	1424	1425	1427	1429	1430	85
86	1432	1434	1435	1437	1439	1440	1442	1444	1445	1447	86
87	1449	1450	1452	1454	1455	1457	1459	1460	1462	1464	87
88	1465	1467	1469	1470	1472	1474	1475	1477	1479	1480	88
89	1482	1484	1485	1487	1489	1490	1492	1494	1495	1497	89
90	1498	1500	1502	1503	1505	1507	1508	1510	1512	1513	90
91	1515	1517	1518	1520	1522	1523	1525	1527	1528	1530	91
92	1532	1533	1535	1537	1538	1540	1542	1543	1545	1547	92
93	1548	1550	1552	1553	1555	1557	1558	1560	1562	1563	93
94	1565	1567	1568	1570	1572	1573	1575	1577	1578	1580	94
95	1582	1583	1585	1587	1588	1590	1592	1593	1595	1597	95
96	1598	1600	1602	1603	1605	1607	1608	1610	1612	1613	96
97	1615	1617	1618	1620	1622	1623	1625	1627	1628	1630	97
98	1632	1633	1635	1637	1638	1640	1642	1643	1645	1647	98
99	1648	1650	1652	1653	1655	1657	1658	1660	1662	1663	99
%	0.0	0.1	0.2	0.3	0.4	0.5	0.6	0.7	0.8	0.9	%

1
%	
0.01	0.1
2	0.2
3	0.3
4	0.4
5	0.5
6	0.6
7	0.7
8	0.8
9	0.9

2
%	
0.01	0.2
2	0.4
3	0.6
4	0.8
5	1.0
6	1.2
7	1.4
8	1.6
9	1.8

Al$_2$O$_3$ 101.94

Weight % 0.0 —20.9

%	0.0	0.1	0.2	0.3	0.4	0.5	0.6	0.7	0.8	0.9	%
0	0	1	2	3	4	5	6	7	8	9	0
1	10	11	12	13	14	15	16	17	18	19	1
2	20	21	22	23	24	25	26	26	27	28	2
3	29	30	31	32	33	34	35	36	37	38	3
4	39	40	41	42	43	44	45	46	47	48	4
5	49	50	51	52	53	54	55	56	57	58	5
6	59	60	61	62	63	64	65	66	67	68	6
7	69	70	71	72	73	74	75	76	77	77	7
8	78	79	80	81	82	83	84	85	86	87	8
9	88	89	90	91	92	93	94	95	96	97	9
10	98	99	100	101	102	103	104	105	106	107	10
11	108	109	110	111	112	113	114	115	116	117	11
12	118	119	120	121	122	123	124	125	126	127	12
13	128	128	129	130	131	132	133	134	135	136	13
14	137	138	139	140	141	142	143	144	145	146	14
15	147	148	149	150	151	152	153	154	155	156	15
16	157	158	159	160	161	162	163	164	165	166	16
17	167	168	169	170	171	172	173	174	175	176	17
18	177	178	179	180	180	181	182	183	184	185	18
19	186	187	188	189	190	191	192	193	194	195	19
20	196	197	198	199	200	201	202	203	204	205	20
%	0.0	0.1	0.2	0.3	0.4	0.5	0.6	0.7	0.8	0.9	%

1 %	
0.01	0.1
2	0.2
3	0.3
4	0.4
5	0.5
6	0.6
7	0.7
8	0.8
9	0.9

Al$_2$O$_3$ 101.94

Weight % 20.0—40.9

%	0.0	0.1	0.2	0.3	0.4	0.5	0.6	0.7	0.8	0.9	%
20	196	197	198	199	200	201	202	203	204	205	20
21	206	207	208	209	210	211	212	213	214	215	21
22	216	217	218	219	220	221	222	223	224	225	22
23	226	227	228	229	230	231	231	232	233	234	23
24	235	236	237	238	239	240	241	242	243	244	24
25	245	246	247	248	249	250	251	252	253	254	25
26	255	256	257	258	259	260	261	262	263	264	26
27	265	266	267	268	269	270	271	272	273	274	27
28	275	276	277	278	279	280	281	282	282	283	28
29	284	285	286	287	288	289	290	291	292	293	29
30	294	295	296	297	298	299	300	301	302	303	30
31	304	305	306	307	308	309	310	311	312	313	31
32	314	315	316	317	318	319	320	321	322	323	32
33	324	325	326	327	328	329	330	331	332	333	33
34	334	334	335	336	337	338	339	340	341	342	34
35	343	344	345	346	347	348	349	350	351	352	35
36	353	354	355	356	357	358	359	360	361	362	36
37	363	364	365	366	367	368	369	370	371	372	37
38	373	374	375	376	377	378	379	380	381	382	38
39	383	384	385	385	386	387	388	389	390	391	39
40	392	393	394	395	396	397	398	399	400	401	40
%	0.0	0.1	0.2	0.3	0.4	0.5	0.6	0.7	0.8	0.9	%

	1
%	
0.01	0.1
2	0.2
3	0.3
4	0.4
5	0.5
6	0.6
7	0.7
8	0.8
9	0.9

½Al₂O₃ 50.97

Weight % 0.0 – 20.9

%	0.0	0.1	0.2	0.3	0.4	0.5	0.6	0.7	0.8	0.9	%
0	0	2	4	6	8	10	12	14	16	18	0
1	20	22	24	26	27	29	31	33	35	37	1
2	39	41	43	45	47	49	51	53	55	57	2
3	59	61	63	65	67	69	71	73	75	77	3
4	78	80	82	84	86	88	90	92	94	96	4
5	98	100	102	104	106	108	110	112	114	116	5
6	118	120	122	124	126	128	129	131	133	135	6
7	137	139	141	143	145	147	149	151	153	155	7
8	157	159	161	163	165	167	169	171	173	175	8
9	177	179	180	182	184	186	188	190	192	194	9
10	196	198	200	202	204	206	208	210	212	214	10
11	216	218	220	222	224	226	228	230	232	233	11
12	235	237	239	241	243	245	247	249	251	253	12
13	255	257	259	261	263	265	267	269	271	273	13
14	275	277	279	281	283	284	286	288	290	292	14
15	294	296	298	300	302	304	306	308	310	312	15
16	314	316	318	320	322	324	326	328	330	332	16
17	334	335	337	339	341	343	345	347	349	351	17
18	353	355	357	359	361	363	365	367	369	371	18
19	373	375	377	379	381	383	385	386	388	390	19
20	392	394	396	398	400	402	404	406	408	410	20
%	0.0	0.1	0.2	0.3	0.4	0.5	0.6	0.7	0.8	0.9	%

1 %	
0.01	0.1
2	0.2
3	0.3
4	0.4
5	0.5
6	0.6
7	0.7
8	0.8
9	0.9

2 %	
0.01	0.2
2	0.4
3	0.6
4	0.8
5	1.0
6	1.2
7	1.4
8	1.6
9	1.8

½Al₂O₃ 50.97

Weight % 20.0—40.9

%	0.0	0.1	0.2	0.3	0.4	0.5	0.6	0.7	0.8	0.9	%
20	392	394	396	398	400	402	404	406	408	410	20
21	412	414	416	418	420	422	424	426	428	430	21
22	432	434	436	438	439	441	443	445	447	449	22
23	451	453	455	457	459	461	463	465	467	469	23
24	471	473	475	477	479	481	483	485	487	489	24
25	490	492	494	496	498	500	502	504	506	508	25
26	510	512	514	516	518	520	522	524	526	528	26
27	530	532	534	536	538	540	541	543	545	547	27
28	549	551	553	555	557	559	561	563	565	567	28
29	569	571	573	575	577	579	581	583	585	587	29
30	589	591	592	594	596	598	600	602	604	606	30
31	608	610	612	614	616	618	620	622	624	626	31
32	628	630	632	634	636	638	640	642	644	645	32
33	647	649	651	653	655	657	659	661	663	665	33
34	667	669	671	673	675	677	679	681	683	685	34
35	687	689	691	693	695	696	698	700	702	704	35
36	706	708	710	712	714	716	718	720	722	724	36
37	726	728	730	732	734	736	738	740	742	744	37
38	746	747	749	751	753	755	757	759	761	763	38
39	765	767	769	771	773	775	777	779	781	783	39
40	785	787	789	791	793	795	797	798	800	802	40
%	0.0	0.1	0.2	0.3	0.4	0.5	0.6	0.7	0.8	0.9	%

1
%	
0.01	0.1
2	0.2
3	0.3
4	0.4
5	0.5
6	0.6
7	0.7
8	0.8
9	0.9

2
%	
0.01	0.2
2	0.4
3	0.6
4	0.8
5	1.0
6	1.2
7	1.4
8	1.6
9	1.8

Fe$_2$O$_3$ 159.68

Weight % 0.0—20.9

%	0.0	0.1	0.2	0.3	0.4	0.5	0.6	0.7	0.8	0.9	%
0	0	1	1	2	3	3	4	4	5	6	0
1	6	7	8	8	9	9	10	11	11	12	1
2	13	13	14	14	15	16	16	17	18	18	2
3	19	19	20	21	21	22	23	23	24	24	3
4	25	26	26	27	28	28	29	29	30	31	4
5	31	32	33	33	34	34	35	36	36	37	5
6	38	38	39	39	40	41	41	42	43	43	6
7	44	44	45	46	46	47	48	48	49	49	7
8	50	51	51	52	53	53	54	54	55	56	8
9	56	57	58	58	59	59	60	61	61	62	9
10	63	63	64	64	65	66	66	67	68	68	10
11	69	70	70	71	71	72	73	73	74	75	11
12	75	76	76	77	78	78	79	80	80	81	12
13	81	82	83	83	84	85	85	86	86	87	13
14	88	88	89	90	90	91	91	92	93	93	14
15	94	95	95	96	96	97	98	98	99	100	15
16	100	101	101	102	103	103	104	105	105	106	16
17	106	107	108	108	109	110	110	111	111	112	17
18	113	113	114	115	115	116	116	117	118	118	18
19	119	120	120	121	121	122	123	123	124	125	19
20	125	126	126	127	128	128	129	130	130	131	20
%	0.0	0.1	0.2	0.3	0.4	0.5	0.6	0.7	0.8	0.9	%

1	
%	
0.01	0.1
2	0.2
3	0.3
4	0.4
5	0.5
6	0.6
7	0.7
8	0.8
9	0.9

Fe$_2$O$_3$ 159.68

Weight % 20.0—40.9

%	0.0	0.1	0.2	0.3	0.4	0.5	0.6	0.7	0.8	0.9	%
20	125	126	126	127	128	128	129	130	130	131	20
21	132	132	133	133	134	135	135	136	137	137	21
22	138	138	139	140	140	141	142	142	143	143	22
23	144	145	145	146	147	147	148	148	149	150	23
24	150	151	152	152	153	153	154	155	155	156	24
25	157	157	158	158	159	160	160	161	162	162	25
26	163	163	164	165	165	166	167	167	168	168	26
27	169	170	170	171	172	172	173	173	174	175	27
28	175	176	177	177	178	178	179	180	180	181	28
29	182	182	183	183	184	185	185	186	187	187	29
30	188	188	189	190	190	191	192	192	193	193	30
31	194	195	195	196	197	197	198	199	199	200	31
32	200	201	202	202	203	204	204	205	205	206	32
33	207	207	208	209	209	210	210	211	212	212	33
34	213	214	214	215	215	216	217	217	218	219	34
35	219	220	220	221	222	222	223	224	224	225	35
36	225	226	227	227	228	229	229	230	230	231	36
37	232	232	233	234	234	235	235	236	237	237	37
38	238	239	239	240	240	241	242	242	243	244	38
39	244	245	245	246	247	247	248	249	249	250	39
40	250	251	252	252	253	254	254	255	255	256	40
%	0.0	0.1	0.2	0.3	0.4	0.5	0.6	0.7	0.8	0.9	%

1	
%	
0.01	0.1
2	0.2
3	0.3
4	0.4
5	0.5
6	0.6
7	0.7
8	0.8
9	0.9

½Fe₂O₃ 79.84

Weight % 0.0—20.9

%	0.0	0.1	0.2	0.3	0.4	0.5	0.6	0.7	0.8	0.9	%
0	0	1	3	4	5	6	8	9	10	11	0
1	13	14	15	16	18	19	20	21	23	24	1
2	25	26	28	29	30	31	33	34	35	36	2
3	38	39	40	41	43	44	45	46	48	49	3
4	50	51	53	54	55	56	58	59	60	61	4
5	63	64	65	66	68	69	70	71	73	74	5
6	75	76	78	79	80	81	83	84	85	86	6
7	88	89	90	91	93	94	95	96	98	99	7
8	100	101	103	104	105	106	108	109	110	111	8
9	113	114	115	116	118	119	120	121	123	124	9
10	125	127	128	129	130	132	133	134	135	137	10
11	138	139	140	142	143	144	145	147	148	149	11
12	150	152	153	154	155	157	158	159	160	162	12
13	163	164	165	167	168	169	170	172	173	174	13
14	175	177	178	179	180	182	183	184	185	187	14
15	188	189	190	192	193	194	195	197	198	199	15
16	200	202	203	204	205	207	208	209	210	212	16
17	213	214	215	217	218	219	220	222	223	224	17
18	225	227	228	229	230	232	233	234	235	237	18
19	238	239	240	242	243	244	245	247	248	249	19
20	250	252	253	254	256	257	258	259	261	262	20
%	0.0	0.1	0.2	0.3	0.4	0.5	0.6	0.7	0.8	0.9	%

1 %	
0.01	0.1
2	0.2
3	0.3
4	0.4
5	0.5
6	0.6
7	0.7
8	0.8
9	0.9

2 %	
0.01	0.2
2	0.4
3	0.6
4	0.8
5	1.0
6	1.2
7	1.4
8	1.6
9	1.8

$1/2 Fe_2O_3$ 79.84

Weight % 20.0—40.9

%	0.0	0.1	0.2	0.3	0.4	0.5	0.6	0.7	0.8	0.9	%
20	250	252	253	254	256	257	258	259	261	262	20
21	263	264	266	267	268	269	271	272	273	274	21
22	276	277	278	279	281	282	283	284	286	287	22
23	288	289	291	292	293	294	296	297	298	299	23
24	301	302	303	304	306	307	308	309	311	312	24
25	313	314	316	317	318	319	321	322	323	324	25
26	326	327	328	329	331	332	333	334	336	337	26
27	338	339	341	342	343	344	346	347	348	349	27
28	351	352	353	354	356	357	358	359	361	362	28
29	363	364	366	367	368	369	371	372	373	374	29
30	376	377	378	380	381	382	383	385	386	387	30
31	388	390	391	392	393	395	396	397	398	400	31
32	401	402	403	405	406	407	408	410	411	412	32
33	413	415	416	417	418	420	421	422	423	425	33
34	426	427	428	430	431	432	433	435	436	437	34
35	438	440	441	442	443	445	446	447	448	450	35
36	451	452	453	455	456	457	458	460	461	462	36
37	463	465	466	467	468	470	471	472	473	475	37
38	476	477	478	480	481	482	483	485	486	487	38
39	488	490	491	492	493	495	496	497	498	500	39
40	501	502	504	505	506	507	509	510	511	512	40
%	0.0	0.1	0.2	0.3	0.4	0.5	0.6	0.7	0.8	0.9	%

1	
%	
0.01	0.1
2	0.2
3	0.3
4	0.4
5	0.5
6	0.6
7	0.7
8	0.8
9	0.9

2	
%	
0.01	0.2
2	0.4
3	0.6
4	0.8
5	1.0
6	1.2
7	1.4
8	1.6
9	1.8

FeO 71.84

Weight % 0.0—20.9

%	0.0	0.1	0.2	0.3	0.4	0.5	0.6	0.7	0.8	0.9	%
0	0	1	3	4	6	7	8	10	11	13	0
1	14	15	17	18	19	21	22	24	25	26	1
2	28	29	31	32	33	35	36	38	39	40	2
3	42	43	45	46	47	49	50	52	53	54	3
4	56	57	58	60	61	63	64	65	67	68	4
5	70	71	72	74	75	77	78	79	81	82	5
6	84	85	86	88	89	90	92	93	95	96	6
7	97	99	100	102	103	104	106	107	109	110	7
8	111	113	114	116	117	118	120	121	122	124	8
9	125	127	128	129	131	132	134	135	136	138	9
10	139	141	142	143	145	146	148	149	150	152	10
11	153	155	156	157	159	160	161	163	164	166	11
12	167	168	170	171	173	174	175	177	178	180	12
13	181	182	184	185	187	188	189	191	192	193	13
14	195	196	198	199	200	202	203	205	206	207	14
15	209	210	212	213	214	216	217	219	220	221	15
16	223	224	225	227	228	230	231	232	234	235	16
17	237	238	239	241	242	244	245	246	248	249	17
18	251	252	253	255	256	258	259	260	262	263	18
19	264	266	267	269	270	271	273	274	276	277	19
20	278	280	281	283	284	285	287	288	290	291	20
%	0.0	0.1	0.2	0.3	0.4	0.5	0.6	0.7	0.8	0.9	%

1 %	
0.01	0.1
2	0.2
3	0.3
4	0.4
5	0.5
6	0.6
7	0.7
8	0.8
9	0.9

2 %	
0.01	0.2
2	0.4
3	0.6
4	0.8
5	1.0
6	1.2
7	1.4
8	1.6
9	1.8

FeO 71.84

Weight % 20.0—40.9

%	0.0	0.1	0.2	0.3	0.4	0.5	0.6	0.7	0.8	0.9	%
20	278	280	281	283	284	285	287	288	290	291	20
21	292	294	295	296	298	299	301	302	303	305	21
22	306	308	309	310	312	313	315	316	317	319	22
23	320	322	323	324	326	327	328	330	331	333	23
24	334	335	337	338	340	341	342	344	345	347	24
25	348	349	351	352	354	355	356	358	359	361	25
26	362	363	365	366	367	369	370	372	373	374	26
27	376	377	379	380	381	383	384	386	387	388	27
28	390	391	393	394	395	397	398	399	401	402	28
29	404	405	406	408	409	411	412	413	415	416	29
30	418	419	420	422	423	425	426	427	429	430	30
31	431	433	434	436	437	438	440	441	443	444	31
32	445	447	448	450	451	452	454	455	457	458	32
33	459	461	462	464	465	466	468	469	470	472	33
34	473	475	476	477	479	480	482	483	484	486	34
35	487	489	490	491	493	494	496	497	498	500	35
36	501	502	504	505	507	508	509	511	512	514	36
37	515	516	518	519	521	522	523	525	526	528	37
38	529	530	532	533	534	536	537	539	540	541	38
39	543	544	546	547	548	550	551	553	554	555	39
40	557	558	560	561	562	564	565	567	568	569	40
%	0.0	0.1	0.2	0.3	0.4	0.5	0.6	0.7	0.8	0.9	%

1 %	
0,01	0.1
2	0.2
3	0.3
4	0.4
5	0.5
6	0.6
7	0.7
8	0.8
9	0.9

2 %	
0,01	0.2
2	0.4
3	0.6
4	0.8
5	1.0
6	1.2
7	1.4
8	1.6
9	1.8

MnO 70.93

Weight % 0.0—20.9

%	0.0	0.1	0.2	0.3	0.4	0.5	0.6	0.7	0.8	0.9	%
0	0	1	3	4	6	7	8	10	11	13	0
1	14	16	17	18	20	21	23	24	25	27	1
2	28	30	31	32	34	35	37	38	39	41	2
3	42	44	45	47	48	49	51	52	54	55	3
4	56	58	59	61	62	63	65	66	68	69	4
5	70	72	73	75	76	78	79	80	82	83	5
6	85	86	87	89	90	92	93	94	96	97	6
7	99	100	102	103	104	106	107	109	110	111	7
8	113	114	116	117	118	120	121	123	124	125	8
9	127	128	130	131	133	134	135	137	138	140	9
10	141	142	144	145	147	148	149	151	152	154	10
11	155	156	158	159	161	162	164	165	166	168	11
12	169	171	172	173	175	176	178	179	180	182	12
13	183	185	186	187	189	190	192	193	195	196	13
14	197	199	200	202	203	204	206	207	209	210	14
15	211	213	214	216	217	219	220	221	223	224	15
16	226	227	228	230	231	233	234	235	237	238	16
17	240	241	242	244	245	247	248	250	251	252	17
18	254	255	257	258	259	261	262	264	265	266	18
19	268	269	271	272	273	275	276	278	279	281	19
20	282	283	285	286	288	289	290	292	293	295	20
%	0.0	0.1	0.2	0.3	0.4	0.5	0.6	0.7	0.8	0.9	%

1 %	
0.01	0.1
2	0.2
3	0.3
4	0.4
5	0.5
6	0.6
7	0.7
8	0.8
9	0.9

2 %	
0.01	0.2
2	0.4
3	0.6
4	0.8
5	1.0
6	1.2
7	1.4
8	1.6
9	1.8

MgO 40.32

Weight % 0.0—20.9

%	0.0	0.1	0.2	0.3	0.4	0.5	0.6	0.7	0.8	0.9	%
0	0	2	5	7	10	12	15	17	20	22	0
1	25	27	30	32	35	37	40	42	45	47	1
2	50	52	55	57	60	62	64	67	69	72	2
3	74	77	79	82	84	87	89	92	94	97	3
4	99	102	104	107	109	112	114	117	119	122	4
5	124	126	129	131	134	136	139	141	144	146	5
6	149	151	154	156	159	161	164	166	169	171	6
7	174	176	179	181	184	186	188	191	193	196	7
8	198	201	203	206	208	211	213	216	218	221	8
9	223	226	228	231	233	236	238	241	243	246	9
10	248	250	253	255	258	260	263	265	268	270	10
11	273	275	278	280	283	285	288	290	293	295	11
12	298	300	303	305	308	310	312	315	317	320	12
13	322	325	327	330	332	335	337	340	342	345	13
14	347	350	352	355	357	360	362	365	367	370	14
15	372	375	377	379	382	384	387	389	392	394	15
16	397	399	402	404	407	409	412	414	417	419	16
17	422	424	427	429	432	434	436	439	441	444	17
18	446	449	451	454	456	459	461	464	466	469	18
19	471	474	476	479	481	484	486	489	491	494	19
20	496	499	501	503	506	508	511	513	516	518	20
%	0.0	0.1	0.2	0.3	0.4	0.5	0.6	0.7	0.8	0.9	%

2
%	
0.01	0.2
2	0.4
3	0.6
4	0.8
5	1.0
6	1.2
7	1.4
8	1.6
9	1.8

3
%	
0.01	0.3
2	0.6
3	0.9
4	1.2
5	1.5
6	1.8
7	2.1
8	2.4
9	2.7

MgO 40.32

Weight % 20.0—40.9

%	0.0	0.1	0.2	0.3	0.4	0.5	0.6	0.7	0.8	0.9	%
20	496	499	501	503	506	508	511	513	516	518	20
21	521	523	526	528	531	533	536	538	541	543	21
22	546	548	551	553	556	558	561	563	565	568	22
23	570	573	575	578	580	583	585	588	590	593	23
24	595	598	600	603	605	608	610	613	615	618	24
25	620	623	625	627	630	632	635	637	640	642	25
26	645	647	650	652	655	657	660	662	665	667	26
27	670	672	675	677	680	682	685	687	689	692	27
28	694	697	699	702	704	707	709	712	714	717	28
29	719	722	724	727	729	732	734	737	739	742	29
30	744	747	749	751	754	756	759	761	764	766	30
31	769	771	774	776	779	781	784	786	789	791	31
32	794	796	799	801	804	806	809	811	813	816	32
33	818	821	823	826	828	831	833	836	838	841	33
34	843	846	848	851	853	856	858	861	863	866	34
35	868	871	873	875	878	880	883	885	888	890	35
36	893	895	898	900	903	905	908	910	913	915	36
37	918	920	923	925	928	930	933	935	937	940	37
38	942	945	947	950	952	955	957	960	962	965	38
39	967	970	972	975	977	980	982	985	987	990	39
40	992	995	997	999	1002	1004	1007	1009	1012	1014	40
%	0.0	0.1	0.2	0.3	0.4	0.5	0.6	0.7	0.8	0.9	%

2 %

0.01	0.2
2	0.4
3	0.6
4	0.8
5	1.0
6	1.2
7	1.4
8	1.6
9	1.8

3 %

0.01	0.3
2	0.6
3	0.9
4	1.2
5	1.5
6	1.8
7	2.1
8	2.4
9	2.7

MgO 40.32

Weight % 40.0—60.9

%	0.0	0.1	0.2	0.3	0.4	0.5	0.6	0.7	0.8	0.9	%
40	992	995	997	999	1002	1004	1007	1009	1012	1014	40
41	1017	1019	1022	1024	1027	1029	1032	1034	1037	1039	41
42	1042	1044	1047	1049	1052	1054	1057	1059	1061	1064	42
43	1066	1069	1071	1074	1076	1079	1081	1084	1086	1089	43
44	1091	1094	1096	1099	1101	1104	1106	1109	1111	1114	44
45	1116	1119	1121	1123	1126	1128	1131	1133	1136	1138	45
46	1141	1143	1146	1148	1151	1153	1156	1158	1161	1163	46
47	1166	1168	1171	1173	1176	1178	1181	1183	1185	1188	47
48	1190	1193	1195	1198	1200	1203	1205	1208	1210	1213	48
49	1215	1218	1220	1223	1225	1228	1230	1233	1235	1238	49
50	1240	1243	1245	1247	1250	1252	1255	1257	1260	1262	50
51	1265	1267	1270	1272	1275	1277	1280	1282	1285	1287	51
52	1290	1292	1295	1297	1300	1302	1305	1307	1309	1312	52
53	1314	1317	1319	1322	1324	1327	1329	1332	1334	1337	53
54	1339	1342	1344	1347	1349	1352	1354	1357	1359	1362	54
55	1364	1367	1369	1371	1374	1376	1379	1381	1384	1386	55
56	1389	1391	1394	1396	1399	1401	1404	1406	1409	1411	56
57	1414	1416	1419	1421	1424	1426	1429	1431	1433	1436	57
58	1438	1441	1443	1446	1448	1451	1453	1456	1458	1461	58
59	1463	1466	1468	1471	1473	1476	1478	1481	1483	1486	59
60	1488	1491	1493	1496	1498	1500	1503	1505	1508	1510	60
%	0.0	0.1	0.2	0.3	0.4	0.5	0.6	0.7	0.8	0.9	%

2 %	
0.01	0.2
2	0.4
3	0.6
4	0.8
5	1.0
6	1.2
7	1.4
8	1.6
9	1.8

3 %	
0.01	0.3
2	0.6
3	0.9
4	1.2
5	1.5
6	1.8
7	2.1
8	2.4
9	2.7

CaO 56.08

Weight % 0.0—20.9

%	0.0	0.1	0.2	0.3	0.4	0.5	0.6	0.7	0.8	0.9	%
0	0	2	4	5	7	9	11	12	14	16	0
1	18	20	21	23	25	27	29	30	32	34	1
2	36	37	39	41	43	45	46	48	50	52	2
3	53	55	57	59	61	62	64	66	68	70	3
4	71	73	75	77	78	80	82	84	86	87	4
5	89	91	93	95	96	98	100	102	103	105	5
6	107	109	111	112	114	116	118	119	121	123	6
7	125	127	128	130	132	134	136	137	139	141	7
8	143	144	146	148	150	152	153	155	157	159	8
9	160	162	164	166	168	169	171	173	175	177	9
10	178	180	182	184	185	187	189	191	193	194	10
11	196	198	200	201	203	205	207	209	210	212	11
12	214	216	218	219	221	223	225	226	228	230	12
13	232	234	235	237	239	241	243	244	246	248	13
14	250	251	253	255	257	259	260	262	264	266	14
15	267	269	271	273	275	276	278	280	282	284	15
16	285	287	289	291	292	294	296	298	300	301	16
17	303	305	307	308	310	312	314	316	317	319	17
18	321	323	325	326	328	330	332	333	335	337	18
19	339	341	342	344	346	348	349	351	353	355	19
20	357	358	360	362	364	366	367	369	371	373	20
%	0.0	0.1	0.2	0.3	0.4	0.5	0.6	0.7	0.8	0.9	%

1

%	
0.01	0.1
2	0.2
3	0.3
4	0.4
5	0.5
6	0.6
7	0.7
8	0.8
9	0.9

2

%	
0.01	0.2
2	0.4
3	0.6
4	0.8
5	1.0
6	1.2
7	1.4
8	1.6
9	1.8

CaO 56.08

Weight % 20.0—40.9

%	0.0	0.1	0.2	0.3	0.4	0.5	0.6	0.7	0.8	0.9	%
20	357	358	360	362	364	366	367	369	371	373	20
21	374	376	378	380	382	383	385	387	389	390	21
22	392	394	396	398	399	401	403	405	407	408	22
23	410	412	414	415	417	419	421	423	424	426	23
24	428	430	432	433	435	437	439	440	442	444	24
25	446	448	449	451	453	455	456	458	460	462	25
26	464	465	467	469	471	473	474	476	478	480	26
27	481	483	485	487	489	490	492	494	496	497	27
28	499	501	503	505	506	508	510	512	514	515	28
29	517	519	521	522	524	526	528	530	531	533	29
30	535	537	538	540	542	544	546	547	549	551	30
31	553	555	556	558	560	562	563	565	567	569	31
32	571	572	574	576	578	580	581	583	585	587	32
33	588	590	592	594	596	597	599	601	603	604	33
34	606	608	610	612	613	615	617	619	621	622	34
35	624	626	628	629	631	633	635	637	638	640	35
36	642	644	645	647	649	651	653	654	656	658	36
37	660	662	663	665	667	669	671	672	674	676	37
38	678	680	681	683	685	687	688	690	692	694	38
39	696	697	699	701	703	704	706	708	710	711	39
40	713	715	717	719	720	722	724	726	728	729	40
%	0.0	0.1	0.2	0.3	0.4	0.5	0.6	0.7	0.8	0.9	%

1

%	
0.01	0.1
2	0.2
3	0.3
4	0.4
5	0.5
6	0.6
7	0.7
8	0.8
9	0.9

2

%	
0.01	0.2
2	0.4
3	0.6
4	0.8
5	1.0
6	1.2
7	1.4
8	1.6
9	1.8

CaO 56.08

Weight % 40.0—60.9

%	0.0	0.1	0.2	0.3	0.4	0.5	0.6	0.7	0.8	0.9	%
40	713	715	717	719	720	722	724	726	728	729	40
41	731	733	735	736	738	740	742	744	745	747	41
42	749	751	752	754	756	758	760	761	763	765	42
43	767	769	770	772	774	776	777	779	781	783	43
44	785	786	788	790	792	793	795	797	799	801	44
45	802	804	806	808	810	811	813	815	817	818	45
46	820	822	824	826	827	829	831	833	834	836	46
47	838	840	842	843	845	847	849	851	852	854	47
48	856	858	859	861	863	865	867	868	870	872	48
49	874	876	877	879	881	883	884	886	888	890	49
50	892	893	895	897	899	900	902	904	906	908	50
51	909	911	913	915	917	918	920	922	924	925	51
52	927	929	931	933	934	936	938	940	941	943	52
53	945	947	949	950	952	954	956	958	959	961	53
54	963	965	966	968	970	972	974	975	977	979	54
55	981	982	984	986	988	990	991	993	995	997	55
56	999	1000	1002	1004	1006	1007	1009	1011	1013	1015	56
57	1016	1018	1020	1022	1023	1025	1027	1029	1031	1032	57
58	1034	1036	1038	1040	1041	1043	1045	1047	1048	1050	58
59	1052	1054	1056	1057	1059	1061	1063	1065	1066	1068	59
60	1070	1072	1074	1075	1077	1079	1081	1082	1084	1086	60
%	0.0	0.1	0.2	0.3	0.4	0.5	0.6	0.7	0.8	0.9	%

1

%	
0.01	0.1
2	0.2
3	0.3
4	0.4
5	0.5
6	0.6
7	0.7
8	0.8
9	0.9

2

%	
0.01	0.2
2	0.4
3	0.6
4	0.8
5	1.0
6	1.2
7	1.4
8	1.6
9	1.8

Na₂O 61.994

Weight % 0.0—20.9

%	0.0	0.1	0.2	0.3	0.4	0.5	0.6	0.7	0.8	0.9	%
0	0	2	3	5	6	8	10	11	13	15	0
1	16	18	19	21	23	24	26	27	29	31	1
2	32	34	35	37	39	40	42	44	45	47	2
3	48	50	52	53	55	56	58	60	61	63	3
4	65	66	68	69	71	73	74	76	77	79	4
5	81	82	84	85	87	89	90	92	94	95	5
6	97	98	100	102	103	105	106	108	110	111	6
7	113	115	116	118	119	121	123	124	126	127	7
8	129	131	132	134	135	137	139	140	142	144	8
9	145	147	148	150	152	153	155	156	158	160	9
10	161	163	165	166	168	169	171	173	174	176	10
11	177	179	181	182	184	185	187	189	190	192	11
12	194	195	197	198	200	202	203	205	206	208	12
13	210	211	213	215	216	218	219	221	223	224	13
14	226	227	229	231	232	234	235	237	239	240	14
15	242	244	245	247	248	250	252	253	255	256	15
16	258	260	261	263	265	266	268	269	271	273	16
17	274	276	277	279	281	282	284	286	287	289	17
18	290	292	294	295	297	298	300	302	303	305	18
19	306	308	310	311	313	315	316	318	319	321	19
20	323	324	326	327	329	331	332	334	336	337	20
%	0.0	0.1	0.2	0.3	0.4	0.5	0.6	0.7	0.8	0.9	%

1 %	
0.01	0.1
2	0.2
3	0.3
4	0.4
5	0.5
6	0.6
7	0.7
8	0.8
9	0.9

2 %	
0.01	0.2
2	0.4
3	0.6
4	0.8
5	1.0
6	1.2
7	1.4
8	1.6
9	1.8

½Na₂O 30.997

Weight % 0.0—20.9

%	0.0	0.1	0.2	0.3	0.4	0.5	0.6	0.7	0.8	0.9	%
0	0	3	6	10	13	16	19	23	26	29	0
1	32	35	39	42	45	48	52	55	58	61	1
2	65	68	71	74	77	81	84	87	90	94	2
3	97	100	103	106	110	113	116	119	123	126	3
4	129	132	135	139	142	145	148	152	155	158	4
5	161	165	168	171	174	177	181	184	187	190	5
6	194	197	200	203	206	210	213	216	219	223	6
7	226	229	232	235	239	242	245	248	252	255	7
8	258	261	265	268	271	274	277	281	284	287	8
9	290	294	297	300	303	306	310	313	316	319	9
10	323	326	329	332	336	339	342	345	348	352	10
11	355	358	361	365	368	371	374	377	381	384	11
12	387	390	394	397	400	403	406	410	413	416	12
13	419	423	426	429	432	436	439	442	445	448	13
14	452	455	458	461	465	468	471	474	477	481	14
15	484	487	490	494	497	500	503	506	510	513	15
16	516	519	523	526	529	532	536	539	542	545	16
17	548	552	555	558	561	565	568	571	574	577	17
18	581	584	587	590	594	597	600	603	606	610	18
19	613	616	619	623	626	629	632	636	639	642	19
20	645	648	652	655	658	661	665	668	671	674	20
%	0.0	0.1	0.2	0.3	0.4	0.5	0.6	0.7	0.8	0.9	%

3

%	
0.01	0.3
2	0.6
3	0.9
4	1.2
5	1.5
6	1.8
7	2.1
8	2.4
9	2.7

4

%	
0.01	0.4
2	0.8
3	1.2
4	1.6
5	2.0
6	2.4
7	2.8
8	3.2
9	3.6

K₂O 94.192

Weight % 0.0—20.9

%	0.0	0.1	0.2	0.3	0.4	0.5	0.6	0.7	0.8	0.9	%
0	0	1	2	3	4	5	6	7	8	10	0
1	11	12	13	14	15	16	17	18	19	20	1
2	21	22	23	24	25	27	28	29	30	31	2
3	32	33	34	35	36	37	38	39	40	41	3
4	42	44	45	46	47	48	49	50	51	52	4
5	53	54	55	56	57	58	59	60	62	63	5
6	64	65	66	67	68	69	70	71	72	73	6
7	74	75	76	77	79	80	81	82	83	84	7
8	85	86	87	88	89	90	91	92	93	94	8
9	95	97	98	99	100	101	102	103	104	105	9
10	106	107	108	109	110	111	112	114	115	116	10
11	117	118	119	120	121	122	123	124	125	126	11
12	127	128	129	131	132	133	134	135	136	137	12
13	138	139	140	141	142	143	144	145	146	147	13
14	149	150	151	152	153	154	155	156	157	158	14
15	159	160	161	162	163	164	166	167	168	169	15
16	170	171	172	173	174	175	176	177	178	179	16
17	180	181	182	184	185	186	187	188	189	190	17
18	191	192	193	194	195	196	197	198	199	201	18
19	202	203	204	205	206	207	208	209	210	211	19
20	212	213	214	215	216	218	219	220	221	222	20
%	0.0	0.1	0.2	0.3	0.4	0.5	0.6	0.7	0.8	0.9	%

1	
%	
0.01	0.1
2	0.2
3	0.3
4	0.4
5	0.5
6	0.6
7	0.7
8	0.8
9	0.9

2	
%	
0.01	0.2
2	0.4
3	0.6
4	0.8
5	1.0
6	1.2
7	1.4
8	1.6
9	1.8

½K₂O 47.10

Weight % 0.0—20.9

%	0.0	0.1	0.2	0.3	0.4	0.5	0.6	0.7	0.8	0.9	%
0	0	2	4	6	8	11	13	15	17	19	0
1	21	23	25	28	30	32	34	36	38	40	1
2	42	45	47	49	51	53	55	57	59	62	2
3	64	66	68	70	72	74	76	79	81	83	3
4	85	87	89	91	93	96	98	100	102	104	4
5	106	108	110	113	115	117	119	121	123	125	5
6	127	130	132	134	136	138	140	142	144	146	6
7	149	151	153	155	157	159	161	163	166	168	7
8	170	172	174	176	178	180	183	185	187	189	8
9	191	193	195	197	200	202	204	206	208	210	9
10	212	214	217	219	221	223	225	227	229	231	10
11	234	236	238	240	242	244	246	248	251	253	11
12	255	257	259	261	263	265	267	270	272	274	12
13	276	278	280	282	284	287	289	291	293	295	13
14	297	299	301	304	306	308	310	312	314	316	14
15	318	321	323	325	327	329	331	333	335	338	15
16	340	342	344	346	348	350	352	355	357	359	16
17	361	363	365	367	369	372	374	376	378	380	17
18	382	384	386	389	391	393	395	397	399	401	18
19	403	405	408	410	412	414	416	418	420	422	19
20	425	427	429	431	433	435	437	439	442	444	20
%	0.0	0.1	0.2	0.3	0.4	0.5	0.6	0.7	0.8	0.9	%

2 %	
0,01	0.2
2	0.4
3	0.6
4	0.8
5	1.0
6	1.2
7	1.4
8	1.6
9	1.8

3 %	
0,01	0.3
2	0.6
3	0.9
4	1.2
5	1.5
6	1.8
7	2.1
8	2.4
9	2.7

TiO$_2$ 79.90

Weight % 0.0—20.9

%	0.0	0.1	0.2	0.3	0.4	0.5	0.6	0.7	0.8	0.9	%
0	0	1	3	4	5	6	8	9	10	11	0
1	13	14	15	16	18	19	20	21	23	24	1
2	25	26	28	29	30	31	33	34	35	36	2
3	38	39	40	41	43	44	45	46	48	49	3
4	50	51	53	54	55	56	58	59	60	61	4
5	63	64	65	66	68	69	70	71	73	74	5
6	75	76	78	79	80	81	83	84	85	86	6
7	88	89	90	91	93	94	95	96	98	99	7
8	100	101	103	104	105	106	108	109	110	111	8
9	113	114	115	116	118	119	120	121	123	124	9
10	125	126	128	129	130	131	133	134	135	136	10
11	138	139	140	141	143	144	145	146	148	149	11
12	150	151	153	154	155	156	158	159	160	161	12
13	163	164	165	166	168	169	170	171	173	174	13
14	175	176	178	179	180	181	183	184	185	186	14
15	188	189	190	191	193	194	195	196	198	199	15
16	200	201	203	204	205	206	208	209	210	212	16
17	213	214	215	217	218	219	220	222	223	224	17
18	225	227	228	229	230	232	233	234	235	237	18
19	238	239	240	242	243	244	245	247	248	249	19
20	250	252	253	254	255	257	258	259	260	262	20
%	0.0	0.1	0.2	0.3	0.4	0.5	0.6	0.7	0.8	0.9	%

1
%	
0.01	0.1
2	0.2
3	0.3
4	0.4
5	0.5
6	0.6
7	0.7
8	0.8
9	0.9

2
%	
0.01	0.2
2	0.4
3	0.6
4	0.8
5	1.0
6	1.2
7	1.4
8	1.6
9	1.8

P₂O₅ 141.96

Weight .% 0.0—20.9

%	0.0	0.1	0.2	0.3	0.4	0.5	0.6	0.7	0.8	0.9	%
0	0	1	1	2	3	4	4	5	6	6	0
1	7	8	8	9	10	11	11	12	13	13	1
2	14	15	15	16	17	18	18	19	20	20	2
3	21	22	23	23	24	25	25	26	27	27	3
4	28	29	30	30	31	32	32	33	34	35	4
5	35	36	37	37	38	39	39	40	41	42	5
6	42	43	44	44	45	46	46	47	48	49	6
7	49	50	51	51	52	53	54	54	55	56	7
8	56	57	58	58	59	60	61	61	62	63	8
9	63	64	65	66	66	67	68	68	69	70	9
10	70	71	72	73	73	74	75	75	76	77	10
11	77	78	79	80	80	81	82	82	83	84	11
12	85	85	86	87	87	88	89	89	90	91	12
13	92	92	93	94	94	95	96	97	97	98	13
14	99	99	100	101	101	102	103	104	104	105	14
15	106	106	107	108	108	109	110	111	111	112	15
16	113	113	114	115	116	116	117	118	118	119	16
17	120	120	121	122	123	123	124	125	125	126	17
18	127	127	128	129	130	130	131	132	132	133	18
19	134	135	135	136	137	137	138	139	139	140	19
20	141	142	142	143	144	144	145	146	147	147	20
%	0.0	0.1	0.2	0.3	0.4	0.5	0.6	0.7	0.8	0.9	%

1 %	
0.01	0.1
2	0.2
3	0.3
4	0.4
5	0.5
6	0.6
7	0.7
8	0.8
9	0.9

$1/2 P_2O_5$ 70.98

Weight % 0.0—20.9

%	0.0	0.1	0.2	0.3	0.4	0.5	0.6	0.7	0.8	0.9	%
0	0	1	3	4	6	7	8	10	11	13	0
1	14	15	17	18	20	21	23	24	25	27	1
2	28	30	31	32	34	35	37	38	39	41	2
3	42	44	45	46	48	49	51	52	54	55	3
4	56	58	59	61	62	63	65	66	68	69	4
5	70	72	73	75	76	77	79	80	82	83	5
6	85	86	87	89	90	92	93	94	96	97	6
7	99	100	101	103	104	106	107	108	110	111	7
8	113	114	116	117	118	120	121	123	124	125	8
9	127	128	130	131	132	134	135	137	138	139	9
10	141	142	144	145	147	148	149	151	152	154	10
11	155	156	158	159	161	162	163	165	166	168	11
12	169	170	172	173	175	176	178	179	180	182	12
13	183	185	186	187	189	190	192	193	194	196	13
14	197	199	200	201	203	204	206	207	209	210	14
15	211	213	214	216	217	218	220	221	223	224	15
16	225	227	228	230	231	232	234	235	237	238	16
17	239	241	242	244	245	247	248	249	251	252	17
18	254	255	256	258	259	261	262	263	265	266	18
19	268	269	270	272	273	275	276	278	279	280	19
20	282	283	285	286	287	289	290	292	293	294	20
%	0.0	0.1	0.2	0.3	0.4	0.5	0.6	0.7	0.8	0.9	%

1

%	
0.01	0.1
2	0.2
3	0.3
4	0.4
5	0.5
6	0.6
7	0.7
8	0.8
9	0.9

2

%	
0.01	0.2
2	0.4
3	0.6
4	0.8
5	1.0
6	1.2
7	1.4
8	1.6
9	1.8

CO₂ 44.01

Weight % 0.0—20.9

%	0.0	0.1	0.2	0.3	0.4	0.5	0.6	0.7	0.8	0.9	%
0	0	2	5	7	9	11	14	16	18	20	0
1	23	25	27	30	32	34	36	39	41	43	1
2	45	48	50	52	55	57	59	61	64	66	2
3	68	70	73	75	77	80	82	84	86	89	3
4	91	93	95	98	100	102	105	107	109	111	4
5	114	116	118	120	123	125	127	130	132	134	5
6	136	139	141	143	145	148	150	152	155	157	6
7	159	161	164	166	168	170	173	175	177	180	7
8	182	184	186	189	191	193	195	198	200	202	8
9	204	207	209	211	214	216	218	220	223	225	9
10	227	229	232	234	236	239	241	243	245	248	10
11	250	252	254	257	259	261	264	266	268	270	11
12	273	275	277	279	282	284	286	289	291	293	12
13	295	298	300	302	304	307	309	311	314	316	13
14	318	320	323	325	327	329	332	334	336	339	14
15	341	343	345	348	350	352	354	357	359	361	15
16	364	366	368	370	373	375	377	379	382	384	16
17	386	389	391	393	395	398	400	402	404	407	17
18	409	411	413	416	418	420	423	425	427	429	18
19	432	434	436	439	441	443	445	448	450	452	19
20	454	457	459	461	464	466	468	470	473	475	20
%	0.0	0.1	0.2	0.3	0.4	0.5	0.6	0.7	0.8	0.9	%

2 %	
0.01	0.2
2	0.4
3	0.6
4	0.8
5	1.0
6	1.2
7	1.4
8	1.6
9	1.8

3 %	
0.01	0.3
2	0.6
3	0.9
4	1.2
5	1.5
6	1.8
7	2.1
8	2.4
9	2.7

CO_2 44.01

Weight % 20.0—40.9

%	0.0	0.1	0.2	0.3	0.4	0.5	0.6	0.7	0.8	0.9	%
20	454	457	459	461	464	466	468	470	473	475	20
21	477	479	482	484	486	489	491	493	495	498	21
22	500	502	504	507	509	511	514	516	518	520	22
23	523	525	527	529	532	534	536	539	541	543	23
24	545	548	550	552	554	557	559	561	564	566	24
25	568	570	573	575	577	579	582	584	586	588	25
26	591	593	595	598	600	602	604	607	609	611	26
27	613	616	618	620	623	625	627	629	632	634	27
28	636	638	641	643	645	648	650	652	654	657	28
29	659	661	663	666	668	670	673	675	677	679	29
30	682	684	686	688	691	693	695	698	700	702	30
31	704	707	709	711	713	716	718	720	723	725	31
32	727	729	732	734	736	738	741	743	745	748	32
33	750	752	754	757	759	761	763	766	768	770	33
34	773	775	777	779	782	784	786	788	791	793	34
35	795	798	800	802	804	807	809	811	813	816	35
36	818	820	823	825	827	829	832	834	836	838	36
37	841	843	845	848	850	852	854	857	859	861	37
38	863	866	868	870	872	875	877	879	882	884	38
39	886	888	891	893	895	898	900	902	904	907	39
40	909	911	913	916	918	920	923	925	927	929	40
%	0.0	0.1	0.2	0.3	0.4	0.5	0.6	0.7	0.8	0.9	%

2

%	
0.01	0.2
2	0.4
3	0.6
4	0.8
5	1.0
6	1.2
7	1.4
8	1.6
9	1.8

3

%	
0.01	0.3
2	0.6
3	0.9
4	1.2
5	1.5
6	1.8
7	2.1
8	2.4
9	2.7

H_2O 18.016

Weight % 0.0—20.9

%	0.0	0.1	0.2	0.3	0.4	0.5	0.6	0.7	0.8	0.9	%
0	0	6	11	17	22	28	33	39	44	50	0
1	56	61	67	72	78	83	89	94	100	105	1
2	111	117	122	128	133	139	144	150	155	161	2
3	166	172	178	183	189	194	200	205	211	216	3
4	222	228	233	239	244	250	255	261	266	272	4
5	278	283	289	294	300	305	311	316	322	327	5
6	333	339	344	350	355	361	366	372	377	383	6
7	388	394	400	405	411	416	422	427	433	438	7
8	444	450	455	461	466	472	477	483	488	494	8
9	500	505	511	516	522	527	533	538	544	549	9
10	555	561	566	572	577	583	588	594	599	605	10
11	610	616	622	627	633	638	644	649	655	660	11
12	666	672	677	683	688	694	699	705	710	716	12
13	722	727	733	738	744	749	755	760	766	771	13
14	777	783	788	794	799	805	810	816	821	827	14
15	832	838	844	849	855	860	866	871	877	882	15
16	888	894	899	905	910	916	921	927	932	938	16
17	944	949	955	960	966	971	977	982	988	993	17
18	999	1005	1010	1016	1021	1027	1032	1038	1043	1049	18
19	1054	1060	1066	1071	1077	1082	1088	1093	1099	1104	19
20	1110	1116	1121	1127	1132	1138	1143	1149	1154	1160	20
%	0.0	0.1	0.2	0.3	0.4	0.5	0.6	0.7	0.8	0.9	%

5 %

%	
0.01	0.5
2	1.0
3	1.5
4	2.0
5	2.5
6	3.0
7	3.5
8	4.0
9	4.5

6 %

%	
0.01	0.6
2	1.2
3	1.8
4	2.4
5	3.0
6	3.6
7	4.2
8	4.8
9	5.4

SO₃ 80.06

Weight % 0.0—20.9

%	0.0	0.1	0.2	0.3	0.4	0.5	0.6	0.7	0.8	0.9	%
0	0	1	2	4	5	6	7	9	10	11	0
1	12	14	15	16	17	19	20	21	22	24	1
2	25	26	27	29	30	31	32	34	35	36	2
3	37	39	40	41	42	44	45	46	47	49	3
4	50	51	52	54	55	56	57	59	60	61	4
5	62	64	65	66	67	69	70	71	72	74	5
6	75	76	77	79	80	81	82	84	85	86	6
7	87	89	90	91	92	94	95	96	97	99	7
8	100	101	102	104	105	106	107	109	110	111	8
9	112	114	115	116	117	119	120	121	122	124	9
10	125	126	127	129	130	131	132	134	135	136	10
11	137	139	140	141	142	144	145	146	147	149	11
12	150	151	152	154	155	156	157	159	160	161	12
13	162	164	165	166	167	169	170	171	172	174	13
14	175	176	177	179	180	181	182	184	185	186	14
15	187	189	190	191	192	194	195	196	197	199	15
16	200	201	202	204	205	206	207	209	210	211	16
17	212	214	215	216	217	219	220	221	222	224	17
18	225	226	227	229	230	231	232	234	235	236	18
19	237	239	240	241	242	244	245	246	247	249	19
20	250	251	252	254	255	256	257	259	260	261	20
%	0.0	0.1	0.2	0.3	0.4	0.5	0.6	0.7	0.8	0.9	%

1
%	
0.01	0.1
2	0.2
3	0.3
4	0.4
5	0.5
6	0.6
7	0.7
8	0.8
9	0.9

2
%	
0,01	0.2
2	0.4
3	0.6
4	0.8
5	1.0
6	1.2
7	1.4
8	1.6
9	1.8

Cl 35.457 Weight % 0.0—20.9

%	0.0	0.1	0.2	0.3	0.4	0.5	0.6	0.7	0.8	0.9	%
0	0	3	6	8	11	14	17	20	23	25	0
1	28	31	34	37	39	42	45	48	51	54	1
2	56	59	62	65	68	70	73	76	79	82	2
3	85	87	90	93	96	99	102	104	107	110	3
4	113	116	118	121	124	127	130	133	135	138	4
5	141	144	147	149	152	155	158	161	164	166	5
6	169	172	175	178	180	183	186	189	192	195	6
7	197	200	203	206	209	212	214	217	220	223	7
8	226	228	231	234	237	240	243	245	248	251	8
9	254	257	259	262	265	268	271	274	276	279	9
10	282	285	288	290	293	296	299	302	305	307	10
11	310	313	316	319	321	324	327	330	333	336	11
12	338	341	344	347	350	352	355	358	361	364	12
13	367	369	372	375	378	381	384	386	389	392	13
14	395	398	400	403	406	409	412	415	417	420	14
15	423	426	429	431	434	437	440	443	446	448	15
16	451	454	457	460	462	465	468	471	474	477	16
17	479	482	485	488	491	493	496	499	502	505	17
18	508	510	513	516	519	522	525	527	530	533	18
19	536	539	541	544	547	550	553	556	558	561	19
20	564	567	570	572	575	578	581	584	587	589	20
%	0.0	0.1	0.2	0.3	0.4	0.5	0.6	0.7	0.8	0.9	%

2
%	
0.01	0.2
2	0.4
3	0.6
4	0.8
5	1.0
6	1.2
7	1.4
8	1.6
9	1.8

3
%	
0.01	0.3
2	0.6
3	0.9
4	1.2
5	1.5
6	1.8
7	2.1
8	2.4
9	2.7

If the equivalent numbers for Cl_2 are desired, the table values should be halved.

F 19.00 Weight % 0.0—10.9

%	0.0	0.1	0.2	0.3	0.4	0.5	0.6	0.7	0.8	0.9	%
0	0	5	11	16	21	26	32	37	42	47	0
1	53	58	63	68	74	79	84	89	95	100	1
2	105	111	116	121	126	132	137	142	147	153	2
3	158	163	168	174	179	184	189	195	200	205	3
4	211	216	221	226	232	237	242	247	253	258	4
5	263	268	274	279	284	289	295	300	305	311	5
6	316	321	326	332	337	342	347	353	358	363	6
7	368	374	379	384	389	395	400	405	411	416	7
8	421	426	432	437	442	447	453	458	463	468	8
9	474	479	484	489	495	500	505	511	516	521	9
10	526	532	537	542	547	553	558	563	568	574	10
%	0.0	0.1	0.2	0.3	0.4	0.5	0.6	0.7	0.8	0.9	%

5
%	
0.01	0.5
2	1.0
3	1.5
4	2.0
5	2.5
6	3.0
7	3.5
8	4.0
9	4.5

If the equivalent numbers for F_2 are desired, the table values should be halved.

S 32.06

Weight % 0.0—10.9

%	0.0	0.1	0.2	0.3	0.4	0.5	0.6	0.7	0.8	0.9	%
0	0	3	6	9	12	16	19	22	25	28	0
1	31	34	37	41	44	47	50	53	56	59	1
2	62	66	69	72	75	78	81	84	87	90	2
3	94	97	100	103	106	109	112	115	119	122	3
4	125	128	131	134	137	140	143	147	150	153	4
5	156	159	162	165	168	172	175	178	181	184	5
6	187	190	193	197	200	203	206	209	212	215	6
7	218	221	225	228	231	234	237	240	243	246	7
8	250	253	256	259	262	265	268	271	274	278	8
9	281	284	287	290	293	296	299	303	306	309	9
10	312	315	318	321	324	328	331	334	337	340	10
%	0.0	0.1	0.2	0.3	0.4	0.5	0.6	0.7	0.8	0.9	%

3 %	
0.01	0.3
2	0.6
3	0.9
4	1.2
5	1.5
6	1.8
7	2.1
8	2.4
9	2.7

½Cr$_2$O$_3$ 76.01

Weight % 0.0—10.9

%	0.0	0.1	0.2	0.3	0.4	0.5	0.6	0.7	0.8	0.9	%
0	0	1	3	4	5	7	8	9	11	12	0
1	13	14	16	17	18	20	21	22	24	25	1
2	26	28	29	30	32	33	34	36	37	38	2
3	39	41	42	43	45	46	47	49	50	51	3
4	53	54	55	57	58	59	61	62	63	64	4
5	66	67	68	70	71	72	74	75	76	78	5
6	79	80	82	83	84	86	87	88	89	91	6
7	92	93	95	96	97	99	100	101	103	104	7
8	105	107	108	109	111	112	113	114	116	117	8
9	118	120	121	122	124	125	126	128	129	130	9
10	132	133	134	136	137	138	139	141	142	143	10
%	0.0	0.1	0.2	0.3	0.4	0.5	0.6	0.7	0.8	0.9	%

1 %	
0.01	0.1
2	0.2
3	0.3
4	0.4
5	0.5
6	0.6
7	0.7
8	0.8
9	0.9

NiO 74.69

Weight % 0.0—5.9

%	0.0	0.1	0.2	0.3	0.4	0.5	0.6	0.7	0.8	0.9	%
0	0	1	3	4	5	7	8	9	11	12	0
1	13	15	16	17	19	20	21	23	24	25	1
2	27	28	29	31	32	33	35	36	37	39	2
3	40	42	43	44	46	47	48	50	51	52	3
4	54	55	56	58	59	60	62	63	64	66	4
5	67	68	70	71	72	74	75	76	78	79	5
%	0.0	0.1	0.2	0.3	0.4	0.5	0.6	0.7	0.8	0.9	%

2 %	
0.01	0.2
2	0.4
3	0.6
4	0.8
5	1.0
6	1.2
7	1.4
8	1.6
9	1.8

BaO 153.36

Weight % 0.0—5.9

%	0.0	0.1	0.2	0.3	0.4	0.5	0.6	0.7	0.8	0.9	%
0	0	1	1	2	3	3	4	5	5	6	0
1	7	7	8	8	9	10	10	11	12	12	1
2	13	14	14	15	16	16	17	18	18	19	2
3	20	20	21	22	22	23	23	24	25	25	3
4	26	27	27	28	29	29	30	31	31	32	4
5	33	33	34	35	35	36	37	37	38	38	5
%	0.0	0.1	0.2	0.3	0.4	0.5	0.6	0.7	0.8	0.9	%

1 %	
0.01	0.1
2	0.2
3	0.3
4	0.4
5	0.5
6	0.6
7	0.7
8	0.8
9	0.9

SrO 103.63

Weight % 0.0—5.9

%	0.0	0.1	0.2	0.3	0.4	0.5	0.6	0.7	0.8	0.9	%
0	0	1	2	3	4	5	6	7	8	9	0
1	10	11	12	13	14	14	15	16	17	18	1
2	19	20	21	22	23	24	25	26	27	28	2
3	29	30	31	32	33	34	35	36	37	38	3
4	39	40	41	41	42	43	44	45	46	47	4
5	48	49	50	51	52	53	54	55	56	57	5
%	0.0	0.1	0.2	0.3	0.4	0.5	0.6	0.7	0.8	0.9	%

½Li₂O 14.94

Weight % 0.0—5.9

%	0.0	0.1	0.2	0.3	0.4	0.5	0.6	0.7	0.8	0.9	%
0	0	7	13	20	27	33	40	47	54	60	0
1	67	74	80	87	94	100	107	114	120	127	1
2	134	141	147	154	161	167	174	181	187	194	2
3	201	207	214	221	228	234	241	248	254	261	3
4	268	274	281	288	295	301	308	315	321	328	4
5	335	341	348	355	361	368	375	382	388	395	5
%	0.0	0.1	0.2	0.3	0.4	0.5	0.6	0.7	0.8	0.9	%

6	
%	
0.01	0.6
2	1.2
3	1.8
4	2.4
5	3.0
6	3.6
7	4.2
8	4.8
9	5.4

ZrO₂ 123.22

Weight % 0.0—5.9

%	0.0	0.1	0.2	0.3	0.4	0.5	0.6	0.7	0.8	0.9	%
0	0	1	2	2	3	4	5	6	6	7	0
1	8	9	10	11	11	12	13	14	15	15	1
2	16	17	18	19	19	20	21	22	23	24	2
3	24	25	26	27	28	28	29	30	31	32	3
4	32	33	34	35	36	37	37	38	39	40	4
5	41	41	42	43	44	45	45	46	47	48	5
%	0.0	0.1	0.2	0.3	0.4	0.5	0.6	0.7	0.8	0.9	%

1	
%	
0.01	0.1
2	0.2
3	0.3
4	0.4
5	0.5
6	0.6
7	0.7
8	0.8
9	90.

½V₂O₅ 90.95

Weight % 0.0—5.9

%	0.0	0.1	0.2	0.3	0.4	0.5	0.6	0.7	0.8	0.9	%
0	0	1	2	3	4	5	7	8	9	10	0
1	11	12	13	14	15	16	18	19	20	21	1
2	22	23	24	25	26	27	29	30	31	32	2
3	33	34	35	36	37	38	40	41	42	43	3
4	44	45	46	47	48	49	51	52	53	54	4
5	55	56	57	58	59	60	62	63	64	65	5
%	0.0	0.1	0.2	0.3	0.4	0.5	0.6	0.7	0.8	0.9	%

2	
%	
0.01	0.2
2	0.4
3	0.6
4	0.8
5	1.0
6	1.2
7	1.4
8	1.6
9	1.8

E. BIBLIOGRAPHY

318 AHRENS, L. H. (1954): Quantitative Spectrochemical Analysis of Silicates. London, Pergamon Press.
— (1954 a): A Note on the Relationship between the Precision of Classical Methods of Rock Analysis and the Concentration of each Constituent. Min. Mag. 30, 467–470.
BACON, CH. (1926): Moldanubische Orthogneise des niederösterreichischen Waldviertels östlich vom Gföhlergneis. Tscherm. Mitt. 37, 126–172.
BARTH, T. F. W. (1948): The Distribution of Oxygen in the Lithosphere. J. Geol. 56, 41–49. Vgl. a. ibid. 57 (1949), 423–425.
— (1948 a): Oxygen in Rocks: A Basis for Petrographic Calculations. J. Geol. 56, 50–60. Vgl. a. ibid. 57 (1949), 415–417.
— (1949): Frequency Distribution of the Minerals in two Petrographic Provinces. J. Geol. 57, 55–61.
— (1952): Theoretical Petrology. New York u. London, Wiley.
— (1955): Presentation of Rock Analyses. J. Geol. 63, 348–363.
BAILEY, E. B., CLOUGH, C. T., WRIGHT, W. B., RICHEY, J. E., WILSON, G. V., etc. (1924): Tertiary and post-Tertiary Geology of Mull, Loch Aline, and Oban. Mem. Geol. Survey Scotland. Edinburgh.
BECKE, F. (1912): Chemische Analysen von krystallinen Gesteinen aus der Zentralkette der Ostalpen. Denkschr. Akad. Wiss. Wien, Math. Natw. Kl. 75, 153–229.
— (1925): Graphische Darstellungen von Gesteinsanalysen. Tscherm. Mitt. 37, 27–56.
BILTZ, W. (1934): Raumchemie der festen Stoffe. Leipzig.
BOEKE, H. E. (1914): Zur chemischen Zusammensetzung der tonerdehaltigen Augite, eine Anwendung quaternärer graphischer Darstellungen auf mineralogische Fragen. Z. Kristallogr. 53, 445–462.
— (1915): Die alkalifreien Aluminiumaugite. C. B. f. Min. etc., 422–431.
— (1916): Die Grenzen der Mischkristallbildung in Muskowit und Biotit. N. Jb. f. Min. etc. 1916 I, 83–117.
— (1916 a): Über die allgemeine Verwendung des gleichseitigen Tetraeders für die Darstellung von Vierstoffsystemen, mit einer Anwendung auf alkali- und tonerdehaltige Hornblenden. N. Jb. f. Min. etc. 1916 I, 118–125.
— (1923): Grundlagen der physikalisch-chemischen Petrographie. 2. Aufl. bearbeitet von W. Eitel. Berlin, Bornträger.
BRANNOCK, W. W. und BERTHOLD, S. M. (1953): The Determination of Sodium and Potassium in Silicate Rocks by Flame Photometry. U.S. Geol. Surv. Bull. 992, 1–14.
BRÖGGER, W. C. (1920): Die Eruptivgesteine des Kristianiagebietes, IV. Das Fen-Gebiet in Telemark, Norwegen. Vidensk. Selsk. Skr. I. Mat. Nat. Kl. No. 9.
BURRI, C. (1926): Die chemischen und provinzialen Verhältnisse der jungen Eruptivgesteine des Pazifischen Ozeans und seiner Umrandung. Schweiz. Min. Petr. Mitt. 6, 115–199 u. Ausz. Inaug. Diss. Univ. Zürich.
— (1928): Zur Petrographie der Natronsyenite von Alter Pedroso (Prov. Alemtejo, Portugal) und ihrer basischen Differentiate. Schweiz. Min. Petr. Mitt. 18, 374–436.

319 BURRI, C. (1944): Über logarithmische Rechenmittel in Mineralogie und Petrographie. Schweiz. Min. Petr. Mitt. 24, 302–315.
- (1955): Der Chemismus der Gesteine. Vierteljahrsschr. Natf. Ges. Zürich, 100, 45–57.
- (1956): Bemerkungen zur Anwendung der Niggli-Werte. Schweiz. Min. Petr. Mitt. 36, 29–48.
- (1959): Petrochemie der Capverden und Vergleich des Capverdischen Vulkanismus mit demjenigen des Rheinlandes. Schweiz. Min. Petr. Mitt. 39 (i. Druck).
- und NIGGLI, P. (1945): Die jungen Eruptivgesteine des mediterranen Orogens I. Publ. herausgeg. v. d. Stiftung ,,Vulkaninstitut Immanuel Friedlaender" 3, Zürich.
- und PARGA-PONDAL, I. (1936): Neue Beiträge zur Kenntnis des granatführenden Cordieritandesites vom Hoyazo bei Nijar (Provinz Almeria, Spanien). Schweiz. Min. Petr. Mitt. 16, 226–262.

CHAYES, F. (1948): A Petrographic Criterion of the Possible Replacement Origin of Rocks. Amer. J. Sc. 246, 413–425.
- (1949): On Correlation in Petrography. J. Geol. 57, 239–254.

CHOFFAT, P. (1916): Les roches éruptives filoniennes intrusives de la région située au nord du Tage. C. R. Ac. Sc. Paris 163, 152–155.

CROSS, CH. W., IDDINGS, J. P., PIRSSON, L. V. und WASHINGTON, H. S. (1902): A quantitative Chemico-Mineralogical Classification and Nomenclature of Igneous Rocks. J. Geol. 10, 555–690.
- (1903): Quantitative Classification of Igneous Rocks. Chicago, Univ. Press.

DEICHA, G. (1955): Essai d'une methode graphique des calculs petrochimiques. Bull. Soc. géol. France (6) 5, 243–247.

DIEHL, E. A. (1938): Geologisch-petrographische Untersuchung der Zone du Grand Combin im Val d'Ollomont. (Prov. Aosta, Italien). Schweiz. Min. Petr. Mitt. 18, 214–402, u. Prom. Arb. ETH Zürich.

ECKERMANN, H. v. (1936): The Loos-Hamra Region. Geol. För. Förh. 58, 129–336.

ESKOLA, P. (1954): A Proposal for the Presentation of Rock Analyses in Ionic Percentage. Ann. Acad. Fennicae (A) III, 38.

EZEKIEL, M. (1941): Methods of Correlation Analysis. 2nd ed. New York, Wiley.

FAIRBAIRN, H. W. etc. (1951): A cooperative Investigation of Precision and Accuracy in Chemical, Spectrochemical, and Modal Analysis of Silicate Rocks. U.S. Geol. Surv. Bull. 980.
- (1953): Precision and Accuracy of Chemical Analysis of Silicate Rocks. Geoch. Cosmochim. Acta 4, 153–156.
- u. SCHAIRER, J. F. (1952): A Test of the Accuracy of Chemical Analysis of Silicate Rocks. Amer. Min. 37, 744–757.

FÚSTER, J. M. (1954): Aplicación de los métodos de P. Niggli al cálculo de la composición mineralógica de las rocas graníticas. Bol. R. Soc. Española de Hist. Nat. (Festschrift für E. Hernandez-Pacheco), 315–330.

GANSSER, A. (1937): Der Nordrand der Tambo-Decke. Geologische und petrographische Untersuchungen zwischen San Bernardino und Splügenpass. Schweiz. Min. Petr. Mitt. 17, 291–523 u. Inaug. Diss. Univ. Zürich.

GOLDSCHMIDT, V. M. (1911): Die Kontaktmetamorphose im Kristianiagebiet. Vid. Selsk. Skr. I, Mat. Natv. Kl. (1911), No. 11.
- (1920): Geologisch-petrographische Studien im Hochgebirge des südlichen Norwegens. V. Die Injektionsmetamorphose im Stavanger-Gebiet. Vid. Selsk. Skr. I. Mat. Natv. Kl. (1920), No. 10.
- (1928): Über die Raumerfüllung der Atome in Kristallen und über das Wesen der Lithosphäre. N. Jb. f. Min. etc. B.B. 57 A (Festschrift O. Mügge), 1119–1130.

GROVES, A. W. (1951): Silicate Analysis. 2nd ed. London.

GRUBENMANN, U. und NIGGLI, P. (1924): Die Gesteinsmetamorphose I. Berlin, Borntraeger.

HILLEBRAND, W. F. (1910): The Analysis of Silicate and Carbonate Rocks. A revision of Bulletin 305. U.S. Geol. Surv. Bull. 422 (1910).
– und LUNDELL, G. E. F. (1954): Applied Inorganic Analysis. 2nd ed., revised by G. E. F. LUNDELL, H. A. BRIGHT und J. I. HOFFMAN. New York.
HOLMES, A. (1921): Petrographic Methods and Calculations. London, Murby.
JAKOB, J. (1952): Chemische Analyse der Gesteine und silikatischen Mineralien. Basel, Birkhäuser.
JOHANNSEN, A. (1931): A Descriptive Petrography of the Igneous Rocks, I. Chicago, Univ. Press.
– (1932): id. II.
JUNG, H. (1928): Die chemischen und provinzialen Verhältnisse der jungen Eruptivgesteine Deutschlands und Nordböhmens. Chem. d. Erde 3, 137–340.
KARAMATA, ST. (1957): Augitgranite im Granodioritmassiv der Boranja (West-Serbien). Schweiz. Min. Petr. Mitt. 37, 51–63.
LACROIX, A. (1916): Les syénites à riebeckite d'Alter Pedroso (Portugal), leurs formes mésocrates (lusitanites) et leur transformation en léptynites et gneiss. C. R. Ac. Sc. Paris 163, 279–283.
– (1916a): La constitution des roches volcaniques de l'Extrême Nord de Madagacsar e de Nosy bé; les ankaratrites de Madagascar en général. C. R. Ac. Sc. Paris 163, 253–258.
– (1917): La composition chimique de la vaugnérite et la position de cette roche dans la systématique. Bull. Soc. Min. France 40, 158–162.
– (1920): La systématique des roches grenues à plagioclases et feldspatoides. C. R. Ac. Sc. Paris 170, 20–25 (Errata p. 148).
– (1922): Minéralogie de Madagascar II, Paris.
– (1924): Les laves analcimiques de l'Afrique du Nord et, d'une facon générale la classification des laves renfermant de l'analcime (scanoite). C. R. Ac. Sc. Paris 178, 529–531.
– (1933): Classification et nomenclature des roches éruptives. In: Contribution à la connaissance de la composition chimique et minéralogique des roches éruptives de l'Indochine. Bull. Serv. géol. de l'Indochine. Hanoi, 20, Fasc. 3., im bes. 15–36 u. 183–206.
LAFFITTE, P. (1957): Introduction à l'étude des roches métamorphiques et des gîtes métallifères. Paris, Masson.
LINDER, A. (1951): Statistische Methoden für Naturwissenschafter, Mediziner und Ingenieure. 2. Aufl. Basel, Birkhäuser.
MARCHET, A. (1931): Zur Petrographie der vorsarmatischen Ergußgesteine bei Gleichenberg in Oststeiermark. Sitz.-Ber. Akad. Wiss. Wien, Math. Nat. Kl. (I) 140, 461–540.
MASSON, R. (1938): Geologisch-petrographische Untersuchungen im untern Valpelline (Prov. Aosta, Italien). Schweiz. Min. Petr. Mitt. 18, 54–213 u. Inaug. Diss. Univ. Zürich.
MICHEL-LÉVY, A. und LACROIX, A. (1887): Sur le granite à amphibole de Vaugnéray (vaugnérite de Fournet). Bull. Soc. Min. France 10, 27–31.
MITTELHOLZER, A. (1936): Beitrag zur Kenntnis der Metamorphose in der Tessiner Wurzelzone mit besonderer Berücksichtigung des Castionezuges. Schweiz. Min. Petr. Mitt. 16, 19–182 u. Prom.-Arb. ETH Zürich.
MUTUSWAMI, T. N. (1952): Nigglis Principles of Igneous Petrogenesis. Proc. Indian Acad. Sc. 36, Sect. A, 1–40.
NIGGLI, E. (1944): Das westliche Tavetscher Zwischenmassiv und der angrenzende Nordrand des Gotthardmassivs. Schweiz. Min. Petr. Mitt. 24, 58–301 u. Inaug. Diss. Univ. Zürich.
NIGGLI, P. (1912): Die Chloritoidschiefer und die sedimentäre Zone am Nordostrande des Gotthardmassivs. Beitr. Geol. Karte Schweiz N. F. 36 u. Inaug. Diss. Univ. Zürich.

321 Niggli, P. (1919): Petrographische Provinzen der Schweiz. Vierteljahrsschr. Natf. Ges. Zürich 64 (Festschrift Albert Heim) 179–212.
- (1920): Lehrbuch der Mineralogie. 1. Aufl. Berlin, Bornträger.
- (1920 a): Systematik der Eruptivgesteine. C. B. f. Min. usw. (1920), 161–174.
- (1923): Gesteins- und Mineralprovinzen I. Berlin, Bornträger.
- (1923 a): Anwendungen der mathematischen Statistik auf Probleme der Mineralogie und Petrographie. N. Jb. f. Min. usw. B.B. 48, 167–222.
- (1927): Zur Deutung der Eruptivgesteinsanalysen auf Grund der Molekularwerte. Schweiz. Min. Petr. Mitt. 7, 116–133.
- (1929): Die chemisch-mineralogische Charakteristik der metamorphen Paragesteinsprovinz am Südrand des Gotthardmassivs. Schweiz. Min. Petr. Mitt. 9 (1929), 160–187.
- (1931): Die quantitative mineralogische Klassifikation der Eruptivgesteine. Schweiz. Min. Petr. Mitt. 11, 296–364.
- (1936 a): Über Molekularnormen zur Gesteinsberechnung. Schweiz. Min. Petr. Mitt. 16, 295–317.
- (1936 b): Die Magmentypen. Schweiz. Min. Petr. Mitt. 16, 335–399.
- (1937): Über die chemische Zusammensetzung der Alkaliglimmer. Z. Kristallogr. (A) 96, 89–106.
- (1938): Die komplexe gravitative Kristallisationsdifferentiation. Schweiz. Min. Petr. Mitt. 18, 610–664.
- (1941): Gesteinschemismus und Mineralchemismus, I. Das Problem der Koexistenz der Feldspäte in den Eruptivgesteinen. Schweiz. Min. Petr. Mitt. 21, 183–193.
- (1943): Gesteinschemismus und Mineralchemismus, II. Die Pyroxene der magmatischen Erstarrung. Schweiz. Min. Petr. Mitt. 23, 538–607.
- (1946): Gesteinschemismus und Mineralchemismus, II. Die Pyroxene der magmatischen Erstarrung, 2. Teil: Allgemeines über Reaktionsgleichungen und Dreiecksdarstellungen mit Hilfe der Basis. Schweiz. Min. Petr. Mitt. 26, 34–43.
- (1950): On the Presentation of Geochemical. Data. Int. Geol. Congr. Rep. 18th Session, Great Britain 1948, Pt. II, Proc. Sect. A: Problems of Geochemistry, 101–115.
- (1950 a): Some Hornfelses from Saxony and the Problems of Metamorphic Facies. Amer. Min. 35, 867–876.
- (1951): Gesteinschemismus und Magmenlehre. Geol. Rundschau 39, 8–32.
- und Lombaard, B. V. (1933): Das Bushveld als petrographische Provinz. Schweiz. Min. Petr. Mitt. 13, 110–186.
- und Niggli, E. (1948): Gesteine und Minerallagerstätten I. Basel, Birkhäuser. Die englische Ausgabe, P. Niggli (1954): Rocks and Mineral Deposits, San Francisco, Freeman, besorgt durch R. L. Parker, enthält gegenüber der deutschen einige zusätzliche Ausführungen über petrochemische Berechnungsmethoden.
- Quervain, Fr. de und Winterhalter, R. U. (1930): Chemismus Schweizerischer Gesteine. Beitr. Geol. Schweiz, Geotechn. Serie 14.

Oenay, T. S. (1949): Über die Smirgelgesteine Südanatoliens. Schweiz. Min. Petr. Mitt. 29, 357–491 und Prom.-Arb. ETH Zürich.

Palm, Q. A. (1954): Vaugnérites et Amphibolites, deux types de roches amphiboliques dans les Cévennes à la hauteur de Largentière (Ardèche). Bull. Soc. géol. France (6) 4, 627–641.

Pearson, K. (1897): Mathematical Contribution to the Theory of Evolution. – On a Form of Spurious Correlation which may arise when Indices are used in the Measurement of Organs. Proc. R. Soc. London 60, 489–498.

Philipsborn, H. v. (1933): Tabellen zur Berechnung von Mineral- und Gesteinsanalysen. Leipzig.

RITTMANN, A. (1933): Die geologisch bedingte Evolution und Differentiation des Somma-Vesuvmagmas. Z. Vulkanolog. 15, 8–94.
- (1944): Vulcani, attività e genesi. Napoli.
RODE, K. P. (1941): The Geology of the Morcote Peninsula and the Petrochemistry of the Porphyry Magma of Lugano. Schweiz. Min. Petr. Mitt. 21, 194–312 u. Inaug. Diss. Univ. Zürich.
SAWARITZKI, A. N. (1954): Einführung in die Petrochemie der Eruptivgesteine. Berlin, Akademie-Verlag.
SHAPIRO, L. und BRANNOCK, W. W. (1956): Rapid Analysis of Silicate Rocks. A Contribution to Geochemistry. Revised from Circular 165. U. S. Geol. Surv. Bull. 1036–C.
SHEPHERD, E. S. (1938): The Gases in Rocks and some related Problems. Amer. J. Sc. (5) 35 A, 311–351.
SCHLECHT, W. G. (1951): Co-operative Investigation of Precision and Accuracy. Anal. Chem. 23, 1568.
STUTZ, A. H. (1940): Die Gesteine der Arolla-Serie im Valpelline (Prov. Aosta, Italien). Schweiz. Min. Petr. Mitt. 20, 117–246 u. Prom.-Arb. ETH Zürich.
TILLEY, C. E. (1924): Contact-Metamorphism in the Comrie Area of the Pertshire Highlands. Q. J. Geol. Soc. 80, 22–71.
TIPPET, L. H. C. (1952): Methods of Statistics. 4th ed. London u. New York.
TRÖGER, W. E. (1935): Spezielle Petrographie der Eruptivgesteine. Berlin.
TURNER, H. W. (1896): Further Contributions to the Geology of the Sierra Nevada. An. Rep. U. S. Geol. Surv. 17, I, 521–762, im bes. 724.
USSING, N. V. (1911): Geology of the Country around Julianehaab. Medd. Grønland 38.
WALLIS, W. ALLEN und ROBERTS, HARRY, V. (1957): Statistics, a new approach. London, Methuen.
WANG, H. S. (1939): Petrographische Untersuchungen im Gebiet der Zone von Bellinzona. Schweiz. Min. Petr. Mitt. 19, 23–109 u. Inaug. Diss. Univ. Zürich.
WASHINGTON, H. S. (1915): The Correlation of Potassium and Magnesium, Sodium and Iron, in Igneous Rocks. Proc. National Acad. Sc. 1, 575–578.
- (1917): Chemical Analyses of Igneous Rocks published from 1884–1913 inclusive with a Critical Discussion of the Character and Use of Analyses. U. S. Geol. Surv. Prof. Pap. 99.
- (1917a): The Quantitative Classification of Igneous Rocks. U. S. Geol. Surv. Prof. Pap. 99, Appendix 1–3, 1151–1180.
- (1919): Manual of the Chemical Analysis of Rocks. New York.
WENK, E. (1954): Berechnung von Stoffaustauschvorgängen. Schweiz. Min. Petr. Mitt. 34, 309–318.
WILLIAMS, H. (1932): Geology of the Lassen Volcanic National Park, California. Univ. of California Publ. Bull. Dept. Geol. Sc. 21, 195–385.
YULE, G. U. und KENDALL, M. G. (1950): An Introduction to the Theories of Statistics. 14th ed. London.

F. AUTHOR INDEX

AHRENS, L. H., 16, 17

BACELAR, BEBIANO, J., 77
BACON, CH., 44, 79
BAILEY, E. B., 92
BARTH, T. F. W., 14, 19, 260, 262, 264, 267, 269
BECKE, F., 37, 39, 42, 44, 201, 202
BERTHOLD, S. M., 15
BIANCHI, A., 239
BILTZ, W., 179
BOEKE, H. E., 39, 79
BONATTI, ST., 239
BRANNOCK, W. W., 15
BRÖGGER, W. C., 100, 164, 168, 169
BURRI, C., 23, 24, 25, 26, 45, 46, 51, 69, 77, 79, 80, 87, 92, 93, 103, 107, 127, 128, 131, 181, 187, 188, 189, 190, 192, 193, 200, 219, 239

CHAUTARD, J., 177
CHAYES, F., 30, 31, 67, 68
CHOFFAT, F., 157
CLOUGH, T. T., 92
CROSS, CH. W., 100

DAL PIAZ, GB., 239
DEICHA, G., 52
DIEHL, E. A., 194, 239, 241, 244, 250, 252

ECKERMANN, H. V., 67, 68
EITEL, W., 39
ERMERT, H., 77
ESKOLA, P., 14, 19, 202, 238, 260, 262
EZEKIEL, M., 30

FAIRBAIRN, H. W., 15, 17
FORBES, E. A., 242
FÚSTER, J. M., 103

GANSSER, A., 238
GIBBS, J. W., 32, 33
GOLDSCHMIDT, V. M., 131, 267, 269
GRILL, E., 239
GROVES, A. W., 15
GRUBENMANN, U., 201, 202, 238

HARKER, A., 83
HILLEBRAND, W. F., 15
HOLMES, A., 45

IDDINGS, J. P., 83, 100

JAKOB, J., 15, 212
JOHANNSEN, A., 45, 148
JUNG, H., 26

KARAMATA, ST., 269
KENDALL, M. G., 26
KNEŽEVIĆ, V., 260

LACROIX, A., 13, 77, 79, 148, 157, 164, 166
LAFITTE, P., 19
LINDER, A., 26
LOMBAARD, B. V., 103
LUNDELL, G. E. F., 15

MARCHET, A., 44, 79
MARIO DE JESUS, A., 77
MASSON, R., 252, 255
MICHEL-LÉVY, A., 148
MITTELHOLZER, A., 212, 229
MUTHUSWAMI, T. S., 103

NIGGLI, E., 178, 229
NIGGLI, P., 19, 21, 23, 24, 25, 45, 63, 70, 72, 75, 77, 80, 81, 82, 83, 87, 88, 89, 92, 92, 103, 104, 107, 125, 127, 128, 130, 149, 178, 181, 187, 188, 189, 190, 192, 197, 200, 201, 202, 219, 229, 238, 239, 250

ORCEL, J., 242
OSANN, A., 92

PALM, Q. A., 148
PARGA-PONDAL, I., 51, 131
PART, G. M., 77
PEARSON, K., 30, 32
PHILIPSBORN, H. V., 45, 180
PIRSSON, L. V., 100
PISANI, 148

QUERVAIN, FR. DE, 81, 139

RAOULT, F., 142, 157, 164
REINISCH, R., 77
RICHEY, J., 92
RITTMANN, A., 76, 87, 190, 270
ROBERTS, H. V., 26
RODOLICO, FR., 242
RÖER, O., 268
ROOTHAAN, H. PH., 239
ROSSONI, P., 239

SCHAIRER, J. F., 16
SCHLECHT, W. G., 16
SHAPIRO, L., 16
SHEPHERD, F. S., 16
STOCKES, H. N., 51

THOMAS, H. H., 52
TILLEY, C. E., 202, 208, 210, 227, 228, 229, 233, 234
TIPPET, L. H. C., 26

TORRE DE ASSUNÇÃO, F. C., 77
TRÖGER, W. E., 149, 153, 157, 158, 159, 161, 164, 168
TURNER, FR. J., 202
TURNER, H. W., 51

USSING, V., 51

VOGT, TH., 202

WALLIS, W. A., 26
WANG, H. S., 194, 195, 197
WASHINGTON, H. S., 15, 44, 51, 87, 100, 130, 174, 175
WILLIAMS, H. W., 84
WILSON, G. V., 92
WINTER, C., 51
WINTERHALTER, R. U., 81, 239
WRIGHT, B. W., 92

YULE, G. U., 26

G. SUBJECT INDEX*

A

α 183
Accessories 118, 121
Achnahaitic magma 95
Acidity, degree of 76
Acmite 117, 124
Actinolite, average 203
Adamelitic magma 94
Aegirine 117, 124
Akermanite 117
Albite 116, 220
— porphyroblastic schist 267, 268
— sericite schist 227
Albite magma 96
al 49
$al-fm-c-alk-$ tetrahedron 77, 78
al -hornblenditic magma 95
alk 49
alk -issitic magma 98
Alkali excess 74, 122, 198
Alkali-gabbroidal magmas, heteromorphism of 157
Alkali-granite aplitic magma 96
Alkali-granitic magma 96
Alkali-granitic magma group 96
Alkali-hornblende peridotitic magma 98
Alkali-jacupirangitic magma 97
Alkali-mafitic magma 98
Alkali-pyrobolic magma group 98
Alkali-syenite aplitic magma 96

Alkali-poor magmas 93
Alkali-rich magmas 93
Almandine 204, 206, 237
Alnöitic magma 99
Alter Pedroso, ossanite-syenites from 79, 80
Alumina excess 61, 63, 75, 121, 198
Alumina excess, "large" 122
Alumina excess, "small" 122
Aluminum silicates 204, 205
Amesite 203, 215, 216, 237
Amphibole and melilite in QLM-triangle 188
Amphibole-essexite, foid-free 157
Amphibole-theralite 164
Analcite 116
Andalusite 206
Andesinitic magma 95
Andradite 204, 206
Anhydrite 107, 118, 204
Anions, consideration of 260
Ankaratritic magma 98
Anorthite 116
Anorthosite-gabbroidal magma 95
Anorthositic magma 95
Antigorite 203, 216, 237
Antophyllite 203
Antsohitic magma 99
Apatite 118
Aplite-granitic magma 94

---

* The symbols for equivalent-normative and basis components are not given in the Subject Index. They are compiled in alphabetical order in Table 1 of the Appendix (see page 272).

Arfvedsonite-lujavrite, Lille Elv, Kangerdluarsuk, Greenland, wt. % and Niggli-values 51
—, calculation of alumina excess 64
—, calculation of cation %, from Niggli-values 57
—, calculation of cation % from the basis 115
—, calculation of cation % from the standard katanorm 247
—, calculation of cation % from wt. 56
—, calculation of Niggli-values from cation % 59
—, calculation of Niggli-values from the standard katanorm 147
—, calculation of the quartz number qz 63
—, calculation of the basis from cation % 114
—, calculation of the basis from wt. % 111
—, calculation of the excess of alkali over alumina 63
—, calculation of the standard katanorm from cation % 137
—, calculation of the standard katanorm from Niggli-values 142
—, calculation of the standard katanorm from Niggli-values via cation % 143
—, calculation of the standard katanorm from the basis 131
—, calculation of wt. % from Niggli-values 54
—, calculation with regard to the anions, based on wt. % 263
—, graphic determination of alkali-feldspar ratios 75
—, projection in the double-tetrahedron 81
Ariégitic magma 95
Arithmetic mean 26
Arkitic magma 98
Arkitic magma group 98
Asymmetric distributions 26
Atlantic suite 21
Augite 203, 204
Augite and hornblende 117, 119

Augite, from fasinite 168
Augite-granite from Boraja, West Serbia 269
Auxilliary means for calculation 44
Average 26
$A_z°$ 76

B
β 184
BaO, calculation of 48
Barth's standard cell 263, 267
"   "   ", norm of 266
Basis 106
Basis, calculation of from cation % 112, 113, 114
"   , calculation of from wt. % 106, 109, 110, 111
Basis bonds 106, 107
"   ", possible combinations of 122
Basis components 106
Basis-group values QLM 181
"   "   "   ", calculation of from Niggli-values 198
Batukitic magma 95
Belugitic magma 95
Beringitic magma 97
Berondrite 157, 164, 166, 168
"   , biotite-bearing 168
"   , olivine-bearing 168
Berondritic magma 97
Biotite 117
"   -hornblende granogabbro 157
"   -pyroxene quartz-diorite 134
"   -quartz gabbro 153
"   -staurolite-kyanite-chlorite schist 259
Biotite magma 99
Boraja (West Serbia), augite-granite from 269
Borolanitic magma 98
Bostonitic magma 96

C
c 49

c' 64
Ca-aluminate 107, 108
" -orthosilicate 107, 108
" -phosphate 107, 118
Calc-alkali series 93
Calcite 107, 118, 204, 219, 220
Calcium feldspar 116
Calcium mica schist 230
Calcium silicate 204
Calc-silicate fels from Castione near Rellinzond, Tessin (Switzerland), calculation of a calcite variant of the standard katanorm 214
" - " ", calculation of the basis from the wt. % 212
" - " ", calculation of the standard epinorm based on the distribution of Al_2O_3 235
" - " ", calculation of the standard epinorm from the basis 229
" - " ", calculation of the standard katanorm from the basis 213
Calculating machines 45
Cancrinite 117
Cape Verde 76, 192, 193
Cation percentages 55
" " , calculation from Niggli-values 57
" " calculation from the basis 114
" " , calculation from the standard katanorm 136
Carbonate-marialite 117
Carbonate-meionite 117
Carbonates 204, 205
c -gabbro-theralitic magma 97
Clinochlore 216, 258
Clinochlore variants 229
Chlorite 216
Chlorite-sericite-chloritoid phyllite 229
Chloritoid 216, 237
Chloritoid phyllite from Curaglia 229
Chloritoid schist, Gotthard Massif 229
Chloritoid-sericite-albite schist 228
Chlor-marialite 117

Chlor-meionite-117
Chromite 107, 118
cl_2 219
co_2 , calculation of 49
Classical methods of rock analysis 17
c -melteigitic magma 97
c - normal magmas 93
Common hornblende 117, 134
co_2 49
CO_2, calculation of 49, 105
Concentration tetrahedron 32, 38
" " , illustration of 39
" " , projection on hexahedron plane 41
" " , projection on tetrahedron plane 40
Concentration lines 32, 33
Concentration triangle 32, 33
" " , rectangular 37
CoO, calculation of 48
Cordierite 118, 121, 204, 206
Cordierite-andesite, garnet-bearing, Hoyazo near Nijar, Almeria, Spain, calculation by consideration of the anions, based on the Niggli-values 262
" - " , calculation of cation % from Niggli-values 57
" - " , calculation of cation % from standard katanorm 139
" - " , calculation of cation % from the basis 115
" - " , calculation of cation % from wt. % 56
" - " , calculation of Niggli-values from cation % 59
" - ", calculation of Niggli-values from the standard katanorm 146
" - ", calculation of the alumina excess 64
" - ", calculation of the basis from cation % 113
" - " , calculation of the basis from wt. % 110
" - " , calculation of the excess of alumina over alkalis 63

Cordierite-andesite, calculation of the portion of calcium not bound to Al 64
" - " , calculation of the quartz number qz 63
" - " , calculation of the standard epinorm from the basis 226
" - " , calculation of the standard katanorm from cation % 137
" - " , calculation of the standard katanorm from Niggli-values 141
" - " , calculation of the standard katanorm from the basis 131
" - " , calculation of wt. % from Niggli-values 54
" - " , graphic determination of the normative feldspar ratios from Niggli-values 72, 75
" - " , projection of the double-tetrahedron 81
" - " , wt. % and Niggli-values 51
Cordierite-spinel hornfels, Creag, near Iolaire, Glen Lednock, Comrie area, Perthshire, Scotland, calculation of a biotite variant of the standard katanorm 211
" - " ", calculation of a clinochlore variant of the epinorm 229
" - " ", calculation of the basis from wt. % 210
" - " ", calculation of the standard epinorm based on the distribution of Al_2O_3 234
" - " ", calculation of the standard epinorm from the basis 228
" - " ", calculation of the standard katanorm from cation % 211
" - " ", calculation of the standard katanorm from the basis 210
Correlation 29, 30
Correlation coefficient 30
Correlation diagram 29
Correlation, false 32
Correlation of Niggli-values 67
Correlation of Niggli-values, of ratios 30, 31
Corundum 107, 108, 118
Corundum-cordierite-spinel hornfels, Tom a'Mhinn, Glen Lednock, Comrie region, Perthshire, Sctoland, calculation of a magnesite variant 228
" - " - " ", calculation of the basis from wt. % 208
" - " - " ", calculation of the standard epinorm, based on the distribution of Al_2O_3 233
" - " - " ", calculation of the standard epinorm from the basis 227
" - " - " ", calculation of the standard katanorm from cation % 209
" - " - " ", calculation of the standard katanorm from the basis 208
CPIW, -classification 100
CPIW, -norm, 100, 121, 174, 175, 176, 200
—, recalculation to the equivalent norm 175
c - poor magma 95
c -rich magma 93
Cr_2O_3, calculation of 48
CsO_2, calculation of 48
Cumbraitic magma 95
Cumulative curve 20, 22
Cumulative lines 20
Cumulative percent 20, 22
Curvature of distribution curves 27

D

Degree of saturation, in SiO_2 182, 183
"   "   " , with respect to SiO_2 182, 183
Degree of silication 76
Desilication 193, 270
Diallagitic magma 95
Diopside 117
Diorite magma 94
Dioritic magma group 94
Double-tetrahedron, deformed 84

E

Engadinitic magma 94
Enstatite 117
Epidote 215, 237
    " -hornblende-albite schist 239
    " -zoisite group 203, 205
Epi-facies 202
Epinorm 203
    " , calculation of standardized 219
    " , calculation of variants adapted to the modus 236
    " , variants of the calculation modus 230
Epi-standard minerals 215
Epizone 202
Equivalent norm 100, 103
    " " , recalculation to CPIW-norm 175, 176
Equivalent-normative components, of igneous rocks 116
    " - " " of metamorphic rocks 203
Equivalent numbers, atomic 45
    " " , molecular 45, 47
Equivalent percentages, molecular 59
    " " , recalculation to wt. %, 177
    " " , recalculation to wt. % 171
Equivalent volumes 178
    " " , computable 104
Equivalent weight, 171, 172
    " " , computable 104
Essexite-akeritic magma 97
    -dioritic magma group 96
    -foyaitic magma 96
    -gabbrodioritic magma 97
    -grabboidal magma 97
Essexitic magma 97
Essexitic magma group 97
Eucritic magma 95
Evisitic-groruditic magma 96
Evisitic magma group 96
Evisitic-pantelleritic magma 96

F

f_2 49
F_2, calculation of 49
Farsunditic magma 94
Fasinite, from Ambahila, Ampasindava, Madagascar 164, 166
—, augite from 168
—, hornblende-augite variant 168
—, melilite variant 165
—, titanite-monticellite variant 165
—, titanite variant 165
Fayalite 107, 108, 117
Fe-actinolite 203
  -akermanite 117
  -amesite 203, 216
  -antigorite 203, 216
  -biotite 117
  -cordierite 118, 121, 204
  -gehlenite 117
  -ottrelite 203, 216
Feldspars 116, 118
Feldspar, choice of formula unit 70
Feldspar-cordierite hornfels 131
Feldspar-cordierite fels, hypersthene-bearing 253
Feldspar diagram 72
Feldspar ratios, normative 186
Feldspar triangle 186
Feldspathoids 116, 119
Femic magmas 92
Ferrisilicate 107, 108
Fluorite 107, 118
fm 49
fm -gabbroidal magma 95
Foid-olivine gabbro 162
Foreland, Alpine 26
Formula units 19
    " ", agreeable 104
    " ", computable 104
Formula weight 19, 44
    " ", of oxides, round numbers 45
Forsterite 107, 108, 117
Foyaitic magma groups 96
Frequency curve 20

Frequency distribution 26
Frequency polygon 20
Frequency values, percentile 20
Function 28

G

γ 185
Gabbrodioritic magma group 95
Gabbroidal magma 95
Gabbroidal magma group 95
Gabbro-melteigitic magma 97
Gabbro-theralitic magma 95
   "   "   magma group 95
Garnet 237
Garnet-biotite-kyanite-staurolite schist 258
Garnet group 204, 205, 206
Gehlenite 117, 125
Geological oxidation 263
Geological reduction 264
Gibelitic magma 96
Glaucophane 203, 236
Gotthard Massif, chloritoid schists of 229
   "   ", southern margin 82
Grammatite 203
Granitic magma 94
Granitic magma group 94
Granodiorite, mesotypes 149
Granodioritic magma 94
   "   "   group 94
Granosyenitic magma 98
Graphic representation 19
Greenlanditic magma 96
Greenschist 239, 247
Grossularite 204, 206
Groups of components 18

H

h 49
Halite 117, 118
Halogenides 205
Hauyne 117
Hawaiitic magma 95
Hedenbergite 117

Hematite 107, 118, 122, 220
Hercynite 107, 108
Heteromorphic facies, conversion to 131
Heteromorphic possibilities of alkali-gabbroidal magmas 157
Heteromorphic relationships, study of, by means of equivalent norm 147
Heteromorphism 13, 148, 200
Histogram 20, 22
H_2O, calculation of 49, 105, 261, 262
Hornblende 203, 204, 216
Hornblende-chlorite-albite-epidote schist 239, 248
Hornblende, common 117, 134
   "   , from essexitic rocks 162, 166
   "   , from gabbro, average 153
Hornblende-olivine monzonite, nepheline-bearing 155
   "   -peridotitic magma, 95
   "   -quartz monzonite 269
Hornblenditic magma 95
Hornblenditic magma group 95
Hypersthene 117

I

Idealized crystal components, projection points 182
Igneous field 77, 78, 80, 81, 82
Igneous rocks, equivalent-normative components of 116
Ijolitic magma 97
Ijolitic magma group 97
Initial magmatite 238
Injection metamorphism in the Stavanger region 267
Ilmenite 118
Intra-Pacific suite 20
Intermediate alkali-rich magmas 93
Ionic radii, empirical 48
Issitic magma 95
Italy, volcanic rocks 26

J

Jacupirangite, nepheline-rich 164

Jacupirangitic magma 95
Jadeite 117
Jumillitic magma 99
Juvitic magma group 98

K

k 50, 185
k' 185
K-acmite 117, 124
K-aegirine 117, 124
K-ferrisilicate 107, 123
K-metasilicate 107
Kajanitic magma 99
Kaliophilite 106, 107, 116
Kammgranitic magma 98
Kamperitic magma 99
Kaolin 203, 217
Kassaitic magma 96
Kata-facies 202
Katanorm 121, 203
Katazone 202,
Kaulaitic magma 98
k-mg- diagram 87
KNaCa-triangle 186
Koswitic magma 95
Kyanite 204, 206
Kyanite-garnet-sericite-chlorite-phyllite 238

L

Labradorite-felsitic magma 95
Lamprodioritic magma 94
Lamproitic magma groups 99
Lamprosommaitic magma 99
Lamprosyenitic magma 99
Lardalitic magma 96
"Large" alumina excess 122
Larnite 204
Larvikitic magma 96
Larsen Peak, Cal., lavas of 84, 85, 86, 87, 187, 190, 191
Lavas of Young Somma and Vesuvius 270
Lepidolite 117
Leucite 116

Leucite-basanite 153
Leucoevisitic magma 96
Leucogabbroidal magma group 95
Leucogranitic magma group 94
Leucomiharaitic magma 95
Leuconzonitic magma 99
Leucopeléetic magma 94
Leucoquartz dioritic magma 93
Leucosommaitic magma 98
Leucosyenitic magma 98
Leucosyenite-granitic magma group 98
Leucotonalitic magma 94
Li_2O, calculation of 48
Loos-Hamra area 67
Lujauritic magma 96
Lujauritic magma group 96
Luscladite 157, 158, 159, 160, 161
Lusitanitic magma 98

M

μ 185
Maenaitic magma 97
Mafatite-olivine gabbro 154
Mafraite 157, 158, 164
" , attempt at calculation of a modus of 161, 163
" , hornblende variant 163
" , perovskite variant 162
" , Ti-augite-biotite variant 161
Magmas, classification, principles of 93
" , the calc-alkali series 94
" , the potassic series 98
" , the sodic series 96
Magmatite, initial 238
Magma-type, concept of 92
Magma-types, review of 94-99
Magnesite 204, 219, 220
Magnetite 118, 123
Margarite 203, 217
Marialite 217
Material balances of isovolumetric metasomatic processes 267
Mean deviation 27
Mean error 27
Mean value 26

Meionite 117
Melagabbroidal-dioritic-magma 95
Melagranodiorite 149
Melagranogabbro 149
Melaplagioclastic magma 95
Mela-quartz-dioritic magma 94
Melarkitic magma 99
Melanatron-gabbroidal magma group 97
" -syenitic magma 97
Melashonkinite magma group 99
Melatheralitic magma 97
Melatinguaitic magma 98
Melteigitic magma 97
Metal-atom numbers 102
Mesofacies 202
Mesozone 202
Metamorphic rocks, epinorm of 214
" " , equivalent normative component of 203
" " , standard katanorm of 206
" " , usage of the equivalent norm for the study of 200
Median value 26
mg 50, 185
mg' 185
Mg-biotite 117
Mg-ottrelite 203, 216
MgFeCa-triangle 186, 187
Mica 117, 120
Miharaitic magma 195
Mineral combinations of the standard epinorm 225
Mineral composition, simplified, normative 64
Missouritic-alnoitic magma group 99
MnO, calculation of 48
Modlibovite-polzenitic magma 98
Modus 26, 100
Moisture 15
Molecular numbers 47
Molecular percent 47
Molecular proportions 47
Molecular quotients 47
Monmouthitic magma 96

Monticellite 117
Monzonitic-dioritic magma 99
Monzonitic magma 99
Monzonitic magma group 99
Monzonitic-syenitic magma 98
Most frequent value 26
Moyitic magma 94
Mugearitic magma 97
Murcialamproitic magma 99
Muritic magma 97
Muscovite 117, 203, 215, 220
Muscovite-chlorite-chloritoid phyllite 257
Muscovite (phengitic) 117, 203

N

Na-carbonate 107, 118, 204
Na-ferrisilicate 107, 123
Na-gehlenite 117
Na-metasilicate 107, 108
Na-sericite, formation of 250
Natron-engadinitic magma 94
 -gabbroidal magma group 97
 -granite-aplitic-magma 94
 -hornblenditic magma 98
 -lamprosyenitic magma 97
 -rapakivitic magma 94
 -syenitic magma 97
 -syenitic magma group 97
Nepheline 107, 108, 116
Niggli-values 23, 24, 25, 47
" " , calculation from cation % 58
" " , calculation from molecular-equivalent % 59
" " , calculation from the standard katanorm 140, 145
" " , calculation from wt. % 47
" " , correlation 67
" " , limits of application 66
" " , recalculation to basis-group values QLM 198
" " , recalculation to cation % 57

Niggli-values, recalculation to the standard katanorm 140
"   ", recalculation to the standard katanorm via cation % 144
"   ", recalculation to wt. % 52
NiO, calculation of 48
Nordmarkitic magma 96
Norm 100
Normal evistic magma 96
"   gabbrodioritic magma 95, 96
"   missouritic magma 99
Normative feldspars, quantitative relations of 71, 86
Normative mineral composition, derivation from the basis 111
Norm of the standard cell 266
Nosean 117
Nosykombitic magma 97

O

o 51
Okaitic magma 97
Oligoclastic magma 95
Olivine group 117
Olivine-theralite 160
Olivine, augite and melilite in the MgFeCa-triangle 189
Olivine-monzonite 150
Olivine-nepheline gabbro 160
"   -theralite gabbro 158
Omphacite, average 203, 207
Opdalitic magma 94
Ophiolite 238
Orbitic magma 94
Ore-peridotitic magma 96
Ortho-augitic magma 96
Ortho-augitic-perioditic magma group 96
Orthoclase 116
Ossipitic magma 95
Ostraitic magma 94
Oxide equivalents 47
Oxide form 14, 260
Oxide molecule 47
Oxide point 183

P

π 185
p 49, 50
P-space 78
Pacific Provinces 187
Paragonite 203
Parameters, statistical 26
Pedrosite 79
Peléetic magma 94
Peraltemic magmas 92
Peralkalic magmas 93
Peridotitic magma 96
Perovskite 118
Pienaaritic magma 98
Pietre Verdi 238
Pistacite 203, 237
Plagioclastic magma group 95
P_2O_5, calculation of 49
Possible combinations of basis components 122
Potassic-dioritic magma group 99
Potassic-foyaitic magma 98
Potassic-gibelitic magma 98
Potassic-hornblenditic magma 99
Potassic-nepheline 106, 116
Potassic-nordmarkitic magma 98
Potassic-polzenitic magma 99
Potassic-series 93
Potassium feldspar 116
Prasinite 239
"   , mineral composition of 238
"   , selected, basis 240
"   , chemical analysis 239
"   , standard epinorm 240
"   "   , standard katanorm 240
Principal constituents of calc-alkali rocks in the MgFeCa-triangle 190
Principal constituents of calc-alkali rocks in the QLM triangle 189
Pyrite 117, 118
Pyrope 204, 206
Pyrophyllite 203
Pyroxenitic magma 95
Pyroxenitic magma group 95

Pyroxenolithic magma 99
Pyroxene and biotite in the MgFeCa-triangle 188
" " " in the QLM triangle 187
Pyroxene-gabbroidal magma 95
Pyroxene, separation of 129
Pulaskitic magma 96

Q

QLM triangle 181
Quantitative ratio M/L 184
Quartz 107, 109, 116
Quartz-diorite, Spanish Peak, Cal., calculation involving anions, based on wt. % 261
— calculation of a biotite variant of the standard katanorm 133
— calculation of a hornblende-biotite variant of the standard katanorm 134
— calculation of alumina excess 64
— calculation of a simplified mineral composition from Niggli-values 72
— calculation of cation % from Niggli-values 57
— calculation of cation % from the basis 114
— calculation of cation % from the standard katanorm 139
— calculation of cation % from wt. % 55
— calculation of Niggli-values from cation % 58
— calculation of Niggli-values from molecular % 60
— calculation of Niggli-values from the standard katanorm 145
— calculation of Niggli-values from wt. % 50
— calculation of the amount of calcium not bound to Al 64
— calculation of the Barth standard cell, based on cation % 264
— calculation of the Barth standard cell, based on Niggli-values 265

Quartz-diorite, calculation of the basis from cation % 112
— calculation of the basis from wt. % 109
— calculation of the excess of alumina over alkalis 63
— calculation of the norm of the standard cell 266
— calculation of the quartz number qz 62
— calculation of the standard epinorm from the basis 226
— calculation of the standard katanorm from cation % 136
— calculation of the standard katanorm from Niggli-values 141
— calculation of the standard katanorm from the basis 128
— calculation of wt. % from Niggli-values 53
— projection in the double-tetrahedron 81
— projection in the feldspar triangle 186
Quartz-diorite magma 94
Quartz-musvocite phyllite 267, 268
Quartz number qz 21, 60

R

Rapakivitic magma 98
Rapid methods, analytical 16
Rare elements 47
Rb_2O, calculation of 48
Reaction coefficients, sum of 105
Reaction equations 105
Regression line 30
Remainder point, possibilities for the displacement of 196
Remainder triangle 194, 240
Riebekite 117
Rockallitic magma 98
Rock metamorphism 202
Rock salt 107
Rouvillitic magma 96
Rutile 107, 108, 118

S

s 49
S, calculation of 49
Salic magma 92
Salitritic magma 98
Saturation coefficient 183
Saturation line 182
Saturation with respect to SiO_2 20
Scapolite group 117
Semialic magmas 92
Semifemic magmas 92
Separation of pyroxenes 129
Serpentine 216
Sheet silicates 203, 204
Shonkinitic magma 99
Shonkinitic magma group 99
Shonkinitic-missouritic magma 99
si 49
$si°$ 76
Siderite 204, 220
si -camperitic magma 95
si -gabbrodioritic magma 95
si -maenaitic magma 97
si -melaplagioclastic magma 95
si -monzonitic magma 99
si -natron-syenitic magma 97
si -oligoclastic magma 93
si -pyroxenitic magma 95
si -syenite granitic magma 98
Silication status 75
Sillimanite 204, 206
Sillimanite-biotite-cordierite-garnet fels, La Vieille, Comba di Vessona, Valpelline, Prov. Aosta, Italy, biotite variant of katanorm 252
— garnet-biotite variant of katanorm 256
— garnet-free cordierite-biotite sillimanite variant of katanorm 254
— garnet variant of the epinorm 257
— mesonorm, garnet-free variant 259
    " , staurolite variant 258
— standard epinorm 256
— standard katanorm 253
SiLU-triangle 37
SiO_2, distribution among normative leucocratic and melanocratic components 90
Skewness 27
Slide-rule, disk-shaped, logarithmic 46
   "   ", simple,   "   45, 46
"Small" alumina excess 122
so_3 49
So_3, calculation of 49
Sodalite 117
Sodic series 93
Sodium feldspar 116
Sommaitic magma 99
Sommaitic magma group 99
Sommaitic-dioritic magma 99
   " -monzonitic " 99
   " -assipitic   " 99
   " -tonalitic   " 99
Spessartite 204
Spinel 107, 108
Spodumene 117
Spurrite 204, 207
SrO, calculation of 48
Standard cell according to Barth 263
Standard deviation 27
   "   " , relative 29
Standard epinorm, calculation of 219
   "   " , mineral combinations of 225
   "   " , variant of the calculation modus 230
Standard katanorm 121, 122
   "   " , calculation from cation % 136, 138
   "   " , calculation from Niggli-values 140
   "   " , mineral combination of 126
   "   " , of metamorphic rocks 206
   "   " , recalculation from Niggli-values 145
Stavanger region, injection metamorphism 267
Subalfemic magmas 92
Subalic magmas 92
Sub-facies 202
Subfemic magmas 92
Subplagi-foyaitic magma group 96

Sulfate-marialite 117
   " -meionite 117
Sviatonossitic magma 98
Syenitic-granitic magma 98
   "   "   magma group 98
   " -ijolitic magma 97
   "   magma 98
   "   " group 98
Symmetrical distribution 26

T

t 63
T 63
T-space 77
Table system 45
Tahititic magma 96
Talc 203, 216, 236
Tasnagranitic magma 94
Tavitic magma 96
Tephroite 117
Thenardite 107, 118
Theralite 157
   " -gabbroidal magma 97
   " - " magma group 97
Theralitic magma 97
ti 49
Ti-augite 159, 166
Tilleyite 204, 207
Tinguaitic magma 96
TiO_2, calculation of 49
Titanite 118
Tonalitic magma 94
Tremola series 122
Triangular coordinate paper 35
Triangular coordinates 23
Trondjemite-aplitic magma 94
Trondjemitic magma 94
Trondjemitic magma group 94
Tschermak's component 117
Turjaitic magma 97
Typomorphic minerals 202

U

Umpektitic magma 96
Upper facies 202
Urtitic magma 96
Urtitinguaitic magma 96

V

Variance 27
Variants of the standard katanorm 133
Variation coefficient 27
Variation range 26
Vaugnerite, biotite-hornblende variant 152
   "   , " variant 152
Vaugnerite from Vaugneray 148
   "   , hornblende variant 152
Vectors in the concentration tetrahedron 39
   "   "   "   " triangle 38
Vesecite-polzenitic magma 98
Vesbic volcano 76, 86, 87
Vesbitic magma 98
Vesuvianite 204, 207
V_2O_3, calculation of 48
Volume percent, recalculation to equivalent % 179
Vredefortitic magma 99

W

w 50, 52, 187
Water of constitution 15
Websteritic magma 95
Weight percentages, calculation from Niggli-values 52
   "   " , recalculation to equivalent % 170
   "   " , recalculation to volume % 179
Wollastonite-diopside-plagioclase fels 213
Wyoming-lamproitic magma 95

X

Xonotlite, 204, 217

Y

Yogoite-lamproitic magma 99
Yogoitic magma 99
Yosemitite-aplitic magma 94
Yoemitite-granitic magma 94

Z

Zinnwaldite 117
Zircon 107, 109, 118
Zoisite 203, 215
zr 49
ZrO_2, calculation of 49